11th International Workshop on Positron and Positronium Chemistry 2014

(PPC-11)

AA002714

Journal of Physics: Conference Series
Volume 618

Goa, India
9 - 14 November 2014

Editors:

P.K. Pujari
K. Sudarshan
D. Dutta

ISBN: 978-1-5108-0753-2
ISSN: 1742-6588

Printed from e-media with permission by:

Curran Associates, Inc.
57 Morehouse Lane
Red Hook, NY 12571

Some format issues inherent in the e-media version may also appear in this print version.

Printed by Curran Associates, Inc. (2015)

For permission requests, please contact the Institute of Physics
at the address below.

Institute of Physics
Dirac House, Temple Back
Bristol BS1 6BE UK

Phone: 44 1 17 929 7481
Fax: 44 1 17 920 0979

techtracking@iop.org

Additional copies of this publication are available from:

Curran Associates, Inc.
57 Morehouse Lane
Red Hook, NY 12571 USA
Phone: 845-758-0400
Fax: 845-758-2634
Email: curran@proceedings.com
Web: www.proceedings.com

Table of contents

Volume 618

11th International Workshop on Positron and Positronium Chemistry (PPC-11)
9–14 November 2014, Goa, India

Accepted papers received: 29 April 2015
Published online: 15 June 2015

Preface

Papers

In Memoriam

Fundamentals: Experiment and theory

Porous materials and confined systems

Experimental techniques and facilities

Journal of Physics: Conference Series **618** (2015) 011001 doi:10.1088/1742-6596/618/1/011001

Preface

The International Workshop on Positron and Positronium Chemistry (PPC) is a prestigious triennial conference series with a rich history. The 11th meeting in the series (PPC-11) was held at Cidade de Goa, Goa, India during 9-14, November, 2014. It was organized by Bhabha Atomic Research Centre (BARC), Mumbai. The co-organizers were Saha Institute of Nuclear Physics (SINP), Kolkata, Indira Gandhi Centre for Atomic Research (IGCAR), Kalpakkam and Indian Association of Nuclear Chemists and Allied Scientists (IANCAS), Mumbai.

PPC-11 attracted participants both from academic institutions and industries. About 120 participants from 20 countries representing all continents participated in the conference. The conference continued the tradition of excellence in terms of quality of presentations and discussions. There were 33 plenary and invited talks, 39 oral presentations and 40 posters. The conference stood true to its multidisciplinary tag with papers presented in the fields of fundamentals of positron and positronium chemistry, applications in polymers, porous materials, metals/alloys, studies in liquids, biological applications as well as developments in theory and experimental techniques. The enthusiastic participation of senior researchers and young students made the scientific program a grand success. In order to encourage the student participants (twenty) and promote excellence, a committee of senior members evaluated their presentations and the top three contributions were awarded. The positron and positronium community paid homage to the memory of late Profs. J. Kristiak and A.T. Stewart. A brief sketch of their life and work was presented by Profs. Jan Kuriplach and Toshio Hyodo, respectively.

All the papers published in these proceedings have been peer reviewed by the participants of PPC-11. Editors thank all the reviewers for sparing their valuable time and helping us in bringing out the proceedings with 43 contributed articles in the scheduled time. We are grateful to the members of the International scientific committee, members of the organizing committee and advisory committee for their support. We wish to thank the members of the positron group at the Radiochemistry Division, Bhabha Atomic Research Centre for shouldering the responsibility of organizing the conference and making it a memorable event.

We wish to conclude by wishing success to the organizers of PPC-12 and hope to meet you all in Poland.

April 2015

P.K. Pujari
K. Sudarshan
D. Dutta

Editors

Journal of Physics: Conference Series **618** (2015) 011001 doi:10.1088/1742-6596/618/1/011001

11th International Workshop on Positron and Positronium Chemistry

International scientific committee

A. Alam (UK)
C.Q. He (China)
T. Hirade (Japan)
Y.C. Jean (USA)
Y. Kobayashi (Japan)
J. Kuriplach (Czech Republic)
F.H.J. Maurer (Sweden) - *Chair*
Y. Nagashima (Japan)
P.K. Pujari (India)
K. Raetzke (Germany)
D.M. Schrader (USA) - *Emeritus*
S.V. Stepanov (Russia)

Local organizing committee

P.K. Pujari, BARC, Mumbai, *Chairman*
R. Acharya, BARC, Mumbai
S. Sodaye, BARC, Mumbai
K. Sudarshan, BARC, Mumbai, *Secretary*
R. Tripathi, BARC, Mumbai
D. Dutta, BARC, Mumbai
S.K. Sharma, BARC, Mumbai
P. Maheshwari, BARC, Mumbai
S. Mukherjee, BARC, Mumbai
Santosh K Gupta, BARC, Mumbai
S.K. Rath, Naval Materials Research Lab, Mumbai

Journal of Physics: Conference Series **618** (2015) 011001 doi:10.1088/1742-6596/618/1/011001

Organizing committee

A. Goswami, BARC, Mumbai, *Chairman*
P.K. Pujari, BARC, Mumbai, *Convener*
A.V.R. Reddy, BARC, Mumbai
B.S. Tomar, BARC, Mumbai
P. Singh, BARC, Mumbai
G.K. Dey, BARC, Mumbai
M.G.R. Rajan, BARC, Mumbai
D. Srivastava, BARC, Mumbai
S.K. Gupta (TPD), BARC, Mumbai
S.K. Gupta (IADD), BARC , Mumbai
I. Samajdar, IITB, Mumbai
C. Ranganathaiah, University of Mysore
V. Ravindrachary, Mangalore University
Y.K. Vijay, University of Rajasthan
G. Amarendra, IGCAR, Kalpakkam
B. Ganguly, SINP, Kolkata
P.M.G. Nambissan, SINP, Kolkata
A.V. Salker, Goa University
K. Madangopal, BARC, Mumbai
K. Sivaji, University of Madras
C.S. Bal, All India Institute of Medical Sciences
M. Patri, Naval Materials Research Laboratory, Mumbai

Advisory committee

R.K. Sinha, Chairman, Atomic Energy Commission, India
S. Basu, BARC, Mumbai
P.R. Vasudeva Rao, IGCAR, Kalpakkam
K.L. Ramakumar, BARC, Mumbai
D. Srivastava, VECC, Kolkata
M.K. Sanyal, SINP, Kolkata
S.K. Aggarwal, BARC, Mumbai
B.N. Jagatap, BARC, Mumbai
S. R. Shetye, Goa University, Goa
C.S. Sundar, IGCAR, Kalpakkam
S.L. Chaplot, BARC, Mumbai
S.G. Markandeya, Mumbai
N. Ramamoorthy, BARC, Mumbai

Journal of Physics: Conference Series **618** (2015) 011001 doi:10.1088/1742-6596/618/1/011001

Participants of the 11th International Workshop on Positron and Positronium Chemistry (PPC-11) held at Goa, India during 9-14, November 2014

Journal of Physics: Conference Series **618** (2015) 011002 doi:10.1088/1742-6596/618/1/011002

Peer review statement

All papers published in this volume of *Journal of Physics: Conference Series* have been peer reviewed through processes administered by the proceedings Editors. Reviews were conducted by expert referees to the professional and scientific standards expected of a proceedings journal published by IOP Publishing.

Journal of Physics: Conference Series **618** (2015) 012001 doi:10.1088/1742-6596/618/1/012001

Jozef Krištiak – In memoriam

Jan Kuriplach

Department of Low Temperature Physics, Faculty of Mathematics and Physics, Charles University in Prague, V Holešovičkách 2, CZ-180 00 Prague, Czech Republic

E-mail: jan.kuriplach@mff.cuni.cz

It is my sad duty to give a memorial address for Dr. Jozef Krištiak who was a well known and respected member of the positron and positronium community since 1980's. It is especially disturbing to remind him at this occasion because he was the organizer of the previous meeting of the PPC series which took place in Slovakia in 2011, and many of the attendees of the current PPC workshop might remember this event very well.

Jozef Krištiak was born in a small town called Turzovka in the north-western part of Slovakia on December 18, 1942. After graduating from the Czech Technical University, Prague (Faculty of Nuclear Sciences and Physical Engineering) in 1966, he started to work at the Institute of Nuclear Research in Řež near Prague and later on he continued at the Institute of Physics, the Slovak Academy of Sciences in Bratislava. The beginning of his scientific career was devoted to nuclear physics, and he received his PhD degree in 1974.

At the beginning of 1980's, Jozef became interested in positron annihilation and started positron studies – being a scientific leader of a small group – of metallic glasses [1], high temperature superconductors [2] and fullerenes [3]. During this period a Doppler broadening spectrometer and afterwards a positron lifetime apparatus were built in his lab in Bratislava.

Later on, Jozef jumped to the field of organic materials and pursued positronium studies of amorphous polymers [4] and molecular materials [5] combining also positrons with other experimental methods. At the same time, he was actively developing the free volume concept. Since 2007, confined systems became the subject of study for the Jozef's group as well [6].

Though I have not cooperated with Jozef frequently, I personally liked his ability to grasp simple principles behind complex processes occurring in studied systems, which – in my honest opinion – was also appreciated by others.

Jozef Krištiak attended regularly PPC meetings and together with his group he was selected to organize the 10th PPC workshop in Slovakia. This event happened in the Smolenice Castle near Bratislava at the beginning of September in 2011 and was evaluated as a highly successful event by the community.

Jozef suffered from an unpleasant health handicap which limited his activities considerably. Yet he was fighting these problems bravely and was looking optimistically at the future. Jozef Krištiak passed away on November 23, 2012. His ideas are further developed and continued by his group, and the positron community will remember him.

I appreciate very much the cooperation with Ondrej Šauša, Jozef's long time collaborator, and Katarína Mülerová, Jozef's daughter, who helped me by supplying the material for this brief note.

[1] Krištiaková K, Tumbev G and Krištiak J 1990 *Phys. Stat. Sol. (a)* **117** K5
[2] Krištiaková K, Šauša O, Krištiak J and Jánoš Š 1989 *Z. Phys. B* **71** 197
[3] Krištiak J, Krištiaková K and Šauša O 1994 *Phys. Rev. B* **50** 2792
[4] Krištiak J, Bartoš J, Krištiaková K, Šauša O and Bandžuch P 1994 *Phys. Rev. B* **49** 6601
[5] Bartoš J, Šauša O, Köhler M, Švajdlenková H, Lunkenheimer P, Krištiak J and Loidl A 2011 *J. Non-Cryst. Solids* **357** 376
[6] Illeková E, Krištiak J, Macová, Maťko I and Šauša O 2013 *J. Therm. Anal. Calorim.* **113** 1187

Alec T. Stewart (1925-2014) and Positronium

Toshio Hyodo
Institute of Materials Structure Science
High Energy Accelerator Research Organization
1-1 Oho, Tsukuba, 305-0801 Japan

E-mail: toshio.hyodo@kek.jp

Abstract. In this article I highlight the positronium-related work of the late Alec Stewart whose valuable contributions helped to establish the field of positron annihilation studies.

It is a great loss to us, both personally and professionally, that Professor Alec T. Stewart of Queen's University, Kingston, Canada, passed away on July 28th, 2014. He began his research life with neutron scattering experiments, working with Professor Bertram Brockhouse at the Chalk River Laboratories in Canada [1-3]. He changed his research field to that of positron annihilation and subsequently established a method of angular correlation of annihilation radiation (ACAR) measurement to investigate positron and positronium (Ps) interactions with matter.

The one dimensional angular correlation of annihilation radiation (1D-ACAR) curve is a projection of the photon-pair momentum density, $\rho_{2\gamma}(\vec{p})$, onto one direction. For, say the x-direction, the curve, N(p), can be expressed as

$$N(p_x) \propto \iint \rho_{2\gamma}(\vec{p}) dp_y\, dp_z\,.$$

Due to momentum conservation during the 2γ annihilation, $\rho_{2\gamma}(\vec{p})$ is, at the same time, the momentum density of the fully-interacting, annihilating spin-singlet electron-positron pair. Stewart studied the momentum dependent enhancement of the annihilation rate resulting in the bulge of the parabolic 1D-ACAR curves for alkali metals, and the effective mass of the positron due to the positron-electron and positron-phonon interactions which can be deduced from the smearing of the Fermi cut-off of the 1D-ACAR. I will not go into detail about this aspect of his work here, but rather concentrate on his interest in Ps.

The ACAR data for the annihilation of a valence electron of the medium with a positron without forming Ps (free positron annihilation) or with a positron bound in Ps (pickoff annihilation) is approximated by the sum of the momenta of the annihilating electron and positron. In the case of the self-annihilation of para-positronium (p-Ps), on the other hand, the relative momenta of the electron and the positron cancel and the ACAR data directly represents the translational total momentum, or the centre of mass momentum, of the Ps.

Ps may exist in either a non-localized or a localized state, depending on its interactions with the medium. When the Ps is in a localized state, $\Phi_{\rm L}(\vec{r}_{\rm e}, \vec{r}_{\rm p})$, with $\vec{r}_{\rm e}$ and $\vec{r}_{\rm p}$ being the coordinates of the electron and the positron in the Ps, the momentum density is determined by the Fourier transform of the electron-positron contact amplitude of the wave function, $\Phi_{\rm L}(\vec{r}, \vec{r})$, such that

$$\rho_{2\gamma}(\vec{p}) \propto \left| \int \Phi_{\rm L}(\vec{r}, \vec{r}) e^{-\vec{p}\cdot\vec{r}} d^3r \right|^2.$$

The greater the localization, the wider the ACAR curve. It was suggested by Ferrell [4] in 1957 that Ps in liquid He was localized in a cavity which Ps itself created because of a strong repulsive exchange force between Ps and a He molecule. Later, Roellig and Kelly [5] estimated the size of the bubble in high density gaseous He from the pickoff annihilation rate of o-Ps. Soon after this, Stewart, with Briscoe and Choi performed a direct measurement of the bubble size in liquid He and other liquids using 1D-ACAR curves [6]. The curves had two components. The broad component was attributed to annihilation of the electrons of the liquid's atoms and molecules and reflected their momentum distributions. The narrow component was attributed to the self-annihilation of *p*-Ps localized in a bubble. Stewart, Briscoe and Steimbacher later added further data which aided in forming a more comprehensive account of the observation [7].

If it is assumed that Ps is localized in a spherical square well of radius R, the total energy of the bubble state, $E_{\rm tot}$, is written as

$$E_{\rm tot} = E_{\rm cm}(R) + \frac{4}{3}\pi R^3 p + 4\pi R^2 \sigma,$$

where $E_{\rm cm}(R)$ is the centre of mass zero-point energy of the localized Ps, the second term corresponds to the work required to remove atoms from the bubble volume in a liquid of pressure p, and the last term is the bubble surface energy with specific surface energy σ. If we assume an infinite well depth, the electron-positron contact amplitude of the Ps is

$$\Phi_{\rm L}(\vec{r},\vec{r}) = \frac{A \sin\sqrt{\frac{2mE_{\rm cm}}{\hbar^2}}\, r}{r}$$

where

$$E_{\rm cm} = \frac{1}{2m}\left(\frac{h\pi}{R}\right)^2$$

It is a straightforward procedure to calculate the 1D-ACAR curve $N(p_y, p_z)$ with $\rho_{2\gamma}(\vec{p})$ for this contact amplitude. The full-width-at-half-maximum $\theta_{\rm FWHM}$ in milliradians has a relation with the radius of the cavity as

$$R = \frac{1.659}{\theta_{\rm FWHM}}\ {\rm nm}\,.$$

The experimentally obtained value was $\theta_{\rm FWHM}$ = 0.86 milliradians and thus R=1.93±0.13 nm. The equilibrium bubble radius estimated by minimizing $E_{\rm tot}$ with respect to the bubble radius gives an equation for R, solving which, with a surface tension of 0.096 dyn cm^{-1} , gives the value R=1.88 nm. (The numbers cited here are from ref. [7] rather than [6].) The good agreement indicates that the method used for measuring the bubble size is feasible.

The analysis of the long positron lifetime [5] verified the formation of bubbles by long-lived *o*-Ps. The ACAR study showed more directly that even a short-lived *p*-Ps atom creates a bubble around it before annihilation. This study also showed that the formation of the bubble is completed in a time shorter than 100ps.

When the Ps is in a delocalized, quasi free state, its wave function, $\Phi_{\vec{k}}(\vec{r}_{\rm e}, \vec{r}_{\rm p})$, is characterized by a translational wave vector $\vec{k}$. If we neglect a periodic structure of the medium as in a crystal, the electron-positron contact amplitude is a plane wave so that its Fourier transform is a delta function $\delta(\hbar k\text{-}p)$ and thus

$$\rho_{2\gamma}(\vec{p}) \propto \sum_{\vec{k}} n_{\vec{k}} \left|\int \Phi_{\vec{k}}(\vec{r},\vec{r}) e^{-\vec{p}\cdot\vec{r}} d^3 r\right|^2 = \sum_{\vec{k}} n_{\vec{k}}\, \delta(\hbar k - p) = n_{\vec{p}/\hbar}\,.$$

Since the thermalized Ps atoms follow the Boltzmann distribution, $n_{\vec{p}/\hbar} \propto \exp\left(\frac{p^2}{2m^* k_B T}\right)$, where m^* is the effective mass of the Ps and $k_{\rm B}$ is the Boltzmann constant. Thus, the width of the component is not determined by the width of the electron-positron contact amplitude, but the Boltzmann distribution of the states and hence temperature dependence is expected. When the medium is a crystal, Ps is actually in a Bloch state. The wave function is modulated by the periodicity of the crystal field that the momentum density has a non-zero component at the momenta corresponding to the reciprocal lattice vectors

$$\Phi_{\vec{k}}(\vec{r},\vec{r}) \propto \sum_{\vec{G}} a_{\vec{G}}(\vec{k}) e^{i(\vec{k}+\vec{G})\cdot\vec{r}} ,$$

where $\vec{G}$ is the reciprocal vectors and $a_{\vec{G}}(\vec{k})$ is the amplitude of the higher momentum components. Then the momentum density is written as

$$\rho_{2\gamma}(\vec{p}) \propto \sum_{\vec{G}} n_{(\vec{p}/\hbar-\vec{G})} \left| a_{\vec{G}}(\vec{p}/\hbar - \vec{G}) \right|^2 .$$

Here $\vec{G}$ is the reciprocal lattice vector which projects $\vec{p}/\hbar - \vec{G}$ into the first Brillouin zone and again $n_{(\vec{p}/\hbar-\vec{G})}$ is the Boltzmann distribution. The resulting 1D-ACAR curve is a series of Gaussian curves centred at the projection of $\vec{G}$ onto the x direction as experimentally shown by Brandt *et al.* [8] and Greenberger *et al.* [9]. Stewart and co-workers [10] investigated the Ps components in crystalline quartz and showed that the orthogonality of the electron part of the Ps wave function and the wave function of the valence electrons in the crystal, or the indistinguishability of the electrons explains the magnitude of the higher momentum components.

Following this, Kubica and Stewart measured the 1D-ACAR for quartz at 4.2K [11]. The width was quite small, corresponding to a distribution at 10K if the effective mass was $2m_e$, where m_e is the mass of the electron in vacuum. The effective mass is known to be greater than $2m_e$ [10,12,13] by analysing the temperature dependence of the width. Therefore, it is safe to say that the Ps in quartz at 4.2K is thermalized within the experimental uncertainty. Thus also, in a solid, thermalization time for Ps is shorter than the lifetime of *p*-Ps.

I worked with Alec Stewart for two years (1980-82) at the Queen's University, Canada, where I studied positron-electron and positron-phonon interactions in potassium [14]. Several years before this, I had found that Ps exists in alkali halides in delocalized Bloch states at low temperatures and self-trapped at higher temperatures including room temperature [15]. This explained why the 1D-ACAR curves for alkali halides systematically measured by Stewart and Pope at room temperature did not show distinct narrow peaks. If we look at the curves carefully, they are more triangular in form than typical 1D-ACAR curves for metals and insulators. The broadened narrow component due to the *p*-Ps being localized in a much smaller volume than that in liquid inert gases gives rise to the characteristic shapes. Alec was interested in this phenomenon and we conducted measurements on the temperature dependence of the positron lifetime in NaF and NaCl [16]. The results were consistent with the picture that Ps which were free at low temperatures underwent transition to a self-trapped state at higher temperatures, and that the accompanying deformation was of a dilation type.

After returning to Japan, I conducted some joint research with Alec and others on the temperature dependence of the Ps momentum density in SiO_2, ice and BaF_2 [17], as well as 2D-ACAR studies of alkali metals [18] and a Compton profile study on Li [19].

I also collaborated with Alec Stewart on the ACAR study of the thermalisation of Ps in gases, which evolved from my experience in Kingston. On considering the relationship between the momentum density and the shape of 1D-ACAR curve, I noticed that an isotropic momentum density always gave a convexed 1D-ACAR curve and the extreme case was a distribution only on a thin spherical shell giving a rectangular curve. Thus, when I measured the 1D-ACAR curve of silica aerogel and found that it had a rather flat top, I reasoned that it must consist of a Gaussian-like broad component from the valence electrons of amorphous silica and a more or less rectangular component from *p*-Ps emitted isotropically in all directions from the grain with an energy determined by the negative work function [20]. These initial findings were then followed by our collaboration with Alec on the thermalisation of positronium via collisions with the silica aerogel grain surfaces and a variety of gas molecules [21].

Prof. Alec Stewart's enthusiasm for physics was undeniable and his approach to understanding physical phenomena was to seek the most straightforward and essential models. Doing research with him was immensely pleasurable and I learned much from him during our conversations together.

References

[1] Brockhouse B N and Stewart A T 1958 *Rev. Mod. Phys.* **30** 236
[2] Brockhouse B N and Stewart A T 1958 *Rev. Mod. Phys.* **30** 250
[3] Brockhouse B N 1994 *Nobel Lecture*
[4] Ferrell R A 1957 *Phys. Rev.* **108** 167
[5] Roellig L O and Kelly T M 1967 *Phys. Rev. Lett.* **18** 387
[6] Briscoe C V, Choi S -I and Stewart A T 1968 *Phys. Rev. Lett.* **20** 493
[7] Stewart A T, Briscoe C V and Steimbacher J J 1990 *Can J. Phys.* **68** 1362
[8] Brandt W, Coussot G and Paulin R 1969 *Phys. Rev. Lett.* **23** 522
[9] Greenberger A, Mills A P, Thompson A and Berko S 1970 *Phys. Lett.* A **32** 72
[10] Hodge C H, McKee B T A, Trifthuser W and Stewart A T 1972 *Can J. Phys.* **50** 103
[11] Kubica P and Stewart A T 1975 *Phys. Rev. Lett.* **34** 852
[12] Fujiwara K and Ikari H 1979 *J. Phys. Soc. Jpn.* 1979 **46** 92
[13] Ikari H 1979 *J. Phys. Soc. Jpn.* 1979 **46** 97
[14] Hyodo T, McMullen and Stewart A T 1986 *Phys. Rev.* B **33** 3050
[15] Hyodo T and Takakusa Y 1977 *J. Phys. Soc. Jpn.* 1979 **42** 1065
[16] Hyodo T and Stewart A T 1984 *Phys. Rev.* B **39** 4164
[17] Nagai Y, Kakimoto M, Hyodo T, Fujiwara K, Ikari H, Elcrup M and Stewart A T 2000 *Phys. Rev.* B **62** 5531
[18] Manuel A A, Oberlity L, Singht A K, Jarlborgt T, Petefi M, Mijnarends P E, Rabous L P L M, Hyodo T and Stewart A T 1993 *J. Phys.: Candens. Matter* **5** 8703
[19] Sakurai Y, Tanaka Y, Bansil A, Kaprzyk S, Stewart A T, Nagashima Y, Hyodo T, Nanao S, Kawata H and Shiotani N 1995 *Phys. Rev. Lett.* **74** 2252
[20] Chang T B, Deng J K, Akahane T, Chiba T, Kakimoto M and Hyodo 1985, *Positron Annihilation*, ed. by Jain P C, Singru R M and Gopinathan K P (World Scientific) 974
[21] Nagashima Y et al 1995 *Phys. Rev.* A **52** 258

Journal of Physics: Conference Series **618** (2015) 012003 doi:10.1088/1742-6596/618/1/012003

Early processes in positron and positronium chemistry: possible scavenging of epithermal e^+ by nitrate ion in aqueous solutions

Serge V. Stepanov[1,2], Vsevolod M. Byakov[2,3], Gilles Duplâtre[4], Dmitrii S. Zvezhinskiy[2], Petr S. Stepanov[5], Alexandr G. Zaluzhnyi[1,2]

[1] National Research Nuclear University MEPhI (Moscow Engineering Physics Institute), 115409, Moscow, Kashirskoe highway, 31. Russia
[2] Institute of Theoretical and Experimental Physics, 117218, Moscow, Russia
[3] D.Mendeleyev University of Chemical Technology, Miusskaya sq., 9, Moscow 125047, Russia
[4] Institut Pluridisciplinaire Hubert Curien, CNRS/IN2P3, BP 28 67037 Strasbourg, Cedex 2, France
[5] Center for Photochemical Sciences, Bowling Green State University, Bowling Green, OH 43403, USA

E-mail: stepanov@itep.ru

Abstract. Positron ionization slowing down, formation of the positron track, reactions of e^+ with track species and its interaction with a scavenger on a subpicosecond timescale, including the process of the positronium formation process are discussed. Interpretation of the positron annihilation lifetime data on positronium formation in aqueous solutions of NO_3^- anions, known as efficient scavengers of the presolvated track electrons, suggests that these ions may also capture epithermal (presolvated) positrons as well.

1. Introduction

Positrons (e^+) as well as positronium atoms (Ps) are recognized as probes of the early radiolytic physicochemical processes. This is due to the fact that the formation of the positronium atom in molecular media takes place on a very short (picosecond) time scale through interaction of the thermalized positron with one of the electrons in the e^+ track [1]. Therefore, introducing electron scavengers into the medium can modify the Ps formation process in a controllable way.

Obviously, the informative potentiality of positron spectroscopy strongly depends on the reliability of a theory describing processes in the positron track, since it deciphers the information coded in the annihilation spectra. So, realistic models are needed for the e^+ track structure, e^+ ionization slowing down [2], thermalization and transformation of the released energy into heat [3, 4], intratrack reactions (ion-electron recombination, solvation, interaction with scavengers), Ps formation process and finally Ps interaction with chemically active radiolytic species [5].

The usual treatment of the measured annihilation spectra in the case of Positron Annihilation Lifetime Spectroscopy (PALS) is reduced to their deconvolution into a set of decaying time exponentials, or of Gaussians in the case of angular correlation of annihilation radiation and Doppler broadening of annihilation radiation. The outcome of such analyses of the spectra are the lifetimes/widths of these components and the corresponding intensities [6, 7].

However, realistic physicochemical processes within the fast positron track, especially in its terminal part (which is termed a blob), suggest some more complex behavior, particularly when different scavengers are dissolved in the studied medium, capable to react with epithermal (presolvated) particles (positrons and track electrons). The inhomogeneous spatial distribution of the track species and their outdiffusion become also an important factor: diffusion kinetics cannot be expressed in terms of mere exponentials or Gaussians [7].

In this case, the use of reaction rate constants (reaction radii or cross-sections), diffusion coefficients, initial size of the terminal positron blob and Ps contact density as the adjustable parameters is physically much more meaningful than the above mentioned "intensities" and lifetimes of some exponential functions. In the following, possible reactions of the presolvated e^+ and e^- with scavengers, taking as an example aqueous solutions of nitrate ions, are discussed.

2. Formation of the positron track

Positrons, produced in nuclear β^+-decay, have initial energies of several hundreds of keV. Passing through a medium they lose energy via ionization. Within 5-10 ps the positron energy drops below the ionization threshold, Fig. 1.

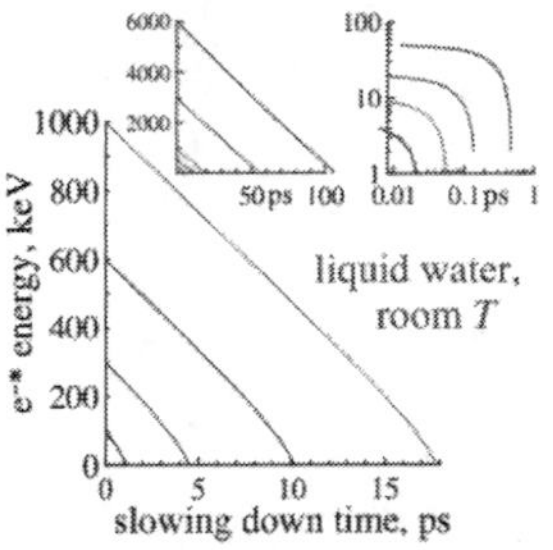

Figure 1. Typical times of ionization slowing down of e^+/e^- vs. its initial kinetic energy in liquid water at room temperature [physics.nist.gov/PhysRefData/Star/Text/ESTAR-u.html].

When the e^+ energy becomes less than the blob formation energy, W_{bl} ($\lesssim$ 0.5 keV or so), the positron creates its terminal blob [1]. The diffusion motion of e^+ in the blob becomes more pronounced: the direction of its momentum changes frequently due to elastic scattering and ionization of surrounding molecules. Mostly all intrablob ionizations are confined within a sphere of radius a_{bl}. The terminal positron blob contains a few tens of ion-electron pairs: $n_0 \approx W_{bl}/W_{iep} \approx 30$, where the average energy W_{iep} required to produce one ion-electron pair is 16-20 eV. The formation of the terminal blob lasts less then 0.01 ps. Further approach towards thermal equilibrium (from about 10 eV down to thermal energies) proceeds primarily via excitations of intra- and intermolecular vibrations and usually takes a few tens of femtoseconds. The ultimate stage of transformation of the projectile positron as well as of secondary track electrons is solvation (in water – hydration). The electron hydration time is about 0.3 ps [8].

3. Reactions of hot electrons and positrons

During the slowing down process subionizing (hot, presolvated) electrons e^{-*} (or positron e^{+*}) can be captured by some solute molecules (or ions): $e^{-*} + S \to S^{-*}$. Subexcitation electrons cannot promote electronic transitions in host molecules, but may excite the solute molecules, this occurring at some definite energy W_* depending on the structure of the excited states S^{-*}, Fig. 2. Below, we shall consider electron trapping, but it may proceed with positrons also.

The fraction φ_e of electrons escaping the resonance capture by the solute (c_S is the solute concentration) depends on the energy spectrum of the subionizing electrons $f(W_0)$ and their stopping power due to excitation of vibrations [2]:

$$\varphi_e(c_S) = \int_0^\infty dW_0 f(W_0) \exp\left[-c_S \int_0^{W_0} \frac{\sigma(W)dW}{|dW/dx|_{vib}}\right]. \quad (1)$$

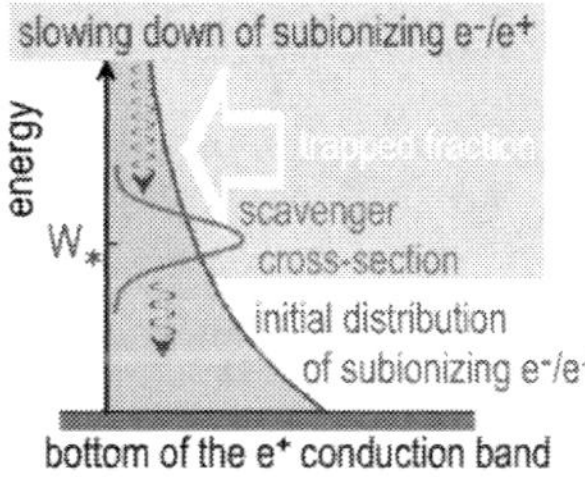

Figure 2. Resonance trapping of e^- or e^+, having the energy $W \approx W_*$, by the scavenger S.

(the decrease in the scavenger concentration c_S inside the blob due to the reaction with hot electrons is neglected here. For simplicity we may assume that the capture cross-section $\sigma(W)$ is nonzero in a rather narrow energy range: $\sigma(W) \approx \sigma_0\delta(W - W_*)$. Substituting this to Eq.1 we easily obtain:

$$\varphi_e(c_S) \approx 1 - F_- + F_- \cdot e^{-c_S/c_{37}^-}, \qquad F_- = \int_{W_*}^{\infty} f(W_0)dW_0, \qquad \frac{1}{c_{37}^-} = \left.\frac{\sigma_0}{|dW/dx|_{vib}}\right|_{W\approx W_*}. \quad (2)$$

Here $1 - F_-$ represents the fraction of track electrons escaping resonance capture even at large S concentration. These electrons are knocked out from molecules with energies below W_* and therefore do not interact with the scavenger.

In a similar way, the solute may also react with subionizing positrons. In this case $\varphi_p = 1 - F_+ + F_+ \exp(-c_S/c_{37}^+)$ would represent the fraction of positrons escaping a resonance capture by S. Fig. 3 illustrates effects of hot electron and positron scavenging.

If the capture cross-section of electrons (positrons) is located on the energy axis near the bottom of the conduction band ($1 - F_- \to 0$), the scavenger is able to capture almost all presolvated (quasifree) e^- (these electrons are precursors of the hydrated electrons). Such a situation is completed when nitrate ions are added into water. In this case the fraction φ_e of the track electrons ending up hydrated decreases exponentially with increasing c_S (the scavenger

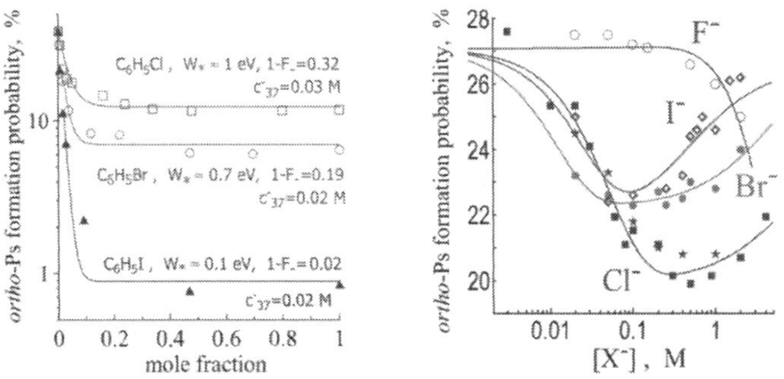

Figure 3. Left: Hot electron scavenging in benzene. Values of ortho-Ps formation probabilities are shown in benzene solutions of C_6H_5Cl (□), C_6H_5Br (○) and C_6H_5I (▲) [Anisimov O A, Molin Yu N Proc. 4th Int. Conf. Positron Annihilation, Helsingør, Danmark, G31, (1976)]. A correlation appears between the inhibition efficiency, F_-, and the energies W_* at which electron trapping by scavengers in a gas phase takes place. Solid lines are calculated according to Eq. 2. Right: Hot positron scavenging in aqueous solutions of alkali halide ions [Duplâtre G et al. 1978 Radiat. Phys. Chem. **11**, 199; Beling C D, Smith F A 1983 Chem.Phys. **81**, 243]. The increase of the Ps yield at higher c_S is due to reactions of halogen ions with H_2O^+.

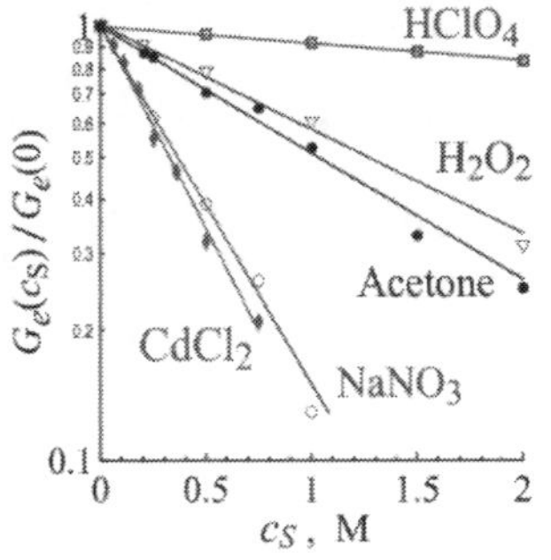

Figure 4. Exponential inhibition of the yield of the hydrated electron in aqueous solutions of various electron scavengers [Wolff R K, Bronskill M J, Hunt J W 1970 J. Chem. Phys. **53(11)**, 4211].

concentration): $\varphi_e \approx \exp(-c_S/c^-_{\rm NO3})$. Therefore, in the same way, the yield of hydrated electrons is decreased: $G_e(c_S) \approx G_e(0)\exp(-c_S/c^-_{\rm NO3})$, where $c^-_{\rm NO3} = 0.53$ M and $G_e(0) \approx 4.3$ electrons per 100 eV is the e^-_{aq} yield in pure water, Fig. 4.

4. Reactions of thermalized quasi-free e^- and e^+. Formation of e^-_{aq}, H_2 and qf-Ps

Now let us consider the next stage – the processes occurring in the e^+ blob with participation of quasi-free (thermalized, but not yet solvated) electrons e^-_{qf}, positrons e^+_{qf} and primary radical-cations (holes, H_2O^+). As we have seen above, the scavenger S (nitrate ions NO_3^-) does not react with thermalized electrons, but with epithermal ones. The most important reactions at this stage are [2, 5]

hydration (solvation): $e^-_{qf} \xrightarrow{\tau_{aq}} e^-_{aq}$, $\quad e^+_{qf} \xrightarrow{\tau_{aq}} e^+_{aq}$,

ion-molecule reaction: $H_2O^+ + H_2O \xrightarrow{\tau_{imr}} H_3O^+ + OH$,

ion-electron recombination: $e^-_{qf} + H_2O^+ \xrightarrow{k_{ie}} H_2O^* \xrightarrow{+H_2O} \begin{cases} H_2 + 2OH \\ H + OH + H_2O, \\ 2H_2O \end{cases}$

qf-Ps formation: $e^-_{qf} + e^+_{qf} \xrightarrow{k_{ep}}$ qf-Ps.

Here $\tau_{aq} \approx 0.3$ ps is the hydration times of e^-_{qf} and e^+_{qf}, assumed to be the same. τ_{imr} is the time of ion-molecule reaction. qf-Ps is the quasi-free positronium atom (precursor of the Ps bubble state), k_{ie}, k_{ep} are recombination reaction rate constants. This scheme assumes that Ps formation proceeds via the reaction of the thermalized positron with one of the track electrons, in competition with the other intratrack processes including hydration and recombination.

The above processes may be described by the following kinetic equations:

$$\frac{\partial c_j(\mathbf{r},t)}{\partial t} = D_j\nabla^2 c_j - \sum_{i(\neq j)} k_{ij}c_ic_j - c_j/\tau_j, \qquad c_j(\mathbf{r},t=0) \propto e^{-r^2/a^2_{bl}}, \quad j=\{i,e,p\}, \quad (3)$$

where c_i, c_e and c_p are the concentrations of H_2O^+, e^-_{qf} and e^+_{qf}, respectively, at a point $\mathbf{r}$ measured from the blob center at time t measured from the end of thermalization. D_j is the corresponding diffusion coefficient and k_{ij} is the reaction rate constant.

For an approximate solution of Eqs. 3, one may use the prescribed diffusion method. It assumes that $c_j(\mathbf{r},t) \approx n_j(t)G_j(\mathbf{r},t,D_j)$, where G_j is the normalized Gaussian function (the Green function of the free diffusion equation) [2, 5]. Integrating Eqs. 3 over the whole volume, we obtain equations for the total numbers n_j of the species in the e^+ blob:

$$\dot n_i = -k_{ie}n_in_e/V_{ie} - n_i/\tau_{imr}, \qquad n_i(0) = n_0, \quad (4)$$

$$\dot n_e = -k_{ie}n_in_e/V_{ie} - n_e/\tau_{aq}, \qquad n_e(0) = n_0\varphi_e, \quad (5)$$

$$\dot{n}_p = -k_{ep} n_e n_p / V_{ep} - n_p/\tau_{aq}, \qquad n_p(0) = \varphi_p, \tag{6}$$

$$V_{ij} = V_0^{ij}(1 + t/\tau_{ij})^{3/2}, \quad V_0^{ij} = [\pi(a_i^2 + a_j^2)]^{3/2}, \quad \tau_{ij} = \frac{a_i^2 + a_j^2}{4(D_i + D_j)}.$$

Here $n_0 \approx$ 30-40 is the initial number of ion-electron pairs in the blob. In accordance with the assumption of the ambipolar diffusion of the blob particles and electrical neutrality of the blob, we adopt that all $D_j \to D_{amb}$, $a_j \approx a_{bl}$ and $V_{ij} \approx V_0(1 + t/\tau)^{3/2}$, where $V_0 = (2\pi a_{bl}^2)^{3/2}$ and $\tau = a_{bl}^2/4D_{amb}$ is a characteristic time of the blob expansion. In water, thermalized e^-, e^+ and H_2O^+ disappear much faster than the blob expands. Therefore this expansion may be neglected: $V_{ij}(t) \approx V_0$.

To calculate the observable formation probabilities of e^-_{aq}, H_2 and qf-Ps, we have to perform the following integrations:

$$P_{\mathrm{H_2O^*}} = \int_0^\infty \frac{k_{ie} n_i n_e dt}{n_0 V_0}, \qquad P_{\mathrm{e_{aq}}} = \int_0^\infty \frac{n_e dt}{n_0 \tau_{aq}}, \qquad P_{\text{qf-Ps}} = \int_0^\infty \frac{k_{ep} n_e n_p dt}{V_0}. \tag{7}$$

Here, we adopt that excited water molecules give molecular hydrogen in approximately one third of all cases. To calculate these integrals we assume that H_2O^+, e^-_{qf} and e^+_{qf} disappear in time exponentially ($\propto e^{-w_j t}$) with the corresponding decay rates

$$w_i \approx \tau_{imr}^{-1}, \quad w_e \approx w_p \approx \tau_{aq}^{-1}, \tag{8}$$

which will be obtained later, using balance equations. Substituting these exponents into Eqs. 7, we easily obtain

$$P_{\mathrm{H_2O^*}} = \frac{\varphi_e w_{ie}}{w_e + w_i}, \qquad P_{\mathrm{e_{aq}}} = \frac{\varphi_e}{\tau_{aq} w_e} = \frac{G_e(c_S)}{G_{\mathrm{H2O}}}, \qquad P_{\text{qf-Ps}} = \frac{\varphi_e \varphi_p w_{ep}}{w_e + w_p}, \tag{9}$$

where $w_{ie} = k_{ie} n_0/V_0$ and $w_{ep} = k_{ep} n_0/V_0$ are the corresponding recombination rate constants (ion-electron and electron-positron) and G_{H2O} is the yield of decomposition of water (= 6.3 molecules per 100 eV) [9]. Now let us take into account the conservation laws for electrons, ions and the positron:

$$\varphi_e = P_{\mathrm{H_2O^*}} + P_{\mathrm{e_{aq}}}, \qquad 1 = P_{\mathrm{H_2O^*}} + \frac{1}{w_i \tau_{imr}}, \qquad \varphi_p = P_{\text{qf-Ps}} + \frac{\varphi_p}{w_p \tau_{aq}}. \tag{10}$$

Substituting the expressions for $P_{\mathrm{H_2O^*}}$, $P_{\mathrm{e_{aq}}}$ and $P_{\text{qf-Ps}}$ to these three equations, we can deduce three unknown quantities w_e, w_i and w_p, solving simple quadratic equations.

To reduce the number of uncertain parameters, we assume that 1) $w_{imr} = 1/\tau_{aq}$; 2) from the experimental values of the e^-_{aq} yield in pure water $G_e(0)/G_{\mathrm{H2O}} \approx 4.3/6.3$ using Eqs. 9 one can find that $w_{ie} \approx 0.625/\tau_{aq}$; 3) from the probability of formation of a quasi-free positronium $P_{\text{qf-Ps}} = 0.46$-0.48 (roughly it is the sum $I_1 + I_3$ of intensities of the shortlived and longlived components of the LT spectrum) at room temperature we obtain $w_{ep} \approx 1.6/\tau_{aq}$.

Note that the further intratrack reactions (qf-Ps transformation into a bubble state, Ps oxidation reactions with chemically active track species (OH, H_3O^+), Ps ortho-para conversion) must be considered by solving the diffusion-recombination equations which describe processes on the nanosecond time scale [5, 7].

5. Results and discussion

Fig. 5 shows how the qf-Ps formation probability in $NaNO_3$ aqueous solutions changes with the concentration of NO_3^- ions at 17 and 39 °C. It is seen that simple extrapolation

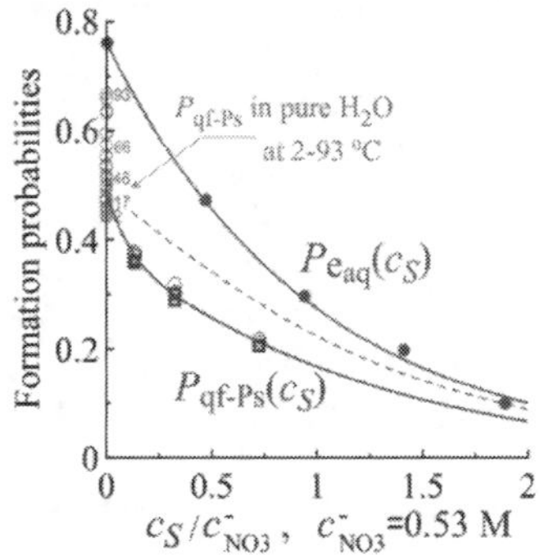

Figure 5. Formation probabilities of e^-_{aq} in NO_3^- aqueous solutions at room T (data are recalculated from Fig. 4) and of qf-Ps at nearly room temperatures (○ – 21 °C, □ – 39 °C [7]). In pure water $P_{\text{qf-Ps}}$ values are shown for a wider T range (2-93 °C). Solid lines are plotted according to Eqs. 9. It was used that $w_{ie}\tau_{aq} = 0.625$, $w_{imr}\tau_{aq} = 1$, $w_{ep}\tau_{aq} = 1.6$, $\varphi_e \approx \exp(-c_S/c^-_{\text{NO3}})$, where $c^-_{\text{NO3}} = 0.53$ M and $\varphi_p = 1 - F_+ + F_+ \exp(-c_S/c^+_{\text{NO3}})$, where $F_+ = 0.25$ and $c^+_{\text{NO3}} = 0.06$ M. Dashed line shows inhibition of qf-Ps formation in case of absence of the trapping of presolvated positrons by NO_3^- ions, i.e. when $\varphi_p = 0$.

($\propto \exp(-c_S/c^-_{\text{NO3}})$) of these data (at non-zero c_S) to zero concentration (at $c_S = 0$) leads to a mismatch with $P_{\text{qf-Ps}}(0)$ values in pure water at the same temperatures. This discrepancy even increases at $T = 60$-90 °C [7]. The dashed line in Fig. 5 shows the theoretically expected dependence for $P_{\text{qf-Ps}}(c_S)$ when trapping of hot positrons by NO_3^- is absent, i.e. if in Eq. 9 we put $\varphi_p = 0$. It is seen that the difference between the dashed and solid red curves is not small. It indicates that nitrate anions capture hot (epithermal) positrons to some extent.

It is also known that the NO_3^- ions react with the hydrated electrons (the rate constant is about $k(NO_3^-) \approx 10^{10}$ $M^{-1}s^{-1}$). Therefore, in a solution at concentration about 0.1 M, the NO_3^- ions capture most of the hydrated electrons in the e^+ blob during 1 ns: $1 - \exp[-k(NO_3^-) \cdot [NO_3^-] \cdot (1\text{ ns})] \approx 2/3$ (the typical concentration of hydrated electrons therein is about 0.01 M). However, treatment of the AMOC and LT spectra in pure water and nitrate aqueous solutions has shown that Ps inhibition is well described by the theory, neglecting Ps formation as a result of combination of e^-_{aq} and the positron. Capture of the hydrated electrons by NO_3^- ions (which definitely takes place in the e^+ blob) does not affect the qf-Ps formation probability [7]. It follows that the hydrated e^- is not a Ps precursor, and the hydrated positron does not react with e^-_{aq}. Most likely, this is due to the low mobility of the hydrated species and to the minute energy gain in this reaction: the energy of the Ps bubble is only slightly below the sum of the energies of the hydrated e^+ and e^-, whereas significant rearrangement of the surrounding molecules is needed in the reaction. Anyway, further experimental investigations are (especially AMOC measurements) are very welcome in this direction.

[1] Stepanov S V, Byakov V M 2003 *Physical and Radiation Chemistry of the Positron and Positronium*, in the book *Principles and Applications of Positron and Positronium Chemistry*. Eds. Y.C. Jean, P.E. Mallone, and D.M. Schrader (World Scientific Publications, Singapore, 2003); Chapter 5, pp.117-149
[2] Byakov V M, Stepanov S V 1996 *Journal of Radioanalytical and Nuclear Chemistry*, Articles, **210(2)**, 371-405
[3] Zvezhinskiy D S, Stepanov S V, Byakov V M, Zgardzinska B 2013 *Materials Science Forum* **733**, 15-18; doi:10.4028/www.scientific.net/MSF.733.15
[4] Stepanov S V, Zvezhinskiy D S, Byakov V M, Duplâtre G, Stepanov P S 2014 *Acta Physica Polonica A*. **125(3)**; 691-695, DOI: 10.12693/APhysPolA.125.691
[5] Stepanov S V, Byakov V M, Zvezhinskiy D S, Duplâtre G, Nurmukhametov R R, Stepanov P S 2012 *Advances in Physical Chemistry*, **2012** 431962; doi:10.1155/2012/431962
[6] Zvezhinskiy D S, Butterling M, Wagner A, Krause-Rehberg R, Stepanov S V 2013 *Journal of Physics: Conference Series*, **443**, 012057; doi:10.1088/1742-6596/443/1/012057
[7] Stepanov S V, Duplâtre G, Byakov V M, Zvezhinskiy D S, Subrahmanyam V S 2014 *Acta Physica Polonica A* **125(3)** 770-774; doi:10.12693/APhysPolA.125.770
[8] Wiesenfeld J M, Ippen E P 1980 *Chem. Phys. Lett.* **73(1)** 47-50; doi:10.1016/0009-2614(80)85199-2
[9] Byakov V M, Grafutin V I 1986 Radiat. Phys. Chem. **28(1)**, 1-18

Positronium bubble oscillation in room temperature ionic liquids -Temperature dependence-

T. Hirade[1]
Nuclear Science and Engineering Center, Japan Atomic Energy Agency,
Tokai, 319-1195 Japan
Graduate School of Science and Engineering, Ibaraki University,
4-12-1 Narusawa, Hitachi, Ibaraki, 316-8511 Japan

E-mail:t.hirade@kurenai.waseda.jp

Abstract. The temperature dependent oscillation of the ortho-positronium pick-off annihilation rate was successfully observed for a room temperature ionic liquid (IL), N,N,N-trimethyl-N-propylammonium bis (trifluoromethanesulfonyl)imide (TMPA-TFSI). The fundamental frequencies at 25C and 30C were 5.85GHz and 4.00GHz, respectively. The decay of the oscillation was faster at higher temperature, 30C. Moreover, the higher harmonic frequencies could explain the change of ortho-positronium pick-off annihilation rate successfully. The macroscopic viscosity of the IL could not explain the appearance of the oscillation. It indicated that the positron annihilation methods were very strong tools to study the properties of IL's in sub-nano meter scale that must be very different from the macroscopic properties.

1. Introduction

Positronium (Ps, a bond state of an electron and a positron) forms mainly by the reaction with an injected positron and one of the excess electrons after the positron thermalization at the end of the track in insulating materials. The detail of the Ps formation in insulating materials, such as liquids or solids, is well explained by the spur reaction model [1]. There are two Ps states because of spins of the electron and the positron. Para-Ps has anti-parallel spins and ortho-Ps has parallel spins. Positrons are anti-particle of electrons and then the mass of an electron and a positron disappears and the energy appears as gamma rays instead. The allowed number of annihilation gamma rays is even for para-Ps and odd for ortho-Ps in the case of the intrinsic annihilation. One gamma ray emitting annihilation is not allowed because of the low of conservation of momentum. Therefore para-Ps gives the fastest intrinsic-annihilation by emitting two gamma rays with lifetime of 125ps, and the ortho-Ps intrinsic-annihilation by emitting three gamma rays has the lifetime of 142ns in vacuum. In condensed matter, the positron in Ps can overlap with electrons on surrounding molecules and pair annihilation with one of those electrons is also possible. This annihilation process is called "pick-off annihilation". Almost all of pick-off annihilations give two gamma rays. The ortho-Ps lifetime in condensed matter is about 1 ~ 10 ns mainly by the pick-off annihilation. Some of the positrons do not form Ps and annihilate with electrons on the surrounding molecules. These positrons are often called "free positrons", because they are free from Ps formation. The lifetime of these positrons is about 400ps in condensed matter.

---

[1] To whom any correspondence should be addressed.

Journal of Physics: Conference Series **618** (2015) 012004 doi:10.1088/1742-6596/618/1/012004

Room temperature ionic liquids (IL's) are liquids consisting of ions. Negligible vapour pressure and other interesting properties are considered to be caused by the ionic nature. There are many considerable applications of ionic liquids, mainly because of their environmentally friendly property.

It was reported that thermalization distance of electrons that has sub-excitation energy is large and that electrons before full solvation in ionic liquids can diffuse [2]. Ps forms before electron and positron solvation at the time range of ~10ps in liquids. Therefore Ps formation reaction is a good tool to study the electrons before full solvation. If these electrons can diffuse for long time, Ps formation is quite possible even at the later positron ages such as 10~100ps and hence the results obtained by the positron annihilation age-momentum correlation (AMOC) experiment for IL could be explained by the delayed Ps formation [3]. AMOC is a combination method of the positron annihilation lifetime and the Doppler broadening of annihilation gamma-rays, i.e. momentum distribution of the annihilating electron. It is possible to discuss the change of the shape of the annihilation gamma-rays energy distribution, i.e. momentum distribution of annihilating electrons, by use of time dependent *S*-parameter applied in [3]. Larger *S* means the sharper line shape. There is the other parameter, *W*-parameter, i.e. wing parameter. It gives the information of annihilation with core electrons. Hence for the change between two positron annihilation processes, the sets of *S* and *W* appear on one straight line on the *S*-*W* plane. However, It was not a straight line for N,N,N-trimethyl-N-propylammonium bis (trifluoromethane-sulfonyl)imide (TMPA-TFSI).

Ps has negative work function in many materials and can create a bubble in liquids. *S*-*W* plots at young ages in TMPA-TFSI did not appear on a straight line and it indicated that the Ps bubble formation would take longer time than other liquids [4]. There is very small space for Ps at very young ages and it is similar to the squeezed Ps [5]. The small p-Ps intrinsic annihilation probability caused by the squeezed state at very young positron ages was successfully observed. [4] On the other hand, there is a good correlation between o-Ps pick-off annihilation rate and the bubble size, which is the well-known Tao-Eldrup formula. [6,7] If the bubble formation is slow in the IL, it is possible to detect the bubble size change by observing the o-Ps pick-off annihilation rate. It was successfully observed as the oscillation of the o-Ps pick-off annihilation rate in TMPA-TFSI at older positron ages than 400-500ps as expected. [8]

Here, we are discussing the oscillation of o-Ps pick-off annihilation rate measured at different temperatures.

2. Experimental

N,N,N-trimethyl-N-propylammonium bis (trifluoromethanesulfonyl)imide (TMPA-TFSI) (Kanto Chemical Co.) was used without any purification in this study. It has the viscosity of 72mPas and the melting temperature of 19C. The oxygen gas in the sample was removed by nitrogen gas bubbling for about 30 min. The sample temperature was controlled by use of a water bath with a temperature controller.

Positron annihilation lifetime (PAL) measurements were performed by use of a fast-fast coincidence method by use of two scintillation detectors. The every detector had a BaF_2 scintillator with the shape of a circular truncated cone with 20mm diameter at the top 30mm diameter at the bottom and 20mm height on the top of a Hamamatsu H3378 photomultiplier tube. The waveforms from these detectors were stored by use of the digital storage oscilloscope (DSO), Wavepro7100A (LeCroy), with 20G sampling/sec. The timing information from every waveform was obtained from the crossing point at the 50% level of the magnitude of the waveform. [4] About 45kBq ^{22}Na positron source sealed with two kapton films whose thickness is about 1.1mg/cm^2 was immersed in IL samples. The time resolution of PAL was about 160ps (fwhm). Every spectrum had about three million counts. The PAL spectra analysis was performed by the PALSfit programme. [9]

3. Results and Discussion

The typical PAL spectrum was shown in Fig.1. Apparently, it is quite usual spectrum. However, PAL spectra measured in IL's gives anomalous long shortest lifetimes. [3] It could be explained by the very

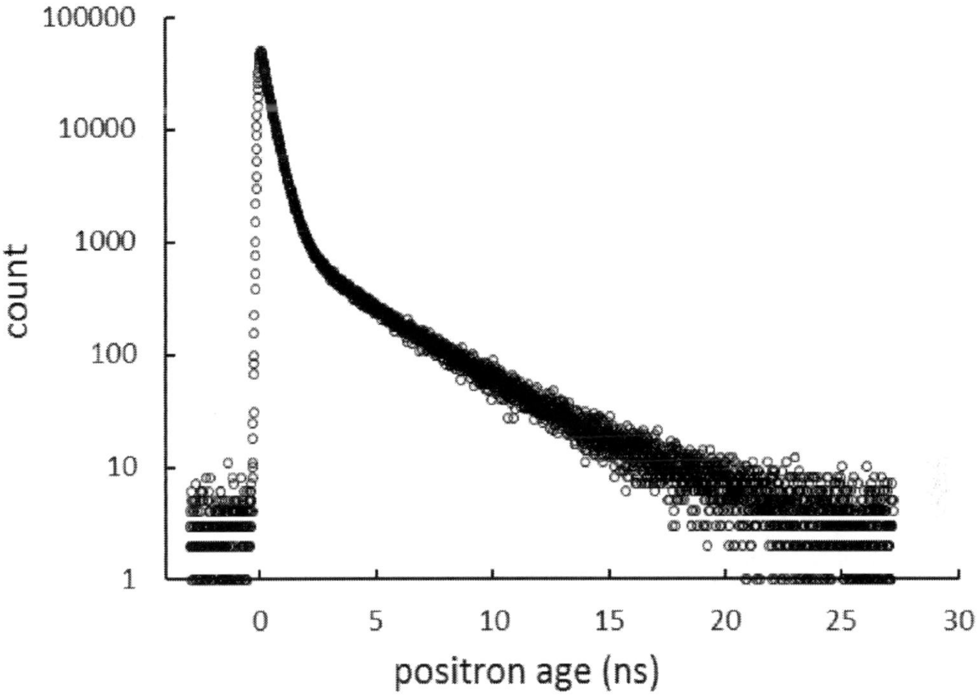

Figure 1. Positron annihilation lifetime spectrum of TMPA-TFSI measured at 25C. Back ground was not subtracted.

slow Ps bubble formation. [4] Moreover, it was indicated that there exist the Ps bubble oscillation until the bubble stabilized. [8] It was needed to extract just the oscillation part from the lifetime spectra indicated in Fig.1. The method used here was exactly same as indicated in [8]. The oscillation part was obtained by just subtracting the simulated lifetime spectrum with four exponentially decaying component fit by use of PALSfit [9] from the measured spectrum. The subtracted part should be just the oscillation of the annihilation counts. The oscillation was caused by the oscillation of the bubble i.e. the oscillation of the o-Ps pick-off annihilation rate. Hence, to obtain the o-Ps pick-off annihilation rate, annihilation counts from o-Ps pick-off annihilation should be divided by the o-Ps population at every positron age. Indeed, the simpler method applied here to standardize just the oscillation part of the annihilation rate was as follows. The oscillation of the annihilation count at every positron age was divided by the o-Ps pick-off annihilation count without the oscillation that is almost proportional to the o-Ps population at the positron age. The o-Ps pick-off annihilation count was calculated by the intensity and the lifetime of the longest lifetime component obtained by the PALSfit analysis.

The oscillation curves of the o-Ps pick-off annihilation rate obtained for TMPA-TFSI at 25C and 30C were indicated in Fig.2 and Fig.3, respectively. The oscillation curve observed at 25C showed beautiful sine curve at the positron ages of 500-900 ps. [8] However, the oscillation curve at 30C was not like the oscillation observed at 25C. The oscillation at 30C was complicated curve. In the case of the oscillation of the gas bubble, there must be the higher harmonics. [10] Therefore it is expected to have higher harmonics even for the Ps bubble oscillation. The solid line indicated in Fig.3 was sum of the second and the third higher harmonics. It successfully reproduced the complicated curve. The fundamental frequency for these higher harmonics was 4.00GHz. Moreover, the decay of the fundamental frequency oscillation was faster at 30C. On the other hand, the higher harmonics was expected even at 25C after the fundamental frequency oscillation decayed. The solid line in Fig.2 was

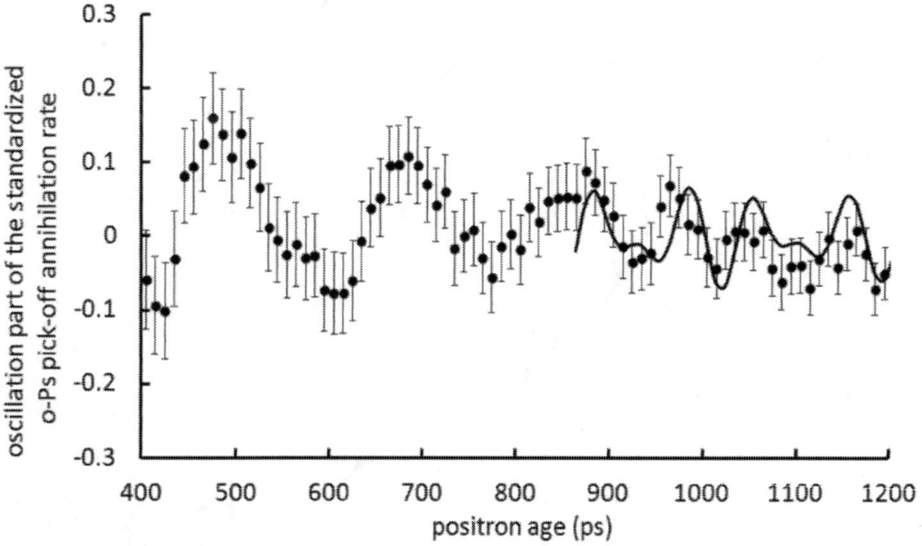

Figure 2. Oscillation part of the standardized o-Ps pick-off annihilation rate in TMPA-TFSI at 25C. The solid line was sum of the second and the third higher harmonics of the fundamental frequency of 5.85GHz.

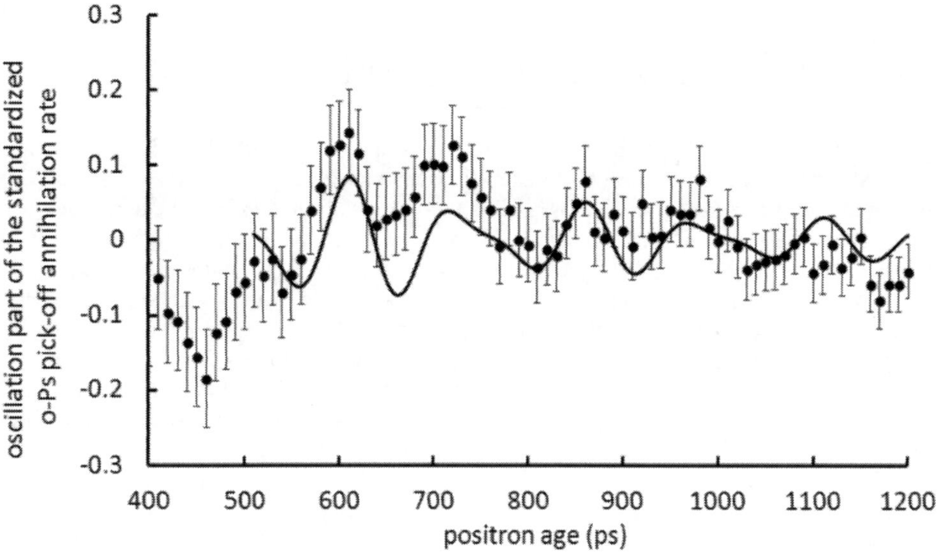

Figure 3. Oscillation part of the standardized o-Ps pick-off annihilation rate in TMPA-TFSI at 30C. The solid line was sum of the second and the third higher harmonics of the fundamental frequency of 4.00GHz.

also obtained by the summation of the second and the third higher harmonics of the fundamental frequency of 5.85GHz that was exactly same frequency for the sine curve appeared at 500-900 ps.

Two important things were obtained here. One was that the decay of the oscillation of the fundamental frequency decayed faster at higher temperatures. The other was that the fundamental frequency was higher at lower temperatures. The faster decay indicated that the energy loss caused by the movement of the molecules were larger at the higher temperatures. However the macroscopic viscosity of TMPA-TFSI is 72mPas that is not so small value. Theoretical studies by Stepanov et al. indicated that the liquids with these kinds of viscosities should not have oscillation. [11] Ps bubble size is sub-nano meter scale and the IL molecular motion in the very small scale around the Ps bubble should be very different from the molecular motion of the usual liquid materials. The higher frequency at the lower temperatures indicated that the modulus was larger.

4. Conclusion

The o-Ps pick-off annihilation rate measured for TMPA-TFSI showed the oscillation at the positron ages of 500-1200ps at 25C and 30C. The oscillation was probably caused by the oscillation of the Ps bubble. The fundamental oscillation could not be observed at 30C. However, the oscillation caused by the higher harmonic frequencies could explain the complicated oscillation of the o-Ps pick-off annihilation rate. Moreover the temperature dependence of the Ps bubble oscillation was observed. The decay of the oscillation was faster at higher temperatures. The frequency was higher at lower temperatures.

5. Acknowledgement

This research was partially supported by a Ministry of Education, Culture, Sports, Science and Technology Grant-in-Aid for Scientific Research (C), 23600011, 2011-2014.

6. References

[1] Mogensen O E 1974 *J. Chem. Phys.* **60** (3) 998
[2] Katoh R, Yoshida Y, Katsumura Y and Takahashi K 2007 *J. Phys. Chem. B* **111** (18) 4770
[3] Hirade T 2009 *Materials Science Forum* **607** 232
[4] Hirade T, Oka T 2013 *Journal of Physics: Conference Series* **443** 012060
[5] Mogensen O E 1994 *Hyperfine Interactions* **84** 377
[6] Tao S J (1972) *J. Chem. Phys.*, 565499
[7] Eldrup M, Lightbody D, Sherwood J N (1981) *Chem. Phys.*, 6351
[8] Hirade T 2014 *Japanese Journal of Applied Physics conference proceedings* **2** to be published
[9] Kirkegaard P, Olsen J V, Eldrup M and Pedersen N J 2009 PALSfit Risø-R-1652(EN)
[10] Fernández D, Maurer P, Martine M, Coey J M D, Möbius M E (2014) *Langmuir* **30** (43) 13065
[11] Stepanov S V, Mikhin,M K, Zvezhinskii D S, Byakov V M (2007) *Radi. Phys.Chem.* **76** 275

Positronium spin conversion during collisions with Xe and its application for measuring the kinetic energy

Kengo Shibuya, Yoshihiro Kawamura and Haruo Saito

Department of Physics, School of Arts and Sciences, The University of Tokyo, Komaba 3-8-1, Meguro, Tokyo 113-8656, Japan

shibuken@gakushikai.jp

Abstract. Positronium (Ps) can undergo ortho–para spin conversion reaction during Ps–Xe collisions due to spin–orbit interaction. We have investigated energy dependence of this reaction rate and found it is nearly proportional to $T^{2.1}$, where T is the temperature (300–623 K), while the pick-off annihilation rate is proportional to T. The strong temperature dependence of the former annihilation rate is attributed to a nature of p-wave scattering because this reaction is forbidden in s-wave scattering. In addition, a new method for measuring Ps kinetic energy has been developed with a high resolution and a high sensitivity by utilizing the strong temperature dependence as a "thermometer." Analyzing the obtained time evolution of Ps kinetic energy, we have determined the momentum transfer cross section during Ps–Xe collisions at a very low energy (40–60 meV) to be $12(2)\times10^{-16}$ cm^2.

1. Introduction

In 2006, a new reaction of positronium (Ps), i.e., Ps spin conversion reaction, was found [1]. Ps has four ground states denoted as $|S,M\rangle$, where S is the total spin and M is the z-projection. The triplet states are $|1,1\rangle$, $|1,0\rangle$ and $|1,-1\rangle$, while the singlet state is $|0,0\rangle$. These four states can be exchanged during Ps scattering with a high-Z atom such as Xe due to spin–orbit interaction. When this reaction occurs, an increase in the two-photon annihilation rate is observed because of a large difference in the lifetime between the triplet states and the singlet state.

Even when a magnetic field is applied, two of the three triplet states, $|1,\pm1\rangle$, remain the same. On the other hand, the last triplet state, $|1,0\rangle$, and the singlet state, $|0,0\rangle$, are mixed each other to make two new ground states, which are denoted as $|+\rangle$ and $|-\rangle$. Under the magnetic field of 0.82 T, the mixture of $|0,0\rangle$ and $|1,0\rangle$ states is 1%, i.e., $|+\rangle=0.994|1,0\rangle+0.111|0,0\rangle$ and $|-\rangle=-0.111|1,0\rangle+0.994|0,0\rangle$ [2]. It should be noted that the Ps lifetime in $|+\rangle$ state is considerably shorter than that in the original triplet states because two-photon annihilation is possible due to the small admixture of the singlet state. Under 0.82 T, the lifetime of $|+\rangle$ state is 9.5 ns.

We used this magnetic effect to evaluate the spin conversion reaction rate discriminating from pick-off annihilation rate. If there is no spin transitions, the lifetime of $|1,\pm1\rangle$ states remains the same when a magnetic field is applied. However, if there are some spin transitions, in other words, the arrows in figure 1 exist, the lifetime of $|1,\pm1\rangle$ states is reduced by the transitions to $|+\rangle$ state. We evaluated the reaction rates from the changes in the lifetime.

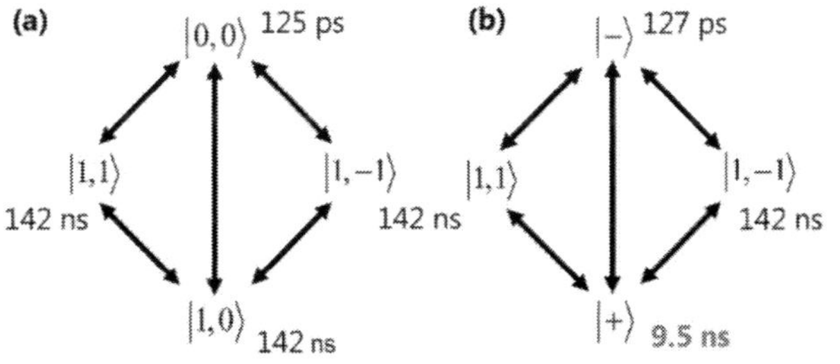

Figure 1. Ground states of Ps and the corresponding lifetimes in vacuum under a magnetic field of (a) 0.00 and (b) 0.82 T. The arrows between the states indicate possible spin transitions.

2. Temperature dependence of spin conversion annihilation

2.1. Experimental

We have two unknown constants: the spin-conversion annihilation rate and the pick-off annihilation rate. To evaluate them separately, we measured the long lifetime components under two magnetic conditions: 0.00 and 0.82 T. The temperature was kept at 300, 423, 540, 588, and 623 K. In addition, the lifetime without Xe gas was also measured to evaluate effect of silica-aerogel (SAG; 0.1 g/cc) used as a Ps creation medium by which positron-source of ^{22}Na (2 μCi) was sandwiched. The SAG decelerated the positron, supplied an electron to the positron, annihilated the free positrons, and took place Ps–Xe interactions in the inter-grain spaces. The measurement system was a digital-oscilloscope-based positron lifetime spectrometer described in our previous reports [3].

2.2. Results and discussion

As shown in figure 2, for example, the measured lifetime spectra exhibited a prompt peak followed by a long lifetime tail. The peak was due to self-annihilation of the singlet Ps and direct annihilations of positrons at the source, the SAG, chamber walls, etc. The tail was due to the triplet Ps and the lifetime was 128.7(3) ns, which was shorter than that in vacuum, i.e., 142.1 ns, because of the pickoff annihilation with the SAG. Next, Xe gas was introduced at the pressure of 250 kPa and the lifetime was reduced to 85.8(3) ns. This reduction was because of both the spin-conversion annihilation and the pick-off annihilation due to the Xe gas. To distinguish them, the third measurement was done under a magnetic field of 0.82 T, and the lifetime was further reduced to 70.4(5) ns. This further reduction was solely because of the spin-conversion from $|1,\pm1\rangle$ state to $|+\rangle$ state.

The followings are simultaneous rate equations for number of Ps in the four ground states, $|+\rangle$, $|-\rangle$, $|1,1\rangle$ and $|1,-1\rangle$:

$$\frac{d}{dt}\begin{pmatrix} N_1 \\ N_2 \\ N_3 \\ N_4 \end{pmatrix} = \begin{pmatrix} -\lambda_+ - k(\xi+2) - \lambda_{po} & k\xi & k\eta & k\zeta \\ k\xi & -\lambda_- - k(\xi+2) - \lambda_{po} & k\zeta & k\eta \\ k\eta & k\zeta & -\lambda_o - 2k - \lambda_{po} & 0 \\ k\zeta & k\eta & 0 & -\lambda_o - 2k - \lambda_{po} \end{pmatrix} \begin{pmatrix} N_1 \\ N_2 \\ N_3 \\ N_4 \end{pmatrix}, \quad (1)$$

where the off-diagonal elements correspond to the spin conversion with the reaction rate of k and magnetic field dependent constants of $\xi = 0.952$, $\eta = 0.780$ and $\zeta = 1.22$ at 0.82 T ($\xi = \eta = \xi = 1.00$ at 0.00 T) [1,2], and the diagonal elements also have the terms of the self-annihilation with the rates of

Journal of Physics: Conference Series **618** (2015) 012005 doi:10.1088/1742-6596/618/1/012005

$\lambda_+ = 1.06\times10^8$ s^{-1}, $\lambda_- = 7.94\times10^9$ s^{-1} and $\lambda_o = 7.04\times10^6$ s^{-1} for $|+\rangle$, $|-\rangle$, $|1,\pm1\rangle$ states, respectively, as well as the pickoff annihilation with the rate of λ_{po} for all the states.

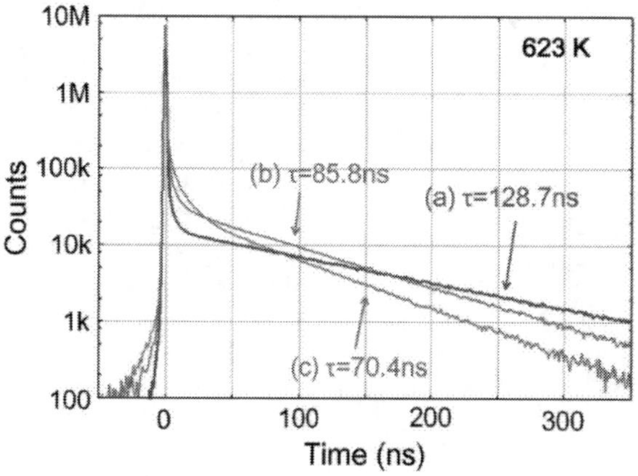

Figure 2. Ps lifetime spectra at 623 K. The measurement conditions of the Xe gas pressure and the magnetic field strength are (a) 0 kPa and 0 T, (b) 250 kPa and 0 T, and (c) 250 kPa and 0.82 T.

The lifetime of the long lifetime component corresponds to the minimum eigenvalue of the four times four matrix in equation (1) that includes two unknown constants of k and λ_{po}. Both can be determined by a maximum likelihood method using two measured lifetimes at 0.00 and 0.82 T. As the results, we obtained the normalized spin-conversion annihilation rate ${}_1Z_{eff}^{sc} = k/4\pi r_0^2 cn = 3.11(14)$ and the normalized pick-off annihilation rate ${}_1Z_{eff}^{po} = \lambda_{po}/4\pi r_0^2 cn = 1.36(17)$ at 623 K, where r_0 is the classical electron radius, c is the velocity of light, and n is the number density of the Xe gas atoms. Similarly we determined the annihilation rates at other four temperatures as shown in figure 3. ${}_1Z_{eff}^{po}$ increased nearly proportional to the temperature (T). On the other hand, ${}_1Z_{eff}^{sc}$ increased proportional to $T^{2.1}$. No other gases have exhibited such a strong temperature dependence [4]. This discrepancy can be understood by considering a nature of *p*-wave scattering whose cross section is proportional to T^2 at low energy because *s*-wave scattering is forbidden for the spin-conversion reaction [5].

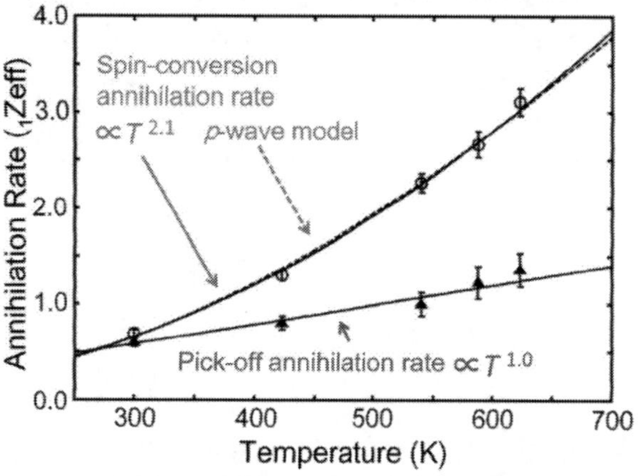

Figure 3. Temperature dependence of the spin-conversion annihilation rate (circles) and the pick-off annihilation rate (triangles) during Ps–Xe collisions. The solid lines are fitting results against power functions of temperature ($300 \le T \le 623$), and the dashed line is calculation result assuming the *p*-wave scattering and the Maxwell-Boltzmann distribution.

Assuming that the spin-conversion occurs only in *p*-wave scattering and that thermalized Ps atoms follows a Maxwell-Boltzmann distribution, we successfully explained the temperature dependence of

${}_1Z_{eff}^{sc}$ as shown in figure 3 with a dashed line. The *p*-wave scattering phase shift (δ_p) is described as $k^3 \cot\delta_p(k) = -1/a + rk^2/2$, where k is the wave number, $a = 6.8$ is the scattering length and $r = -4.1$ is the effective range [5]; then, the *p*-wave scattering cross section (σ_p) is described as $\sigma_p(k) = 12\pi \sin^2\delta_p / k^2$, and the spin conversion annihilation rate (λ_{sc}) is $\lambda_{sc} = nf\int\rho(v)\sigma_p(v)v\,dv / \int\rho(v)dv$, where $f = 5.3\times10^{-4}$ is the spin conversion annihilation rate per collision and $\rho(v)$ is a Maxwell–Boltzmann distribution at T. This model is in a good agreement with the experimental results and shown in figure 3 as *p*-wave model with the dashed line.

3. Application of the spin-conversion reaction for measuring Ps energy

Because of the strong temperature dependence of ${}_1Z_{eff}^{sc}$, we can estimate Ps energy (E) by measuring the two-photon annihilation rate at the time of t ($\lambda_2(t)$). Thermalization of Ps takes a considerably long time because the mass is very small, and it is a major source of uncertainty in precise measurement of Ps hyperfine splitting as a test of quantum electrodynamics [6]. As theoretical calculation is very difficult even for a simple system such as Ps–He scattering, high accuracy data is required [7,8].

However, conventional methods, namely, angular correlation of positron annihilation radiation (ACAR) and Doppler broadening (DB) do not have enough resolution especially at a very low energy near thermalization [8-10]. DB measures energy deviation from 511 keV, and ACAR measures angular deviation from 180 degrees. Both deviations are proportional to $\sqrt{E}$. On the other hand, ${}_1Z_{eff}^{sc}$ is proportional to $T^{2.1} \propto E^{2.1}$ as described in section 2. Then, we can expect the resolution about 80 times higher than those of the conventional methods. In addition, the two conventional methods are sensitive only to the two-photon annihilation events emitting back-to-back ca. 511 keV gamma-rays so that one needs a magnetic field for Zeeman-mixing under which only Ps atoms in $|+\rangle$ state provide effective information. On the other hand, our new method employs both two- and three-photon annihilation events of all the triplet Ps atoms as described later so that we can expect the sensitivity more than six times larger than those of the conventional methods.

The measurement system is similar to a digital age-momentum correlation positron annihilation spectrometer [11] but we use two kinds of digitizers; the triple coincidence events were recorded by a digital oscilloscope (8 bit and 5 GHz sampling) for timing information from two BaF_2 scintillation detectors and by a digitizer (14 bit and 100 MHz sampling) for energy information from a high-purity Ge detector. All waveforms were analyzed after the measurement by a PC.

3.1. Analysis and discussion

Figure 4 (a) is the energy spectrum having a full-energy peak at 511 keV and the Compton edge at about 340 keV. Using this energy information, we make two kinds of lifetime spectrum; One is two-gamma lifetime spectrum ($I_2(t)$) belonging to the full-energy peak (509–515 keV), and the other is three-gamma lifetime spectrum ($I_3(t)$) belonging to the valley between the full-energy peak and the Compton edge (410–473 keV) where two-gamma events are considered less detected as shown in figure 5 (b, c), respectively.

Next, $I_2(t)$ is divided by $I_3(t)$ to obtain a decreasing function that is proportional to $\lambda_2(t)$ as

$$\frac{I_2(t)}{I_3(t)} = \frac{\varepsilon_2 N(t)\lambda_2(t)}{\varepsilon_3 N(t)\lambda_3} = \frac{\varepsilon_2}{\varepsilon_3\lambda_3}\lambda_2(t), \tag{2}$$

where ε_2 (ε_3) is two- (three-) gamma event detection efficiency, $\lambda_3 = 7.04\times10^6\ \text{s}^{-1}$ is three-gamma annihilation rate constant, and $N(t)$ is the total number of the triplet Ps. Because lifetimes of the tails of $I_2(t)$ and $I_3(t)$ are the same value of 101.4 ns, $[I_2(t)/I_3(t)]\lambda_3$ converges at $1.15\times10^7\ \mathrm{s}^{-1}$ and $\lambda_2(t)$ converges at $2.85\times10^6\ \mathrm{s}^{-1}$. Therefore, $\varepsilon_2/\varepsilon_3 = 4.04$ because equation (2) should be kept at any time. As all the constants are fixed, we obtain the absolute value of the time evolution of the two-gamma annihilation rate as $\lambda_2(t) = 1.74\times10^6 \times [I_2(t)/I_3(t)]$, which can be converted into time evolution of Ps average kinetic energy as shown in figure 5 (the outer y-axis), because the relationship between $\lambda_2(t)$ and T has been obtained as described in section 2.

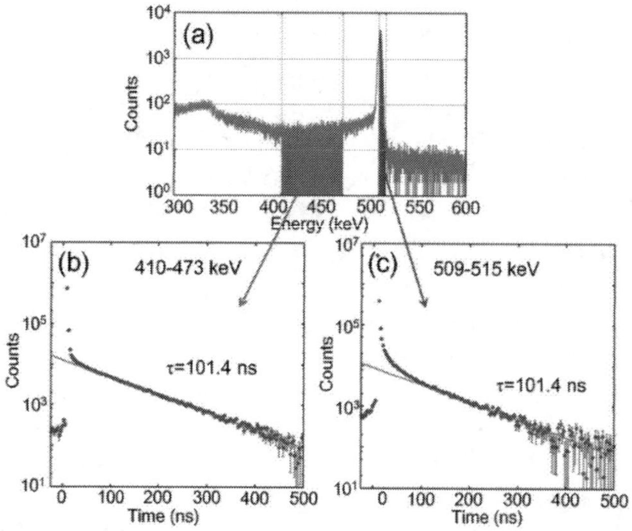

Figure 4. (a) Energy spectrum measured using a high-purity Ge detector, (b) three-gamma lifetime spectrum, and (c) two-gamma lifetime spectrum.

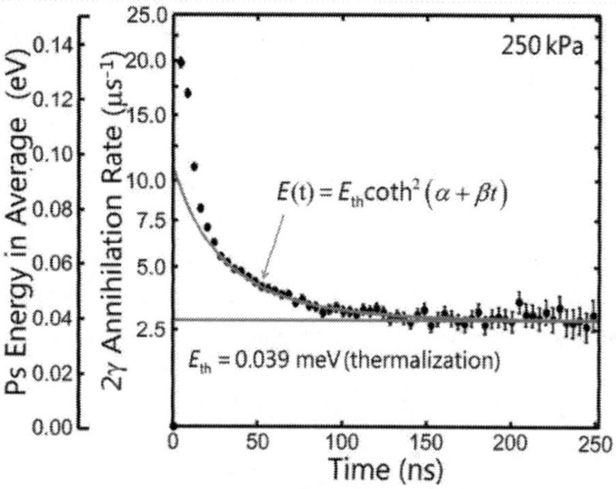

Figure 5. Time evolutions of the two-gamma annihilation rate (the inner y-axis) and the kinetic energy (the outer y-axis). The solid lines are the fitting result and the thermal energy.

The time evolution of Ps kinetic energy ($E(t)$) is described as a hyperbolic faction assuming classical elastic collisions: $E(t) = E_{\rm th}\coth^2(\alpha + \beta t)$, where $E_{\rm th}$ is the thermal energy. α is a parameter corresponds to the vertical range and β is one corresponds to the decay time [10]. β can be divided into two components corresponding to the Xe gas and the SAG as follows:

$$\beta(n) = \beta_{\rm Xe}(n) + \beta_{\rm SAG} = \sigma_{\rm m} n v_{\rm th}\frac{m_{\rm Ps}}{M_{\rm Xe}} + \frac{v_{\rm th}}{L}\frac{m_{\rm Ps}}{M_{\rm SAG}} \quad , \tag{3}$$

where $\sigma_{\rm m}$ is the momentum transfer cross section during Ps-Xe collision, $v_{\rm th}(T)$ is Ps velocity at thermal energy (8.26×10^4 m/s at 300 K), $m_{\rm Ps} = 1.82\times10^{-30}$ kg is the mass of a Ps atom, $M_{\rm Xe} = 2.18\times10^{-25}$ kg is the mass of a Xe atom, $L = 7.0\times10^{-8}$ m is the mean distance between the SAG grains [10], and $M_{\rm SAG}$ is the effective mass of SAG atoms at the collisions. $\sigma_{\rm m}$ and $M_{\rm SAG}$ are unknown constants to be determined.

Least squares fittings for $E(t)$ at two gas densities yield $\beta^{-1} = 89(2)$ ns for $n = 6.06\times10^{25}$ m^{-3} (250 kPa at 300K) and $\beta^{-1} = 121(6)$ ns for $n = 2.42\times10^{25}$ m^{-3} (100 kPa at 300K). Therefore, we obtain $\sigma_{\rm m} = 12(2)\times10^{-16}$ cm^2 and $M_{\rm SAG} = 110(20)$ u in the energy range between 40–60 meV. This $\sigma_{\rm m}$ is not far from the cross section of $\sigma_{\rm w} = \pi r_{\rm w}^2 = 14.7\times10^{-16}$ cm^2, where $r_{\rm w}$ is van der Waals radius.

4. Conclusion

The spin conversion annihilation rate during Ps-Xe collisions exhibits a strong temperature dependence, which can be well explained by a nature of *p*-wave scattering because the reaction is forbidden in *s*-wave scattering. By utilizing this temperature dependence as a method for measuring Ps kinetic energy, we successfully obtained its time evolution and the momentum cross section for Ps-Xe collisions at a very low energy about 40-60 meV with a high sensitivity and a high resolution. The spin conversion reaction would open a new way to study low-energy atomic/molecular collisions [12,13].

References

[1] Saito H and Hyodo T 2006 *Phys. Rev. Lett.* **97** 253402
[2] Nagai Y, Nagashima Y and Hyodo T 1999 *Phys. Rev.* B **60** 7677
[3] Saito H, Nagashima Y, Kurihara T and Hyodo T 2002 *Nucl. Instrum. Methods* A **487** 612
[4] Vallery R S, Leanhardt A E, Skalsey M and Gidley D W 2000 *J. Phys.* B **33** 1047
[5] Mitroy J and Novikov S A 2003 *Phys. Rev. Lett.* **90** 183202
[6] Ishida A, Namba T, Asai S, Kobayashi T, Saito H, Yoshida M, Tanaka K and Yamamoto A 2014 *Phys. Lett.* B **734** 338
[7] Adhilari S K, Biswas P K and Sultanov R A 1999 *Phys. Rev.* A **59** 4829
[8] Engbrecht J J, Erickson M J, Johnson C P, Kolan A J, Legard A E, Lund S P, Nyflot M J and Paulsen J D 2008 *Phys. Rev.* A **77** 012711
[9] Change T, Xu M and Zeng X 1987 *Phys. Lett.* A **126** 189
[10] Nagashima Y, Kamimoto M, Hyodo T, Fujiwara K, Ishikawa A, Chang T, Deng J, Akahane T, Chiba T, Suzuki K, McKee B T A and Stewart A T 1995 *Phys. Rev.* A **52** 258
[11] Inoue K, Nagai Y, Tang Z, Toyama T, Hosoda Y, Tsuto A and Hasegawa M 2011 *Phys. Rev.* B **83** 115459
[12] Shibuya K, Nakayama T, Saito H and Hyodo T 2013 *Phys. Rev.* A **88** 012511
[13] Shibuya K, Kawamura Y and Saito H 2013 *Phys. Rev.* A **88** 042517

Journal of Physics: Conference Series **618** (2015) 012006 doi:10.1088/1742-6596/618/1/012006

Evidence for a positron bound state on the surface of a topological insulator

K Shastry[1], A H Weiss[1,*], B Barbiellini[2], B A Assaf[2], Z H Lim[1], P V Joglekar[1], D Heiman[2]
[1] Physics Department, University of Texas at Arlington, Arlington, TX 76019-0059 USA.
[2] Physics Department, Northeastern University, Boston, MA 02115 USA.
*weiss@uta.edu

Abstract. We describe experiments aimed at probing the sticking of positrons to the surfaces of topological insulators using the Positron Annihilation induced Auger Electron Spectrometer (PAES). A magnetically guided beam was used to deposit positrons at the surface of Bi_2Te_2Se sample at energy of ~2eV. Peaks observed in the energy spectra and intensities of electrons emitted as a result of positron annihilation showed peaks at energies corresponding to Auger peaks in Bi, Te and Se providing clear evidence of Auger emission associated with the annihilation of positrons in a surface bound state. Theoretical estimates of the binding energy of this state are compared with estimates obtained by measuring the incident beam energy threshold for secondary electron emission and the temperature dependence positronium (Ps) emission. The experiments provide strong evidence for the existence of a positron bound state at the surface of Bi_2Te_2Se and indicate the practicality of using positron annihilation to selectively probe the critically important top most layer of topological insulator system.

1. Introduction

Positron annihilation spectroscopies have been shown to have a high degree of surface selectivity in systems in which positrons can become trapped in an image potential surface state at the time of annihilation [1]. Positrons in such states annihilate primarily with atoms in the topmost atomic layers and annihilation related signals convey information regarding the elemental composition and electronic structure of these layers [2, 3]. Topological Insulators (TI) [4] are a novel class of materials that behave as conductors on the surface and insulators in the bulk. In these materials the electronic and spintronic states of interest are confined to the topmost layer of the surface [5], [6].

In this paper we present the results of experiments that provide strong evidence that positrons can be efficiently trapped in surface localized states on TI surfaces. These experiments show that it is likely that positron spectroscopic techniques such as Two Dimensional – Angular Correlated Annihilation Radiation (2D- ACAR), Doppler broadening [7] and Spin-polarized positron spectroscopies [8] can be successful used to selectively probe the momentum distribution and spin properties of the topologically constrained electron surface states on TI materials.

Shastry et al. [9] have reported preliminary results showing positron annihilation induced Auger peaks from a Bi_2Te_2Se sample which provided strong evidence for the localization of the positron in a surface state at the time of annihilation. Subsequently, theoretical calculations by R.Sainz and B. Barbiellini [10] based on the model in which the positron is assumed to be bound to the surface in the

Journal of Physics: Conference Series **618** (2015) 012006 doi:10.1088/1742-6596/618/1/012006

form of a physisorbed Ps atom [11] indicated that the positron would be bound to the surface of Bi_2Te_2Se. They calculated the binding energy of the Ps to be between - .12 eV and -.4 eV.

Here we provide details of the PAES measurements on Bi_2Te_2Se along with the results of two additional experiments, Auger mediated sticking and Ps thermal desorption to confirm the existence of the surface state and help provide limits on the binding energy.

2. Experimental Details

The Bi_2Te_2Se topological insulator sample used in our experiments was grown on a Si (111) substrate using Molecular Beam Epitaxy (MBE) [6]. The positron measurements were carried out using a magnetic bottle analyzer type Time-Of-Flight Positron Annihilation induced Auger Electron Spectrometer (TOF-PAES) at University of Texas at Arlington [12],[13][9]. The positrons are generated via β decay from a ^{22}Na radioactive source and are moderated using a polycrystalline tungsten (W) moderator. A combination of electric and axial magnetic field is used to guide the slow positron beam to the sample. A magnetic field of 4mT was used to efficiently transport the low energy positrons to the sample and the resulting electrons from the sample. The beam energy of the incident positrons is given by

$$E_p = e(V_m - V_s) + \varphi_m^+ \qquad (1)$$

φ_m^+ is the positron work-function of the moderator (W). The values V_m and V_s represent the biases on the moderator and sample respectively, e is the magnitude of the electronic charge. The beam energy for PAES and Ps fraction measurement were 9 eV and 70 eV respectively. The beam energies for the AMS measurement ranged from 2.7 eV to 6.2 eV.

The PAES and AMS measurements were done in the reverse timing scheme where the signal from the Micro Channel Plate is used as the START signal and the delayed BaF_2 signal is the STOP signal to a Time to Amplitude Converter (TAC). The resulting spectrum is a histogram of the time of flight of electrons from the sample to the MCP.A calibration procedure detailed in (Ref .13) is used to reference the energy of the electron with respect to the vacuum band. For the Ps fraction, the gamma spectrum from the NaI detector is passed to a timing amplifier and then to a multi-channel analyzer. A heater button behind the sample is used to raise the temperature of the sample in-situ.

3. Results and Discussion

3.1. Positron Annihilation Induced Auger Electron Spectrum (PAES) Measurement:

The TOF PAES of the Bi_2Te_2Se sample are shown in Fig. 2. A corresponding energy spectra (shown in Fig. 3) was obtained using a transformation [9]. The energy spectrum shows the Auger peaks at 240 eV and 100 eV due to the Bismuth N-VV transition [14], at 30 eV due to the Tellurium N-VV transition, at 42 eV due to the Selenium M-VV transition [15] and a Carbon Auger S-VV peak at 260 eV.

The observation of annihilation induced Auger peaks is a strong indication for the existence of a positron surface state on the surface of Bi_2Te_2Se. It should be noted that the beam energy used to obtain the Auger signal is very low to cause Auger transitions due to impact ionizations. In addition, if the positrons were in a bulk state at the time of annihilation only a tiny fraction of the positrons would annihilate close enough to the surface for an electron to escape without losing energy. This distance is in the order of 10Å for a 60 eV electrons. The Bi_2Te_2Se sample is 40 nm thick; if the Auger signal is from below 10Å the Auger signal intensity would be 100 times weaker than that from the surface. It is therefore clear that for a significant Auger signal to be observed for a given material for an incident beam energy below the core ionization energy, the positrons must be trapped in an image correlation well at the surface at the time of annihilation. Hence the observation of a positron annihilation induced Auger signal for a 9eV positron beam provides strong evidence for the existence of a surface localized positron bound state. A further implication of the observation of the PAES signal from the TI surface is that positron spectroscopies such as 2D-ACAR and Doppler Broadened Annihilation Gamma Spectroscopy can be used to sample the characteristics of the top atomic layers of these materials.

3.2. Auger Mediated Sticking (AMS) Measurement:

We report here the results of a series of measurement aimed at estimating the surface state binding energy by determining the cut-off energy for positron induced secondary electron emission from the Bi_2Te_2Se sample. Previous experiments by Mukherjee et al. [16] have demonstrated that measurements of the secondary electron yield as a function of incident positron beam energy yield estimates of the surface state binding energy in agreement (within the errors) with previous estimates of the surface state binding energy obtained from Ps thermal desorption [20]. Following reference [16], the maximum energy that a secondary electron can escape from the sample for a process in which a positron makes a direct transition from a positive energy scattering state to a surface bound state may be found using:

$$E_{Kmax} = E_p + E_{bss} - \varphi^- \qquad (2)$$

where E_{Kmax} is the maximum energy a secondary electron can escape from the surface, E_p is the kinetic energy with which the positrons hit the sample, E_{bss} is the surface state binding energy, φ^- is the work-function. Figure 4 is the secondary electron spectrum at different beam energies. It may be observed in figure 3 that the area under the secondary electron peak is going to zero as the beam energy decreases.

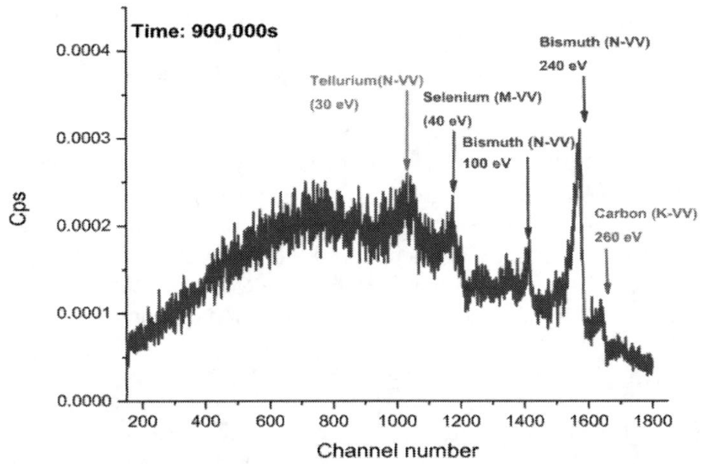

Figure 1. (Color Online)The TOF Auger spectrum for Bi_2Te_2Se sample after sputtering. The Auger transitions due to C (K VV) at 257 eV, Bi (N VV) at 240 eV and (N-VV) at 100 eV, Te (N-VV) at 30 eV, Se (M-VV) at 40 eV are shown.

The emission of secondary electrons is energetically precluded when E_{Kmax}=0. Inserting this value into equation (2) we find;

$$E_{Th} = \varphi^- - E_{bss} \qquad (3)$$

where E_{Th} is the threshold kinetic energy above which it is energetically possible for the positrons to knock out secondary electrons. The data points plotted in Figure 4 show the values of the integrated intensities under the measured positron induced secondary electron energy distributions (shown in figure 3) as a function of incident positron kinetic energy. A straight line fit to the data shown in figure 4 was used to estimate the value, E_{Th}, at which the integrated secondary electron yield would go to zero. Using this method we obtained a value of E_{Th} = 1.8 eV for the Bi_2Te_2Se sample measured. The surface state binding energy was then calculated from the resulting value of E_{Th} using equation (3). Reported values of the electron work-function values for Bi_2T_2Se ranged from 4.32 eV [17] to 5.23 eV [18] were taken from literature. The surface state binding energy was calculated to be between 2.5 eV to 3.4 eV respectively. The initial theoretical calculations based on first principles, for the surface state binding energy varied from 1.61 eV to 2.61 eV.

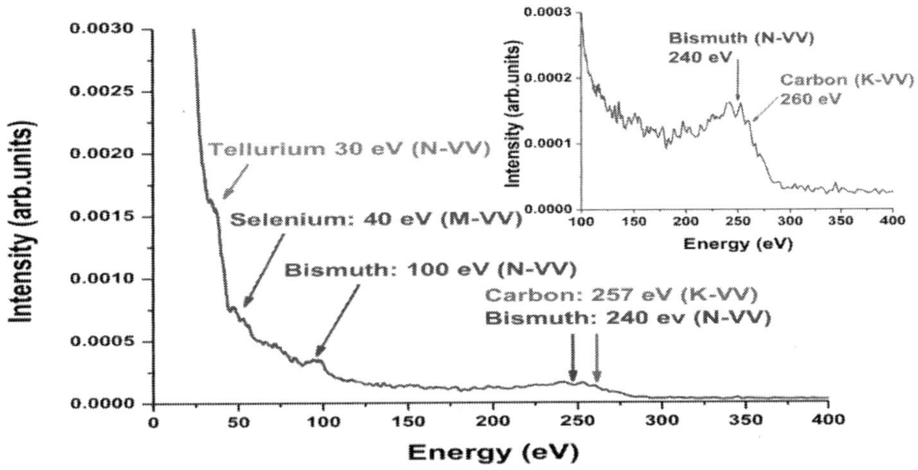

Figure 2. (Color Online) The energy spectrum for Bi_2Te_2Se sample after sputtering. The peaks due to the Auger transitions C (K VV) at 257 eV, Bi (N VV) at 240 eV (N-VV) at 100 eV, Te (N-VV) at 30 eV, Se (M-VV) at 40 eV are indicated by arrows. Inset: magnified region between 200 and 400 eV.

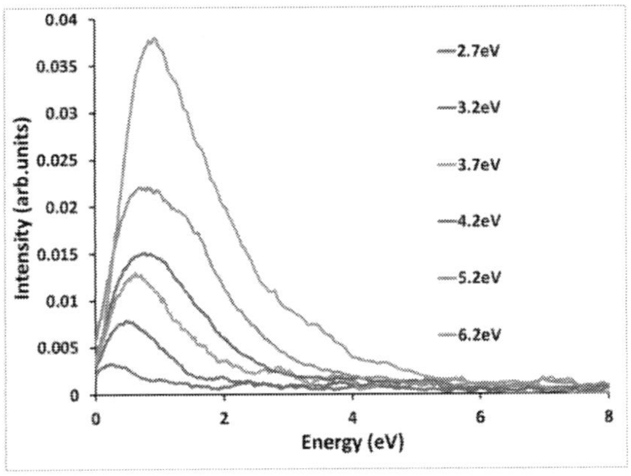

Figure 3.(Color Online) The energy distribution of secondary electrons obtained at a series of incident positron kinetic energies ranging from 2.7eV to 6.2 eV.

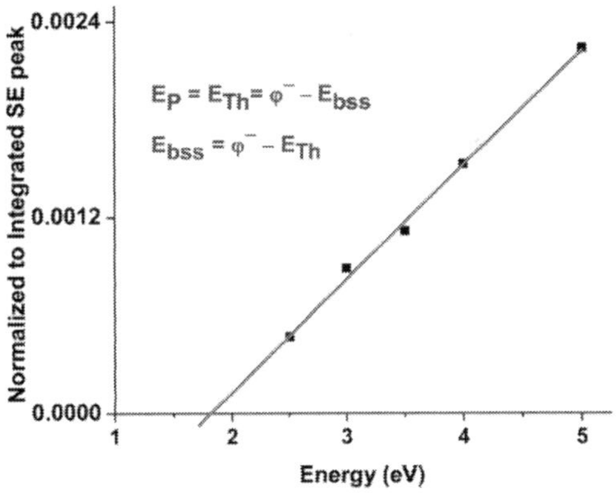

Figure 4. (Color Online) The data points represent the integrated intensity under the secondary electron energy distributions show in figure 3. The line is a fit to the data.

The AMS measurement on Bi_2Te_2Se sample constitute a second piece of evidence for the existence of a positron surface state. If the AMS signal came from the bulk and not from the surface, the maximum energy of the secondary electron would be given by

$$E_{Kmax} = E_p - \varphi^- + \varphi^+ \quad (4)$$

where E_P is the kinetic energy of the positrons incident on the surface, E_{Kmax} is the maximum Kinetic energy of the outgoing electron measured outside the sample surface, φ^- *and* φ^+ are the electron and positron work function respectively. For secondary electrons to be ejected out of the sample (E_k>0eV), E_p should be greater than the difference between the electron and positron work functions. Typically for materials the value of φ^- is ~ 5eV while φ^+ is <1 eV. Therefore, from equation (4) the minimum incident beam energy for secondary electron emission should be > 4 eV, but in our experiment the beam energy was 2 eV. The lower value of E_{Th} observed in our experiment is consistent with the existence of a surface state whose binding energy is of order 2 eV below the bulk state.

3.3. Positronium (Ps) Fraction Measurement:

In the third experiment an estimate of the energy, E_a, required to remove the positron from a surface bound state was estimated from a measurement of the temperature dependence of the fraction of incident positron emitted as positronium, f_{Ps}, from the Bi_2Te_2Se surface using the method of Mills [19, 20]. The positron fraction, f_{Ps}, was determined from the annihilation gamma spectra measured using a NaI (Tl) detector [19] using a calibration procedure detailed in reference 9. The Ps fraction obtained from the Bi_2Te_2Se at room temperature was f_{Ps} = 0.11 which is significantly lower than the value of f_{Ps} obtained from metal surface (e.g. f_{Ps} = 0.51 for Cu (100)[20]). The data points shown indicate the measured values of f_{Ps} for 5 different temperatures. The solid lines are functions obtained using the thermal activation model described in reference 20 for three values of the term E_a, representing the activation barrier for thermal emission. The value of the activation barrier, E_a, can be estimated from the surface state binding energy [20] using equation 5

$$E_a = E_{bss} + \varphi^- - 6.8\,eV \qquad (5)$$

E_{bss} is the surface state binding energy, φ^- is the electron work-function, 6.8 eV is the binding energy of ground state Ps. Based on published values of φ^- and the corresponding values of E_{bss} found from the AMS measurements discussed above, the values of E_a determined from equation 5 range from 0 eV and 1.8 eV. It may be seen in figure 5. That the data are not consistent with models in which E_a = of 0 eV corresponding to spontaneous Ps emission, nor E_a = 1.8eV.

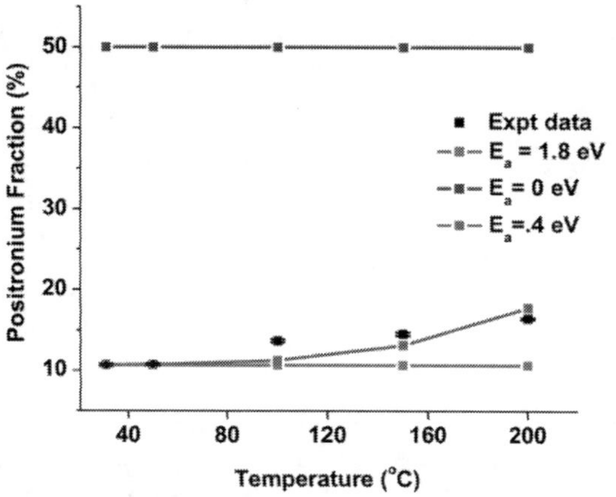

Figure 5. (Color Online)The data points correspond to the measured values of the Ps fraction, f_{Ps}, for 5 different temperatures. The curves represents values obtained using the model given in Ref. 20 for 3 different activation energies.

The curve corresponding to an activation energy of E_a = 0.4eV is consistent with our f_{Ps} vs temperature data. This value of E_a is also consistent with our AMS measurements assuming φ^- = 4.7 eV. Based on the Ps fraction dependence on temperature of Bi_2Te_2Se, the value of E_a that is most consistent with our data is 0.4 eV.

4. Conclusion

We have presented the results of three different measurements which provide strong evidence for a positron bound state on the surface of the topological insulator. The observation of positron annihilation induced Auger electron peaks from the Bi_2Te_2Se surface provide strong evidence that the positrons are localized at the surface at the time of annihilation. Measurements of the dependence of the positron induced secondary electron yield as a function of incident beam energy and the temperature dependence of the fraction of incident positrons forming Ps provided complementary evidence for the existence of the surface state. Latter experiments allowed us to estimate values for the activation energy for Ps thermal emission, $E_a = 0.4$ eV, and a positron surface state binding energy $E_{bss} = 2.7$ eV. These values obtained are within the range of preliminary theoretical calculations [10]. Though the effects observed in this paper are not topological in nature, the results indicate that low energy positrons can be trapped with high efficiency into a surface localized bound state. Consequently, the topological effects could be probed with positrons employing other positron techniques such as polarized positron beam, 2D- ACAR and Doppler broadening spectrocopies, which are of most interest for device applications.

We acknowledge Dr.Moodera for his help in growing the Bi_2Te_2Se samples. We also acknowledge Dr.Sainz and Dr.Barbiellini for their useful discussions. We also acknowledge funding from Welch Grant 1100 and NSF DMR 0907679.

References

[1] Weiss, A H and Coleman, P G Surface Science with Positrons, in Positron Beams and Their Applications, World Scientific Publishing, pp 129-189 (2000).
[2] Weiss, A H et al., Phys.Rev.Lett. 61, 2245 (1988).
[3] Asoka-Kumar, P et al., Phys.Rev.Lett. 77, 2097 (1996).
[4] Kane, C L Mele, E J Phys. Rev. Lett 95, 1456802 (2005).
[5] Moore, J E Nature 464, 194 (2010).
[6] Assaf, B A Ph.D thesis, Northeastern University, (2014).
[7] Coleman, P G Positron beams and applications, World Scientific Publishing, pp 129-189 (2000).
[8] Gidley, D W, Koymen A R et.al Phys.Rev.Lett. 49,24(1982).
[9] Shastry, K et al. Bulletin of the American Physical Society Volume 58, Number 1 M13.00004 (2013) and Shastry, K Ph.D dissertation, December (2014).
[10] Saniz, R et al., Journal of Physics: Conference Series 505 (2014) 012002.
[11] Platzman, P M et al., Phys. Rev. B 33 5900 (1986).
[12] Xie, S Ph.D. thesis, UT Arlington, (2002).
[13] Mukherjee, S F et.al, Rev.Sci.Instrum (under preperation).
[14] Roy Sumalay et al., Phys.Rev.B 90, 155456 (2014).
[15] Zhu, J G et al.,Journal of Applied Physics ,Volume:97 Issue:10.
[16] Mukherjee, S et al., Phys.Rev. Lett. 104, 247403 (2010).
[17] Hao, G Journal of Applied Physics 113, 024306 (2013).
[18] Niesner, D et al.,Journal of Electron Spectroscopy and Related Phenomena 195 (2014).
[19] Mills Jr, A P et al., Solid State Comm, Vol 31, (623-626).
[20] Mills Jr, A P et al., Phys.Rev.Lett, Vol 41, Number 26.

Journal of Physics: Conference Series **618** (2015) 012007 doi:10.1088/1742-6596/618/1/012007

First test of O(α) correction on the energy spectrum in the orthopositronium decay

S Adachi[1], T Yamaji[1], A Ishida[1], T Namba[2], S Asai[1] and T Kobayashi[2]

[1] Department of Physics, Graduate School of Science, The University of Tokyo, 7-3-1 Hongo, Bunkyo, Tokyo 113-0033, Japan

[2] International Center for Elementary Particle Physics, The University of Tokyo, 7-3-1 Hongo, Bunkyo, Tokyo 113-0033, Japan

E-mail: adachi@icepp.s.u-tokyo.ac.jp

Abstract. Positronium (Ps) is an ideal system for the precision test of quantum electrodynamics (QED) in bound state. Orthopositronium (o-Ps) decays into three gamma rays dominantly, and their energy spectrum has detail information on bound-state QED. The purpose of this study is to check the spectrum at the O(α) level accuracy. We constructed a setup for the precise measurement. Ps is created at silica aerogel in a vacuum chamber, and its decay gamma rays are measured with a $LaBr_3$ (Ce) scintillator. In this setup, the materials around the Ps assembly and the detector are well-controlled, and all their responses are coded in a GEANT4 simulator in detail. After the measurement for 100 days, the obtained spectrum is compared with the expected spectrum from the QED calculations and the GEANT4 simulator. The obtained spectrum is consistent with O(α) calculation and excludes the lowest-order calculation at 92% C.L.

Our experimental setup is shown in Figure 1. An aerogel with density 0.13g/cm^3 is used as o-Ps source. A 0.2mm-thick plastic scintillator is placed between the aerogel and a ^{22}Na positron source with approximate strength of 200kBq. The aerogel, plastic scintillator and ^{22}Na are in a vacuum chamber, being evacuated down to lower than 10Pa. The chamber is a cylindrical 30mm-diameter vacuum container made of 1mm-thick aluminium, and it has reflector (125μm-thick polyvinylidene difluoride) inside to work as a light guide for scintillation light. The light from the plastic scintillator is corrected by the PMTs (Hamamatsu H7195) on both sides in order to produce the β^+ trigger. In this experiment, it is necessary to minimize the amount of the materials to reduce the influence of the scattering in the materials, because a measured spectrum is much distorted by the detector response and this influence. Therefore, the aerogel, plastic scintillator and ^{22}Na are fixed by 50μm-thick Mylar and supported by 0.5cm-thick Teflon in the chamber. In addition, the whole setup is lifted up approximately 1m above the floor to reduce the influence of the scattering on the floor. The energy of the gamma rays from the o-Ps decay is measured by a $LaBr_3$ (Ce) scintillator (Saint-Gobain B380). Its crystal size is of ϕ38.1mm $\times$ 50.8mm. This detector is strong for pile up events because it has quick rise time ($\sim$10ns) and fall time ($\sim$50ns). The charge of the signal from the $LaBr_3$ is measured by an ADC (Lecroy 2249W). The data acquisition trigger is the delayed coincidence of the 2μs-width β^+ trigger and $LaBr_3$ signal. The time between the β^+ trigger and $LaBr_3$ signal is obtained by a time-to-digital converter (TDC), which is direct 2GHz-clock-counting type. The total run period in vacuum is approximately 100days. The room temperature was maintained within ±0.5°C to ensure stability of the ADC and the TDC.

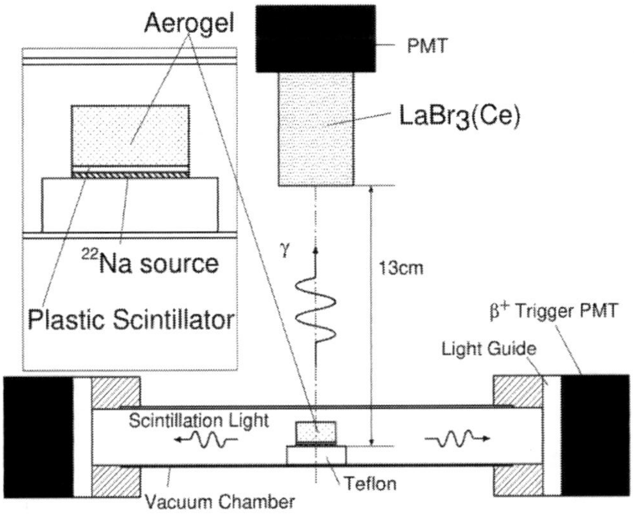

Figure 1. The top view of the experimental setup. Rectangular inset : A magnified view of the source region.

The decay time can be obtained from the time spectrum between the β^+ trigger and $LaBr_3$ signal. It is 129.5 ± 1.5ns. This is the measured decay time and not the o-Ps decay time (142.05 ns [1]) because of the background from the "pickoff" processes in which o-Ps collides with an atomic electron of a material or of a gas to cause an annihilation into 2γ. To estimate the pickoff contributions, we measured the energy spectrum in 1atm oxygen gas with the same setup. The total run period in the oxygen gas is approximately 30 days. In oxygen gas, the quenching rate of the process called "spin conversion" is high. Spin conversion is the conversion of o-Ps into para-positronium (p-Ps), instantly causing 2γ decay of the p-Ps. The measured decay time in the oxygen gas is 29 ± 2ns, which is shortened by the quenching effect of spin conversion. The energy spectrum in the oxygen gas has mainly 2γ-decay events. However, the 511keV peak in 2γ spectrum of spin conversion has narrower shape than that of the pickoff annihilation because the pickoff is influenced by the initial momenta of the colliding atomic electron. Thus, it is necessary to broaden the peak-width in the oxygen spectrum to obtain the correct 2γ spectrum of the pickoff annihilation. The correction amount of the peak-width can be obtained from the difference between the detector resolution and the measured peak-width of the pickoff annihilation. We used the average value ($\Delta\sigma = 1$ keV) in [2] and [3].

We used a GEANT4 Monte Carlo simulation (MC), in which three gamma rays from an o-Ps decay are simultaneously generated and they follow the energy distribution of the QED calculation. All interactions of a gamma ray with materials are taken into account. We generated the MC spectra from the lowest-order QED calculation plus the $O(\alpha)$ correction and from the lowest-order QED calculation, respectively (referred to as "the $O(\alpha)$ MC spectrum" and "the lowest-order MC spectrum"). In order to estimate the accuracy of these MCs, we measured the energy spectra of two monoenergetic gamma sources (^{51}Cr 320keV, ^{137}Cs 662keV) with the exactly same setup. From these measurements, we estimated that the systematic error on the MC spectra of o-Ps decay is less than 0.24% in the energy region between 400keV and 510keV.

By using the lowest-order MC spectrum and the $O(\alpha)$ MC spectrum, we defined "the mixing MC spectrum" with a parameter "F", which represents the mixing degree of the lowest-order

Journal of Physics: Conference Series **618** (2015) 012007 doi:10.1088/1742-6596/618/1/012007

MC spectrum and the O(α) MC spectrum. The count of the mixing MC spectrum at energy E, $N_{MC}(E;F)$ is defined by the equation,

$$N_{MC}(E;F) = S_{norm}\left\{(1-F)N_{L.O.}(E) + FN_{O(\alpha)}(E)\right\}, \tag{1}$$

where $N_{L.O.}(E)$ and $N_{O(\alpha)}(E)$ are the counts of the lowest-order MC spectrum and of the O(α) MC spectrum, respectively. S_{norm} is determined by the total number of the measured spectrum in the energy region of 400–530keV. This mixing MC spectrum represents the lowest-order MC spectrum at $F = 0$ and represents the O(α) MC spectrum at $F = 1$. We verified the O(α) correction by the fit of this mixing MC spectrum to the measured spectrum after the pickoff contributions subtracted. In this fit, we used the energy region of 400–530keV. That's because the O(α) correction is large and the influence of the scattering in the materials is small in this energy region.

The fit result is shown in the Figure 2. The measured spectrum is flat along one. In contrast, the lowest-order MC spectrum is upward to the right. The best-fit values of F is 0.90 ± 0.64 with $\chi^2/\mathrm{NDF} = 14.5/23$ [68.3% confidence level (C.L.) intervals], where NDF is the number of degrees of freedom. Moreover, $F = 0$ is excluded at 92% C.L. In conclusion, the measured spectrum is consistent with the lowest-order QED calculation plus O(α) correction and excludes the lowest-order MC spectrum at 92% C.L.

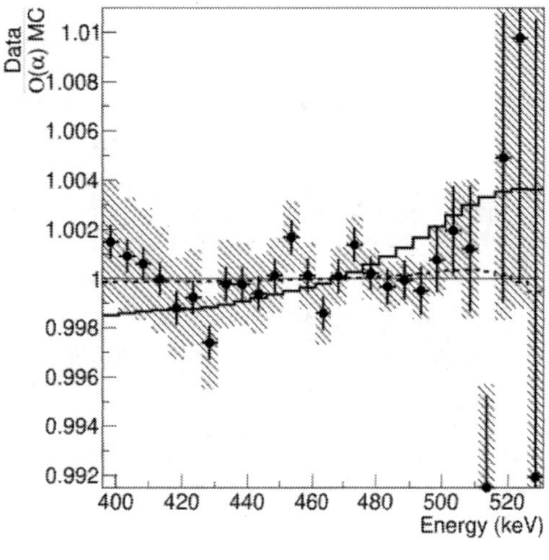

Figure 2. The χ^2-fitting of the o-Ps MC spectrum to the o-Ps measured spectrum. All spectra are divided by the O(α) MC spectrum, which results in that the O(α) MC spectrum is 1 in this graph. The closed circle is the measured spectrum divided by the O(α) MC spectrum and its error bar represents statistical error. The error band represents the full error which includes the systematic error on the measured spectrum and on the MC spectrum. The dashed line is the best fit MC spectrum divided by the O(α) MC spectrum. The solid line is the lowest-order MC spectrum divided by the O(α) MC spectrum.

References

[1] Kataoka Y, Asai S and Kobayashi T 2009 *Phys. Lett. B* **671** 219–223 ISSN 0370-2693

[2] Orito S, Yoshimura K, Haga T, Minowa M and Tsuchiaki M 1989 *Phys. Rev. Lett.* **63**(6) 597–600

[3] Asai S, Orito S, Sanuki T, Yasuda M and Yokoi T 1991 *Phys. Rev. Lett.* **66**(10) 1298–1301

Journal of Physics: Conference Series **618** (2015) 012008 doi:10.1088/1742-6596/618/1/012008

The collision between positronium (Ps) and muonium (Mu)

Hasi Ray[1,2,3] **and Rina De**[4]
[1]Study Center, S-1/407, B. P. Township, Kolkata 700094, India
[2]Department of Physics, New Alipore College, New Alipore, Kolkata 700053, India
[3]National Institute of T.T.T. and Research, Salt Lake City, Kolkata 700106, India
[4]Department of Physics, R.R.R. Mahavidyalaya, Radhanagar, Hoogly 712406, India

E-mail: hasi_ray@yahoo.com

Abstract. The collision between a positronium (Ps) and a muonium (Mu) is studied for the first time using the static-exchange model and considering the system as a four-center Coulomb problem in the center of mass frame. An exact analysis is made to find the s-wave elastic phase-shifts, the scattering-lengths for both singlet and triplet channels, the integrated/total elastic cross section and the quenching cross section due to ortho to para conversion of Ps and the conversion ratio.

1. Introduction

The importance of the studies on positron (e^+) and positronium (Ps) is well known [1,2]. The Ps is an exotic atom which is itself its antiatom and the lightest isotope of hydrogen (H); its binding energy is half and Bohr radius is twice that of H. The muonium (Mu) is a bound system of a positive muon (μ^+) and an electron (e^-) and also a hydrogen-like exotic atom. Its nuclear mass is one-ninth (1/9) the mass of a proton; the ionization potential and the Bohr radius are very close to hydrogen atom. Mu was discovered by Hughes [3] in 1960 through observation of its characteristic Larmor precession in a magnetic field. Since then research on the fundamental properties of Mu has been actively pursued [4-8], as has also the study of muonium collisions in gases, muonium chemistry and muonium in solids [9]. Generally chemical properties of an element are dependent on ionization potential and the Bohr radius. In Mu these are very close to H. However the vibrational frequencies involving Mu are higher than H.

We study the elastic collision between Ps and Mu when both are in ground states. The system is studied in center of mass frame using the static-exchange model (SEM) [10]. The theory is useful to include the non-adiabatic short-range effects due to exchange of electrons. Here the Schrodinger equation of motion is solved exactly considering the system as a four-center Coulomb problem [11] following a coupled-channel methodology and partial wave analysis. Both the direct and exchange amplitudes are calculated exactly. The H-like atomic wavefunctions are used to describe the atoms. There are total six Coulomb interaction terms in the system, from them two are used to describe the atoms. So effectively there are four Coulomb interaction terms between the two H-like atoms and there are total eight different terms in the amplitude to consider both the direct and exchange channels. In addition, there is one off-shell term for the rearrangement channel to include the effect of inelastic flux. We study the singlet (+) and the triplet (-) s-wave elastic phase-shifts (δ_0) and the corresponding scattering lengths utilizing the effective range theory.

Again the annihilation is an important phenomenon in Ps and gas atom collisions. The positrons emitted from a radioactive source into a gas at normal pressure do not suffer annihilation by collision with atomic electrons, before they have a high probability of capturing an electron from a gas atom to form Ps [12]. If the captured e^- has the opposite spin to the e^+ forming para-Ps, the lifetime towards mutual annihilation is only about 10^{-10} sec. But if the spins are parallel, forming ortho-Ps, the lifetime is much longer about 10^{-7} sec. The mutual annihilation of para-Ps is associated with emission of two γ-ray photons each having energy 511 meV. Annihilation may occur due to the exchange collision of Ps with gas atoms by which ortho-Ps transforms into para-Ps, the phenomenon is known as quenching. So it is of interests to study quenching in Ps and Mu collision when both are in ground states. We study the total elastic cross section (σ), the quenching cross section (σ_q) and the conversion ratio for Ps(1s)-Mu(1s) collision. An exact and ab-initio SEM code recently introduced by Ray [11,13] to study a four-body Coulomb problem in the center of mass frame is used.

2. Theory

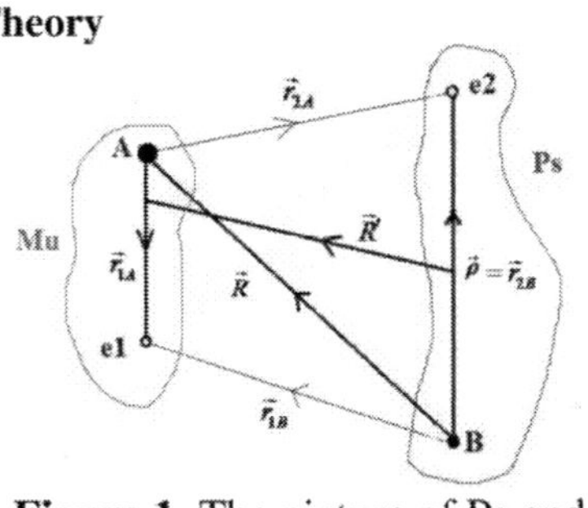

Figure 1. The picture of Ps and Mu system.

The description of the present Ps and Mu system is made in **Figure 1**. The μ^+ of the Mu and the e^+ of Ps are placed at points A and B respectively; e1 and e2 are the two electrons attached to the nuclei A and B which are at a distance r_{1A} and r_{2B} from A and B respectively. Accordingly $\vec{r}_{2A}$ and $\vec{r}_{1B}$ are defined. $\vec{R}$ is the vector joining A and B and $\vec{R}'$ is the vector joining the center of masses of the two atoms so that

$$\vec{R}' = \vec{R} + \frac{m_e}{m_A + m_e}\vec{r}_{1A} - \frac{m_e}{m_B + m_e}\vec{r}_{2B} \quad (2.1),$$

when m_A, m_B and m_e are the masses of μ^+, e^+ and e^- respectively. The initial and final state wavefunctions of the system are defined as

$$\psi_i = \exp(i\vec{k}_i \cdot \vec{R}')\phi_{1s}^A(\vec{r}_{1A})\eta_{1s}^B(\vec{r}_{2B}) \quad (2.2a),$$

$$\psi_f = (1 \pm P_{12})\exp(i\vec{k}_f \cdot \vec{R}_f)\phi_{1s}^A(\vec{r}_{1A})\eta_{1s}^B(\vec{r}_{2B}) \quad (2.2b),$$

so that $\vec{k}_i$ and $\vec{k}_f$ represent the initial and final momenta of the projectile, $\phi_{1s}^A(\vec{r}_{1A})$ and $\eta_{1s}^B(\vec{r}_{2B})$ represent the ground state wavefunctions of Mu and Ps respectively and P_{12} is the exchange or antisymmetry operator; $\vec{R}_f$ is defined as

Journal of Physics: Conference Series **618** (2015) 012008 doi:10.1088/1742-6596/618/1/012008

$$\vec{R}_f = \vec{R} + \frac{m_e}{m_A + m_e}(\vec{r}_{2B} - \vec{R}) - \frac{m_e}{m_B + m_e}(\vec{r}_{1A} + \vec{R}) \quad (3).$$

The Coulomb interaction between the atoms in the direct and exchange channels are defined as

$$V_{Direct}(\vec{R}, \vec{r}_{1A}, \vec{r}_{2B}) = \frac{\mu^+ e^+}{R} - \frac{\mu^+ e_2^-}{|\vec{R} - \vec{r}_{2B}|} - \frac{e^+ e_1^-}{|\vec{R} + \vec{r}_{1A}|} + \frac{e_1^- e_2^-}{|\vec{R} + \vec{r}_{1A} - \vec{r}_{2B}|} \quad (4a),$$

$$V_{Exchange}(\vec{R}, \vec{r}_{1A}, \vec{r}_{2B}) = \frac{\mu^+ e^+}{R} - \frac{\mu^+ e_1^-}{|\vec{r}_{1A}|} - \frac{e^+ e_2^-}{|\vec{r}_{2B}|} + \frac{e_1^- e_2^-}{|\vec{R} + \vec{r}_{1A} - \vec{r}_{2B}|} \quad (4b),$$

respectively. Here the magnitudes of all the Coulomb terms in the numerators in equation (4a) and (4b) are equal to unity in atomic unit (a.u.).

In elastic collision $|k_i| = |k_f|$, so only the direction of final momentum ($\vec{k}_f$) changes due to scattering or collision.

3. Results and Discussion

The present s-wave elastic phase shifts for the singlet and triplet spin-states of the two system electrons are presented in **figure** 2. The corresponding plot of the k^2 versus $k\cot(\delta_0)$ are presented in **figure** 3; following effective range theory one can easily calculate both the singlet and triplet scattering lengths from the corresponding cut-points of the extended curves with the y-axis [11,14] since it is just the inverse of the y-cut multiplied by a negative sign. In Ps-Mu system the singlet scattering length is 7.45 a.u. and the triplet scattering length is 2.50 a.u. whereas in Ps-H system these are 7.25 a.u. and 2.49 a.u. It should be noted that all the curves are almost similar to Ps and H system when both are in ground states. Both the scattering lengths are slightly greater than Ps and H system. It is useful to note that the reduced masses of Ps-H and Ps-Mu systems are 1.9978 a.u. and 1.9807 a.u. respectively.

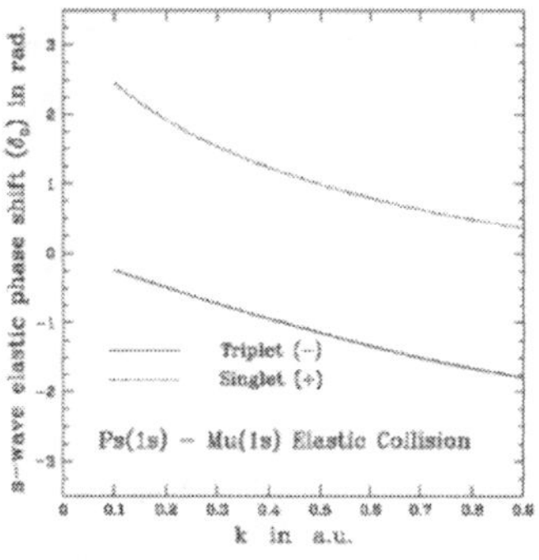

Figure 2. The variation of s-wave elastic phase shifts (δ_0) for the singlet (+) and the triplet (-) spin-states of the electrons.

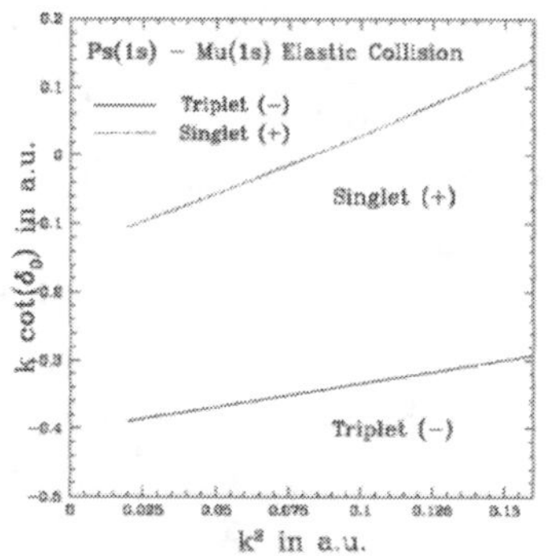

Figure 3. The plot k^2 versus $k\cot(\delta_0)$ to determine the scattering lengths following the effective-range theory.

The integrated or total elastic cross section (σ) and the quenching cross section (σ_q) using the present SEM theory are presented in **figure 4**. An augmented-Born approximation is used to include the effect of higher partial waves more accurately. These data are also very close to Ps-H system. In **figure 5**, the ortho- to para- conversion ratio (σ / σ_q) for Ps-Mu system are compared with Ps-H system. Both almost coincide.

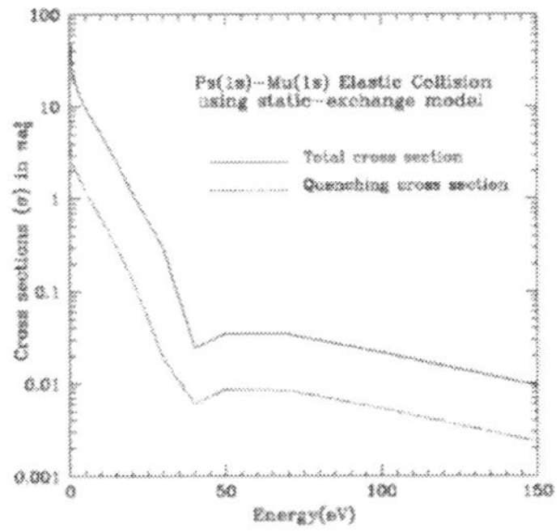

Figure 4. The variation of total or integrated elastic cross section (σ) and the quenching cross section (σ_q) with the incident energy.

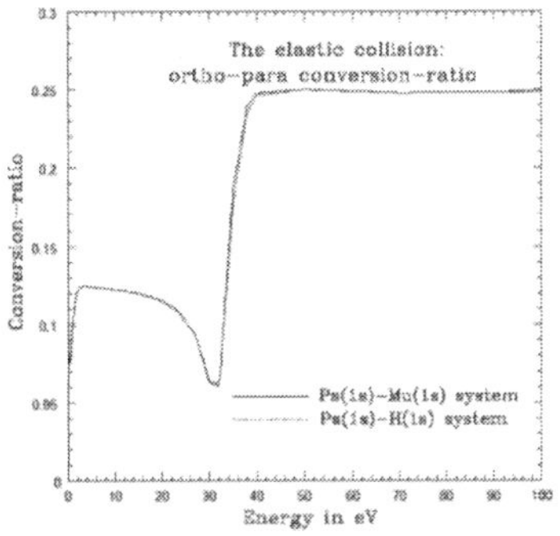

Figure 5. The variation of conversion ratio (σ / σ_q) with the incident energy.

4. Conclusion

We study the Ps and Mu elastic collision using the static-exchange model (SEM) considering the system as a four center Coulomb problem in the center of mass frame. The s-wave elastic phase shifts, the scattering lengths for the singlet and triplet spin-states of the system electrons, the integrated/total elastic cross section, the quenching cross section and ortho-para conversion ratio for Ps(1s)-Mu(1s) elastic scattering are being reported for the first time. It is observed that all the results are very close to Ps-H system and both the scattering lengths are slightly greater than Ps-H system.

References:

[1] Massey H S W and Mohr C B O 1954 *Proc. Phys. Soc.* 67 695-704.
[2] Burke P G Schey H M and Smith K 1963 *Phys. Rev.* **129** 1258
[3] Hughes V W et al 1960 *Phys. Rev. Lett.* **5** 63-65.
[4] Hughes V W and Kinoshita T 1977 *Muon Physics I* Edited by Hughes V W and Wu C S (Academic Press, New York, 1977) 12-199.
[5] Hughes V W and Putlitz G zu 1984 *Comments in Nucl. Part. Phys.* **12** 259 .
[6] Hughes V W 1985 *Ann. Phys. Fr.* **10** 955.
[7] Chu S, Mills A P Jr, Yodh A G, Nagamine K, Miyake Y and Kuga T 1988 *Phys. Rev. Lett.* **60** 101-104.
[8] Mayer V et al 2000 *Phys. Rev. Lett.* 84 1136-1139.
[9] Hughes V W 1989 *Proceedings The Hydrogen Atom* ed. G. F. Bassani et al 171-181 (Springer-Verlag Berlin, Hiedelberg, 1989).
[10] Ray H and Ghosh A S 1996 *J. Phys. B* **29** 5505, 1997 *ibid* **30** 3745.
[11] Ray H 2014 *Pramana* **83** 907-914.
[12] Ghosh A S, Sil N C and Mandal P 1982 *Phys. Rep.* **87** 313-406.
[13] Ray H 2014 *arXiv:*1409.3371.
[14] Ivanov I A, Mitroy J and Varga K 2002 *Phys. Rev. A* **65** 032703.

Journal of Physics: Conference Series **618** (2015) 012009 doi:10.1088/1742-6596/618/1/012009

High-precision calculation of loosely bound states of $LiPs^+$ and $NaPs^+$

Takuma Yamashita and Yasushi Kino*

Department of Chemistry, Tohoku University, Sendai, 980-8578, Japan

E-mail: *y.k@m.tohoku.ac.jp

Abstract. A positronic alkali atom would be the first step to investigate behavior of a positronium (Ps) in an external field from atoms/molecules because the system can be regarded as a simple three-body system using model potentials reflecting electron orbitals of the ion core. In order to precisely determine binding energies and structures of positronic alkali atoms ($LiPs^+$ and $NaPs^+$), we improve the model potential so as to reproduce highly excited atomic energy levels of alkali atoms (Li and Na). The polarization potential included by the model potential is expanded in terms of Gaussian functions to finely determine a short range part of the potential which has been assumed to be a simple form. We find better reproducibility not only of atomic levels of the alkali atoms but also of the dipole polarizability of the core ion than previous works. We construct a model potential between a positron and an ion core based on the model potential between the valence electron and ion core. Binding energies associated with a dissociation of the alkali ion core and positronium, and interparticle distances are recalculated. Our results show slightly deeper bound than other previous studies.

1. Introduction

Behavior of a positron (e^+) in atoms/molecules is quite different from that of an electron because of its difference in charge. In particular, positronium (Ps; a bound state of a positron and an electron) formation is of importance. The Ps lifetime depends on the surrounded atoms/molecules and annihilation γ-rays of Ps reflect electronic states of atoms/molecules. The Ps lifetime spectromery has been applied to investigate nano-structure of materials such as defects of metal and free volumes of polymer. Recently, high-precision experiments have been done such as a precise test of QED for Ps in i-C_4H_{10} gas [1]. Therefore, effects given to a Ps by an external field from atoms/molecules have become more and more important.

A positronic alkali atom (APs^+; "A" means an alkali atom), which is a bound state of an alkali ion core (Li^+ or Na^+) and a Ps, is a good testing ground to investigate the effects of external fields to a Ps. We can treat the APs^+ as a simple three-body system composed of an alkali ion core, a valence electron and a positron by introducing model potentials which consists of a static and polarization potential. The static potential reflects core electron orbital without excitation and is calculated from Hartree-Fock wavefunctions. The polarization potential reflects virtual excitation of the orbitals and consists of short range part and well-known long-range part which is the lowest polarization potential. The short range part reflects higher order of the excitations, but it is usually assumed to be a simple function which only avoids a divergence at the origin. For the availability of precise calculations, a lot of calculations have been done [2–13]. When an alkali atom binds a positron, the valence electron transfers to the positron and forms a Ps halo. Since the Ps is a neutral particle, this state is a loosely bound state and has a large fraction of the Ps-alkali ion configuration. Kubota and Kino [11] have pointed out that the fraction

Journal of Physics: Conference Series **618** (2015) 012009 doi:10.1088/1742-6596/618/1/012009

having the positron-alkali atom configuration is small but plays an indispensable role for the weakly bound system. In this mechanism, the polarization of the alkali ion core by the valence electron and the positron is of importance. The polarization potential consists of a short and long range parts whose asymptotic form is r^{-4}. Contribution of the short range part of the polarization potential should be small because the Coulomb interaction from the nucleus is highly dominant in that region. The short range part, therefore, can not be determined within a single particle approximation. The small binding energy of APs^+ depends on the model potential strongly [10, 12].

In this work, we introduce a more flexible polarization potential having a larger parameter space than other previous studies in which a simple empirical function was used for the inner part of the potential. Binding energies and structures of $LiPs^+$ and $NaPs^+$ are recalculated by a high-precision three-body calculation, based on one of the most accurate calculation [11]. Atomic units (a.u.; $m_e = \hbar = e = 1$) are used throughout this paper except where mentioned otherwise.

2. Theory

2.1. Improvement of model potential

A model potential between the ion core and valence electron $V_e(r_e)$ is composed of static, exchange and polarization potentials, namely,

$$V_e(r_e) = V_{st}(r_e) + V_{exch}(r_e) + V_{pol}(r_e), \tag{1}$$

where the r_e is a distance between the ion core and valence electron. The static potential $V_{st}(r_e)$ and the local exchange potential $V_{exch}(r_e)$ [14] are given by a standard Hartree calculation [15,16] by a computer program [17, 18] written by Albright *et al.* A polarization potential is written as

$$V_{pol}(r_e) = -\frac{\alpha_d}{2r_e^4}\left[1 - f(r_e)\right], \tag{2}$$

where α_d is a dipole polarizability, $\alpha_d = 0.192$ for Li and $\alpha_d = 0.923$ for Na given by Gien [19]. The $f(r_e)$ is a short range function determined numerically and has boundary conditions of $f(r_e \to 0) = 1$ and $f(r_e \to \infty) = 0$. Usually, the $1 - f(r_e)$ is written in an exponential form [18] or a polynomial form [20, 21]. In this work, we expand the polarization potential by Gaussian functions and increace flexibility:

$$V_{pol}(r_e) = \sum_{i=1}^{N} C_i r_e^2 e^{-a_i r_e^2}, \tag{3}$$

where C_i and a_i are optimization parameters. Although the Gaussian function decreases rapidly, a linear conbination of Gaussian functions with small $0.0001 < a_i < 0.1$ can effectively reproduce the asymptotic form of the polarization potential.

The Schrödinger equation about the valence electron can be written as

$$H_A\psi_{nl}, = \varepsilon_{nl}\psi_{nl}, \tag{4}$$

with

$$H_A = -\frac{1}{2}\left\{\frac{1}{r_e}\frac{d^2}{dr_e^2}r_e - \frac{l(l+1)}{r_e^2}\right\} + V_e(r_e), \tag{5}$$

where ψ_{nl} is a wavefunction of the alkali atom, ε_{nl} the eigenenergy of ψ_{nl}, n the principal quantum number ($n \geq 2$ for Li and $n \geq 3$ for Na) and l the orbital angular momentum quantum number. We determine the parameters C_i and a_i in (3) so that the calculated energy levels ε_{nl} can reproduce experimental energy levels [22] of the alkali atom within the boundary conditions.

Journal of Physics: Conference Series **618** (2015) 012009 doi:10.1088/1742-6596/618/1/012009

According to literature calculations [2, 3], the model potential between the ion core and positron is given by

$$V_{\mathrm{p}}(r_{\mathrm{p}}) = -V_{\mathrm{st}}(r_{\mathrm{p}}) + V_{\mathrm{pol}}(r_{\mathrm{p}}), \tag{6}$$

where the r_{p} is a distance between the ion core and the positron. The polarization potential between the ion core and positron is attractive and has the same form as the potential (3) because the polarization effect is quadratic in charge of the projectile [23].

2.2. *Precise calculation of APs^+ systems*

In order to solve the APs^+ three-body system precisely, we employ a Gaussian expansion method (GEM) which has been applied to a variety of few-body systems (see the review article [24] and references therein). In our previous work, precision of the three-body calculation is ensured by a test calculation using the same Peach potential as in Refs. [20, 21]. Total Hamiltonian H_{APs^+} and the three-body wavefunction Ψ_{APs^+} are given in our previous study [11] and satisfy the Schrödinger equation:

$$H_{\mathrm{APs}^+}\Psi_{\mathrm{APs}^+} = E_{\mathrm{APs}^+}\Psi_{\mathrm{APs}^+}. \tag{7}$$

The binding energy associated with the dissociation of A^++ Ps is given by a subtraction of the free Ps energy (-0.25 a.u.) from the three-body energy E_{APs^+}.

3. Results and Discussion

In table 1, calculated energy levels of Li and Na atom are shown together with experimental values. The model potential (1) obtained in this work reproduces observed energy levels of alkali atoms within $|\varepsilon_{nl}^{\mathrm{theo}} - \varepsilon_{nl}^{\mathrm{obs}}| < 0.000\,004$, where $\varepsilon_{nl}^{\mathrm{theo}}$ is calculated energy levels and $\varepsilon_{nl}^{\mathrm{obs}}$ experimental energy levels. This reproducibility is more a precise result than results of the other previous calculations. In our previous work [11], the dipole polarizability α_{d} in (2) was a little different from experimental values. We numerically determine the inner empirical part of the polarization potential by the expansion of Gaussian functions (3) using experimental polarizabilities.

Then we show the results of three-body calculations. In table 2, the convergence of the ground state energy of APs^+ is shown. We obtain the energies in 7-digit accuracy using more than 6000 basis functions. In table 3, the binding energy associated with the A^++ Ps dissociation and the expectation values of interparticle distances are shown together with other theoretical results. The results in this work show slightly deeper bound and larger polarization of a bound Ps than previous results. Since the Ps component in APs^+ is located far away from the ion and the long range parts of $V_{\mathrm{e}}(r_{\mathrm{e}})$ and $V_{\mathrm{p}}(r_{\mathrm{p}})$ cancel each other, small differences among the calculations are mainly caused by their short range parts. In the previous calculations listed in table 3, the short range part of the potentials has a simple analytical form having a few optimization parameters for theoretical convenience. In this work, we expand the part with sufficient number of functions with much more optimization parameters than previous calculations. Moreover some of previous calculations do not include the exchange potential which is effective only in the short range. Thus, we conclude the present result should be more accurate than previous results.

4. Summary

We improve the model polarization potential for precise calculation of loosely bound states of APs^+ with a fine-tuned polarization potential. As a result, we get a good reproducibility up to highly excited states using polarizabilities based on experimental values. These flexible polarization potentials can be applied to other alkali atom systems and high-precision calculations of positron-alkali atom scattering.

The three-body calculation is carried out by GEM using the improved model potential. The binding energies are recalculated in 7-digit accuracy. The geometry of APs^+ system shows slightly larger polarization and deeper bound of Ps but no contradiction with other previous works.

Table 1. Energy levels of Li and Na atoms calculated with various model potentials. The experimental values are taken from the NIST Atomic Spectra Database [22].

State		Theory			Experiment [22]
		Han *et al.* [10]	Kubota *et al.* [11]	This work	
	2s	−0.197 952	−0.198 136	−0.198 142	−0.198 142
	3s	−0.074 224	−0.074 173	−0.074 180	−0.074 182
Li	4s	−0.038 641	−0.038 613	−0.038 617	−0.038 615
	5s		−0.023 636	−0.023 639	−0.023 636
	6s		−0.015 945	−0.015 944	−0.015 945
	3s	−0.188 855	−0.188 859	−0.188 858	−0.188 858
	4s	−0.071 581	−0.071 586	−0.071 577	−0.071 578
Na	5s		−0.037 588	−0.037 588	−0.037 584
	6s		−0.023 134	−0.023 135	−0.023 132
	7s		−0.015 664	−0.015 660	−0.015 662

Table 2. Convergence of the energies of bound states as a function of the number of basis functions for $LiPs^+$ and $NaPs^+$.

Number of basis functions	$LiPs^+$	Number of basis functions	$NaPs^+$
5216	−0.252 578 6	5216	−0.250 479 5
5691	−0.252 578 7	5660	−0.250 479 9
6159	−0.252 578 8	6128	−0.250 480 2
6651	−0.252 578 8	6620	−0.250 480 3

Table 3. Binding energies and expectation values of inter-particle distances $\langle r_e \rangle$, $\langle r_p \rangle$, $\langle r_{ep} \rangle$ for the bound state of $LiPs^+$ and $NaPs^+$. The r_{ep} is a distance between a positon and a valence electron.

	Method	Binding energy	$\langle r_e \rangle$	$\langle r_p \rangle$	$\langle r_{ep} \rangle$
	This work	0.002 579	8.987	9.859	3.403
	GEM [11]	0.002 615	8.950	9.824	3.404
$LiPs^+$	FCSVM [6, 7]	0.002 478 9	9.110 1	9.967 8	3.396 7
	AHM [10]	0.002 455	9.135	9.991	3.393
	FEM [9]	0.002 37	9.01(10)	9.87(10)	3.40(5)
	This work	0.000 480	16.66	17.07	3.165
	GEM [11]	0.000 401	17.87	18.25	3.152
$NaPs^+$	FCSVM [4]	0.000 473	16.818	17.231	3.162
	AHM [10]	0.000 447	17.12	17.52	3.159
	FEM-ITM [12]	0.000 357	18.24	18.62	3.146

Acknowledgments
This work was financially supported by a Grant-in-Aid for Scientific Research from the Ministry of Education, Culture, Sports, Science and Technology (MEXT), Japan. The computation was partially carried out on the supercomputers in Kyushu University and KEK.

References
[1] Ishida A, Namba T, Asai S, Kobayashi T, Saito H, Yoshida M, Tanaka K and Yamamoto A 2014 *Phys. Lett. B* **734** 338
[2] Ryzhikh G and Mitroy J 1997 *Phys. Rev. Lett.* **79**(21) 4124
[3] Ryzhikh G G, Mitroy J and Varga K 1998 *J. Phys. B: At. Mol. Opt. Phys.* **31** 3965
[4] Mitroy J and Ryzhikh G G 1999 *J. Phys. B: At. Mol. Opt. Phys.* **32** L621
[5] Mitroy J, Bromley M W J and Ryzhikh G G 2002 *J. Phys. B: At. Mol. Opt. Phys.* **35** R81

Journal of Physics: Conference Series **618** (2015) 012009 doi:10.1088/1742-6596/618/1/012009

[6] Mitroy J 2004 *Phys. Rev. A* **70**(2) 024502
[7] Mitroy J 2005 *Phys. Rev. Lett.* **94**(3) 033402
[8] Le A T, Bromley M and Lin C D 2005 *Phys. Rev. A* **71**(3) 032713
[9] Shertzer J and Ward S 2006 *Phys. Rev. A* **73**(2) 022504
[10] Han H l, Li Y, Zhang X and Shi T 2008 *J. Chem. Phys.* **128** 244314
[11] Kubota Y and Kino Y 2008 *New J. Phys.* **10** 023038
[12] Shertzer J and Ward S 2010 *Phys. Rev. A* **81**(6) 064505
[13] Yamashita T, Irisawa A and Kino Y 2014 *Jpn. J. Appl. Phys. Conf. Proc.* **011005** 2
[14] Furness J B and McCarthy I E 1973 *J. Phys. B: At. Mol. Phys.* **6** 2280
[15] Hartree D R 1928 *Proc. Camb. Phil. Soc.* **24** 111
[16] Mitroy J 1983 Ph.D. thesis University of Melbourne
[17] Bartschat K 1996 *Computational Atomic Physics Ch. 2* (Berlin: Springer) p 15
[18] Albright B J, Bartschat K and Flicek P R 1993 *J. Phys. B: At. Mol. Opt. Phys.* **26** 337
[19] Gien T 1987 *Phys. Rev. A* **35**(5) 2026
[20] Peach G, Saraph H E and Seaton M J 1988 *J. Phys. B: At. Mol. Opt. Phys.* **21** 3669
[21] Peach G 1982 *Comments At. Mol. Phys.* **11** 101
[22] Kramida A, Yu Ralchenko, Reader J and NIST ASD Team 2014 NIST Atomic Spectra Database (ver. 5.2) `http://physics.nist.gov/asd`. National Institute of Standards and Technology, Gaithersburg, MD.
[23] Charlton M and Humberston J W 2001 *Positron Physics Ch.1* (Cambridge: Cambridge University) p 35
[24] Hiyama E, Kino Y and Kamimura M 2003 *Prog. Part. Nucl. Phys.* **51** 223

Journal of Physics: Conference Series **618** (2015) 012010 doi:10.1088/1742-6596/618/1/012010

Kinetic energy of Ps formed by Ore mechanism in Ar gas

Yosuke Sano[1], Yasushi Kino[1], Toshitaka Oka[1,2], Tsutomu Sekine[2]

[1]Department of Chemistry, Tohoku University, Sendai 980-8578, Japan

[2]Center for the Advancement of Higher Education, Tohoku University, Sendai 980-8576, Japan

E-mail: y.k@m.tohoku.ac.jp

Abstract. In order to investigate kinetic energy of positronium (Ps) formed by Ore mechanism, we performed positron annihilation age-momentum correlation (AMOC) measurements in Ar gas for 5.0 MPa and 7.5 MPa at room temperature. From the time dependence of Doppler broadening of *para*-Ps (*p*-Ps) self-annihilation gramma-ray component, we observed Ps slowing down process. Using a simple slowing down model, we obtained the initial kinetic energy of Ps formed by Ore mechanism and Ps-Ar momentum transfer cross section. The initial kinetic energy was 3.9 eV which was higher than the kinetic energy of Ps formed at the upper limit of Ore gap. The momentum transfer cross section was 0.019 ± 0.010 nm^2 in between 1 eV and 3.9 eV, and was close to the theoretical calculation.

1. Introduction

A positron injected into dense gases has been studied to investigate a correlation between positronium (Ps) and gas molecules [1, 2]. In this case, a Ps is formed during the slowing down of a positron. However, pick-off annihilation of the Ps with gas should depend on the kinetic energy of the Ps. The kinetic energy of Ps was measured to investigate the Ps slowing down in rare gases [3, 4, 5].

In a gas phase, Ps formation is explained by Ore mechanism. An energitic positron picks up one of electrons in an atom (A) and forms a Ps.

$$e^+ + A \rightarrow Ps + A \quad (1)$$

A suitable positron energy for a Ps formation is proposed as follows [6]. The kinetic energy of positrons must exceed the Ps formation threshold energy $E_0 = E_I - E_{Ps}$, where E_I is the ionization potential of the gas atom and $E_{Ps} = 6.8$ eV is the bounding energy of Ps. If E_{ex} is the lowest significant excitation energy for the atom, then for positrons of energy in the range $E_0 < E < E_{ex}$ the only possible inelastic process is Ps formation. This energy range is so-called Ore gap [2], and the initial kinetic energy of Ps can be estimated. Rare gas are suitable to observe the kinetic energy of Ps formed by Ore mechanism because the Ps loses its kinetic energy very slowly only by elastic collisions below 5.1 eV which is the lowest Ps excitation energy. In the case of an Ar gas, Ore gap ranges from 9 eV (Ps formation threshold) to 11.7 eV (1st electronic excitation threshold of Ar). The kinetic energy of Ps formed by Ore mechanism supposed to range from 0 to 2.7 eV.

In this study, we determined the time dependence of the kinetic energy of the Ps and discuss Ore mechanism.

2. Experimental

We performed a positron annihilation age-momentum correlation (AMOC) measurement in Ar gas for 5.0 MPa and 7.5 MPa at room temperature, to observe the initial kinetic energy of Ps and to reveal the Ps slowing down process below 6.8 eV. Time resolution of this system was 300 ps at the full width at half maximum (FWHM). Energy resolution of this system was 1.23 keV at 511 keV at FWHM. The positron source (^{22}Na, 0.5 MBq) was sandwiched between two sheets of 5 μm of the Ni foil (purity > 99 %). The positron source was placed in the center of a cylindrical high-pressure vessel with a diameter of 2.8 cm and a height of 5.0 cm. Detail of the experimental set up of this work was described elsewhere [7, 8].

3. Results and discussion

Since the Doppler broadening of p-Ps (intrinsic lifetime = 0.125 ns) annihilation gamma-ray is narrower than those of any other annihilation gamma-rays [9], we resolved the time dependent Doppler broadening spectrum into two components by using two Gaussian functions. Intensities of the p-Ps and other annihilation components were estimated by the positron annihilation lifetime spectrum. We, therefore, obtained the p-Ps component as a function of positron annihilation time by subtracting the broad component from the Doppler Broadening spectrum. The p-Ps component whose lifetime was 0.125 ns observed from 0 ns to 1 ns. From the Doppler broadening of the p-Ps component, the kinetic energies of Ps were obtained and listed in Table 1 as a function of an annihilation time. Resolution of the annihilation time was 0.2 ns.

In order to determine a momentum transfer cross section, we adopted a Ps slowing down model [3, 10] which was described as,

$$\frac{dE_{\mathrm{Ps}}(t)}{dt} = -\frac{2n_{\mathrm{Ar}}\sigma_m M_{\mathrm{Ar}}}{(M_{\mathrm{Ar}}+m_{\mathrm{Ps}})^2}\sqrt{2m_{\mathrm{Ps}}E_{\mathrm{Ps}}(t)}\left(E_{\mathrm{Ps}}(t)-\frac{3}{2}k_{\mathrm{B}}T\right), \tag{2}$$

where σ_{m} is the Ps-Ar momentum transfer cross-section, M_{Ar} the Ar mass, m_{Ps} the Ps mass, n_{Ar} the number density of Ar, k_{B} the Boltzmann constant, and T the temperature. Since theoretical calculation indicates that the momentum transfer cross section weakly depends on the Ps kinetic energy below the Ps break-up threshold, we assumed the cross section to be a constant value and solved the differential equation of (2). Time-averaged kinetic energy of Ps is given as

$$\overline{E_{\mathrm{Ps}}}(t_1,t_2) = \frac{1}{t_2-t_1}\int_{t_1}^{t_2} E_{\mathrm{Ps}}(t)dt, \tag{3}$$

where t_1 and t_2 are lower and upper limit of the annihilation time range.

The momentum transfer cross section was determined so that the difference between the observed time-averaged kinetic energy of Ps and the calculated one was minimum value. The

Table 1. Observed and calculated time-averaged kinetic energies of Ps, $\overline{E^{\mathrm{obs}}_{\mathrm{Ps}}}(t_1,t_2)$ and $\overline{E_{\mathrm{Ps}}}(t_1,t_2)$, in Ar gas.

Pressure (MPa)	5.0		7.5	
Number density (nm^{-3})	1.3		2.0	
Time range $t_1 \sim t_2$ (ns)	$\overline{E^{\mathrm{obs}}_{\mathrm{Ps}}}$ (eV)	$\overline{E_{\mathrm{Ps}}}$ (eV)	$\overline{E^{\mathrm{obs}}_{\mathrm{Ps}}}$ (eV)	$\overline{E_{\mathrm{Ps}}}$ (eV)
0.0 ~ 0.2	3.7	3.7	3.7	3.5
0.2 ~ 0.6	2.5	3.1	2.0	2.3
0.6 ~ 1.0	1.8	2.0	1.4	1.0

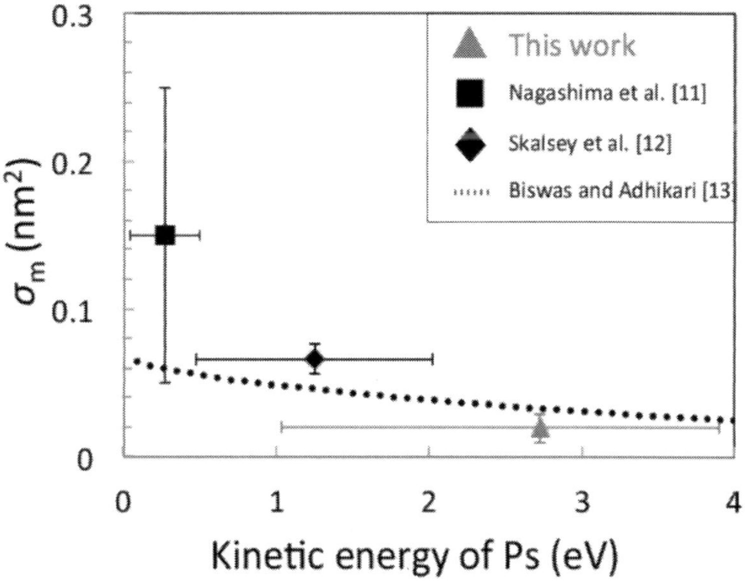

Figure 1. Momentum transfer cross section σ_m of Ps-Ar elastic scattering as a function of the kinetic energy of Ps. Symbols (▲ This work, ■ [11], ♦ [12]) show experimental values. The horizontal and vertical error bars show observed energy range and statistical error, respectively. Dotted line shows theoretical calculation [13]

calculated time-averaged kinetic energies of Ps obtained are shown in Table 1 together with the observed ones. The momentum transfer cross section was determined to be 0.019 ± 0.010 nm^2, and is shown together with literature values in Fig. 1. For the experimental values, the horizontal error bars represent the energy range used for the determination of the cross section. The theoretical calculation [13] was closer to our result than other experimental results [11, 12].

Calculated initial kinetic energy of Ps, $E_{\mathrm{Ps}}(0)$, was 3.9 eV which was higher than the upper limit of the Ore gap energy (2.7 eV). The Ps formation cross section in a positron-Ar collision has a peak around 20 eV which is higher than the Ore gap [14, 15]. During the slowing down of positrons, the Ps can be formed and its kinetic energy would be around 13 eV. The Ps with a kinetic energy more than 6.8 eV would immediately break-up into a positron and an electron by a collision with Ar atoms [16]. In this experimental condition, the density of the gas was high enough to slow Ps down to 6.8 eV within several ps. The Ps whose energy was lower than 6.8 eV would survive and slow down until the annihilation took place in the Ar gas. Thus, the initial distribution of the kinetic energy of Ps would range below 6.8 eV. The value of $E_{\mathrm{Ps}}(0)$ would be an expectation value of the distribution. This results suggested that the initial distribution of the Ps formed by Ore mechanism should change according to the break-up process of the energetic Ps.

4. Summary

Time dependence of Doppler broadening of p-Ps self-annihilation gamma-ray was observed by the AMOC measurement. Initial kinetic energy of Ps formed by Ore mechanism in Ar gas was determined from the Doppler broadening, and was 3.9 eV which was higher than the kinetic energy of Ps formed at the upper limit of the Ore gap. This suggests that the break-up process of the energetic Ps should be considered for the energy distribution of the Ps formed by Ore mechanism which proposed effective range of the initial energy of the Ps formed in gases. From the Ps slowing down model, we obtained momentum transfer cross sections as well as the time dependence of the kinetic energy. The cross section was 0.019 ± 0.010 nm^2 in between 1 eV and 3.9 eV, and was close to the theoretical calculation [13].

Acknowledgments

We would like to thank Dr. T. Hirade (Japan Atomic Energy Agency, JAEA) for his expert advice for the AMOC system. This work was financially supported in part by International Advanced Research and Education Organization, Tohoku University.

References

[1] Tao S J 1970 Phys. Rev. A **1** 1257
[2] Griffith T C and Heyland G R 1978 Phys. Rep. **39** 169
[3] Nagashima Y, Kakimoto M, Hyodo T, Fujiwara K, Ichimura A, Chang T, Deng J, Akahane T, Chiba T, Suzuki K, McKee B T A and Stewart A T 1995 Phys. Rev. A **52** 258
[4] Engbrecht J J, Erickson M J, Johnson C P, Kolan A J, Legard A E, Lund S P, Nyflot M J and Paulsen J D 2008 Phys. Rev. A **77** 012711
[5] Coleman P G, Rayner S, Jacobsen F M, Charlton M and West R N 1994 J. Phys. B **27** 981
[6] Ore A 1949 Univ. Bergen Arbok No.9
[7] Oka T, Sano Y, Kino Y and Sekine T 2014 Eur. Phys. J. D **68** 156
[8] Sano Y, Kino Y, Oka T and Sekine T 2014 Jpn. J. Appl. Phys. Conf. Proc. **2** 011004
[9] Hirade T 2009 Chem. Phys. Lett. **480** 132
[10] Sauder William C 1968 J. Res. Natl. Bur. Stand. **72A** 91
[11] Nagashima Y, Hyodo T, Fujiwara K and Ichimura A 1996 AIP, New York, in The Physics of Electronic and Atomic Collisions, edited by L. J. Dube, B. A. Mitchell, W. McConkey, and C. E. Brion, AIP Conf. Proc. No. 360
[12] Skalsey M, Engbrecht J J, Nakamura C M, Vallery R S and Gidley D W 2003 Phys. Rev. A **67** 022504
[13] Biswas P K and Adhikari S K 2002 Chem. Phys. Lett **317** 129
[14] Surko C M, Gribakin G F and Buckman S J 2005 J. Phys. B **38** R57
[15] Laricchia G, Armitage S, Leslie D E, Szłuińska M and Van Reeth P 2003 Radiat. Phys. Chem. **68** 21
[16] Armitage S, Leslie D E, Garner A J and Laricchia G 2002 Phys. Rev. Lett. **89** 173402

Theoretical study of the thermalization of the positronium interacting with atoms

Omar Morandi and Paul-Antoine Hervieux

Institut de Physique et Chimie des Matériaux de Strasbourg, CNRS and University of Strasbourg, France

E-mail: omar.morandi@ipcms.unistra.fr

Abstract. We have investigated the thermalization properties of a gas made by positronium atoms in interaction with a dense gas. Our theoretical model is based on the kinetic Boltzmann formalism that focuses on the microscopical description of the binary positronium-atom collision. We analyze the evolution of the velocity distribution of the positronium atoms during the thermalization process. Our simulations reveal that due to the collision with the heavy mass atoms, the positronium distribution differs significantly from the expected Maxwell-Boltzmann distribution. In particular, a large depletion of ultra-cold atoms is observed.

1. Introduction

One of the challenging of the nowadays antimatter physics concerns the production of cold atoms made by antimatter. An important effort is devoted to the production of anti-hydrogen and dense positronium. The positronium is a hydrogen-like atom composed by one electron and one positron and is the most easily produced atom formed by one antiparticle. It is likely formed in porous materials or aerogel containing fine grains [1]. By diffusing highly energetic positrons in porous materials, thermalized atoms of positronium are formed in less than few nanoseconds after the positron injection [2]. In order to accelerate the production of thermal positronium, the positronium is produced in the presence of noble or heavy mass gases (sometime denoted as remoderators). The thermalization process proceed faster in the presence of the remoderator because the positronium kinetic energy can be transferred to the atoms of the remoderator by elastic collisions [3]. The investigation of the interaction between the positronium and cold atoms is today an active area of research. Experiments have been performed by diffusing low density positronium gas inside a gas of heavy atoms (typically xenon [4, 5]) or helium [6]. In particular, the role played by the quenching, the spin orbit, the pick-off interaction on the modification of the positronium lifetime have been investigated [7, 8, 9, 10, 11, 12].

Measurements of the positronium decay process indicate that the positronium gas in interaction with a remoderator or a solid could thermalize at room or at cryogenic temperature (around 150 K) before the complete annihilation of the positronium. However, few questions remain open and a further investigation of the atomic dynamics is necessary in order to understand the experiments. As an example, in Ref. [13] it is shown that two positronium populations with a large temperature difference can be emitted by silica sample containing cavities that are arranged in the form a regular array of cylindric channels. Moreover, measurements on the spin-orbit interaction between positronium and xenon revealed that the positronium lifetime is strongly affected by the value of the kinetic energy of the positronium that interacts with the xenon atoms [8].

So far, the theoretical description of the positronium interaction with a solid or a gas is based on the Nagashima's model [3] or on some macroscopic rate equations [10]. One of the basis assumption of these approaches is to consider that the positronium gas is always well described by the equilibrium Maxwell-Boltzmann (MB) distribution. However, since the positronium is typically crated in a far-from-equilibrium condition, this hypothesis need a deeper investigation. With this motivation, on the basis of a theoretical microscopical kinetic model, in the present work we investigate the thermalization of a gas made of positronium atoms interacting with an external gas. With our method, we are able to describe the complete out-of-equilibrium transient evolution of the velocity distribution of the atoms.

2. Results

One of the most efficient way to produce a gas made of positronium is to irradiate a porous material or a aerogel with a highly energetic positron beam. The positrons lose their initial kinetic energies by collisions with the atoms of the solid and the positronium is formed by capturing one electron from one orbital state of the solid. In the case of porous materials, the positronium is mostly formed in the proximity to the internal surfaces of the pores and diffuses inside the solid until it escapes into the void.

Various experiments have established that when the positronium is prepared by using porous materials, the atoms are emitted with a temperature significantly higher than the temperature of the solid. In the case of silica, the positronium atoms are emitted with a characteristic energy that is found to be around 0.5 eV. However, very little is known on the form of the energy distribution of the atoms. It is usually assumed that the positronium thermalize quasi-instantaneously to a MB distribution. Theoretical studies that could confirm such an assumption have not yet been performed. The interest of the present work, is to clarify this point. In a previous study, we showed that the internal positronium thermalization can be rather efficient [14, 15, 16]. By internal thermalization we mean the process through which the total energy of the gas is redistributed among all the atoms leading to the formation of a Maxwell-Boltzmann distribution. Our previous result was derived under the assumption that the positronium is so dense than the probability that two positronium atoms collide is significantly high. In this work, we consider a rarefied positronium gas. Under such condition the bi-positronium collision probability becomes negligible and the only thermalization channel is the collision with the remoderator.

We describe the evolution of the positronium gas interacting with a gas constituted by heavy mass atoms by using the Boltzmann formalism. In this context, the binary heavy-atom positronium collision process is described by the following equation

$$\frac{\partial f(\mathbf{p}_1)}{\partial t} = \frac{2}{(2\pi\hbar)^3}\frac{(M+m)^2}{M^2m^2}\int \sigma\left[f(\mathbf{p}_2)g(\mathbf{P}_2) - f(\mathbf{p}_1)g(\mathbf{P}_1)\right]\delta\left(\mathbf{p}_1 + \mathbf{P}_1 - \mathbf{p}_2 - \mathbf{P}_2\right)$$
$$\times\,\delta\left(\frac{\mathbf{p}_1^2}{2m} + \frac{\mathbf{P}_1^2}{2M} - \frac{\mathbf{p}_2^2}{2m} - \frac{\mathbf{P}_2^2}{2M}\right)\mathrm{d}\mathbf{P}_1\,\mathrm{d}\mathbf{p}_2\,\mathrm{d}\mathbf{P}_2 - \frac{f}{\tau}. \quad (1)$$

Here, f and g denote, respectively, the distribution functions of the positronium and of the external gas. The cross section $\sigma = 4\pi a^2$, a is the scattering length, m is the positronium mass, M is the atomic mass of the external gas and τ is the positronium lifetime. In our simulations, we assume that the scattering cross section is independent from the atomic energy. In the case of highly energetic positrons, some correction to the scattering cross section should be considered [17]. Due to the large mass and density differences, we assume that the remoderator distribution function g is not modified by the presence of the positronium. We assume that g is a constant Maxwell-Boltzmann distribution at room temperature.

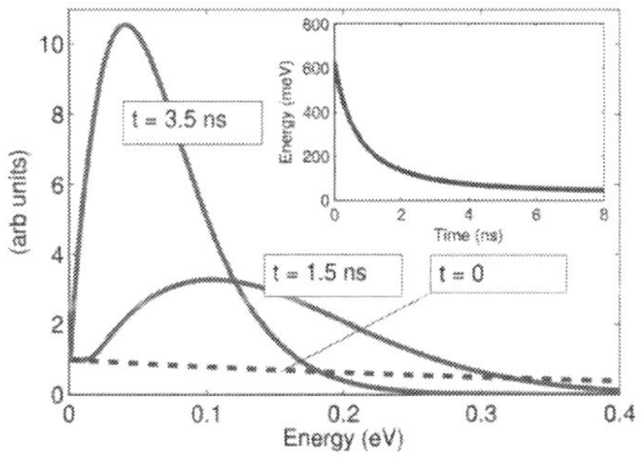

Figure 1. Dynamics of the positronium distribution function induced by collision with a gas made of helium atoms (pressure 1 atm at the room temperature). The dashed-blue line depicts the initial condition (MB distribution at the temperature of 5000 K). The evolution of the mean positronium energy is depicted in the inset.

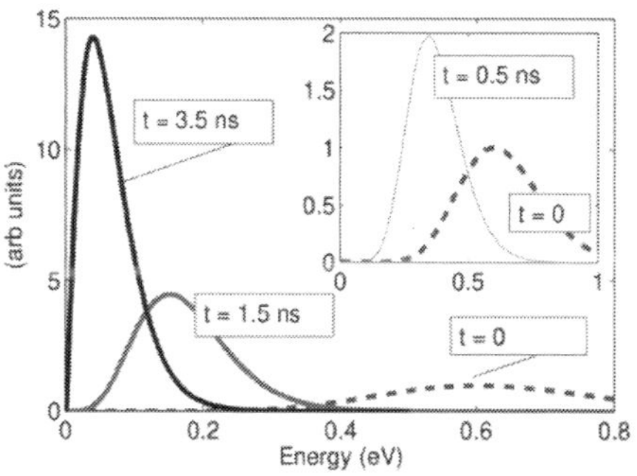

Figure 2. Same case of figure 1 but for an initial gaussian energy distribution of the positronium atoms. Mean energy equal to 0.6 eV and variance equal to 150 meV (dashed-blue line). In the inset we depict the zoom of the initial condition and of the atomic distribution at time $t = 0.5$ ns.

In order to investigate the dynamics of the energetic distribution of the atoms, we consider a low density positronium gas that initially ($t = 0$) is described by a gaussian distribution function with mean energy $\overline{E}$ and energy dispersion ΔE. We investigate the interaction of the positronium with a gas of helium or xenon atoms. Both atoms have already been considered for the experiment of positronium thermalization. In our simulation, we assume that the remoderator density is equal to 6×10^3 nm^{-3} (250 kPa). The temperature is the room temperature. The scattering cross section of these processes have been measured by various Authors [3, 4]. In our simulations we use the scattering length $a = 2a_0$ where a_0 is the Bohr radius.

The main results of our simulations are displayed in figures. 1-3. In order to illustrate the dynamics of the collision between the positronium and the heavy atom gases, we depict the evolution of the positronium distribution at different times and the evolution of the mean positronium energy (insets of figure 1 and figure 3). We have investigated the influence of the form of the initial positronium energetic distribution on the thermalization dynamics. At this aim, in our simulations we used two different initial conditions for the positronium gas. In the first case (figure 1) we have assumed that the positronium is emitted by the porous material according to the thermal Maxwell-Boltzmann (MB) distribution (dashed-blue line of figure 1) at temperature of 5000 K (that corresponds to the mean emission energy of 600 meV). In the right panel (figure 2) we have depicted the numerical results obtained by taking a Gaussian energy distribution (mean energy $\overline{E} = 0.6$ eV and variance $\Delta E = 150$ meV). We have depicted the positronium distribution function at $t = 1.5$ ns and $t = 3.5$ ns after the positronium formation. The comparison between figure 1 and figure 2 indicates that the memory of the initial condition is lost in less than $3 - 3.5$ ns (note the difference on the energy scale of the two panels). Moreover, our simulations show that the interaction with the gas is able to equilibrate the energy of the positronium with mean room energy (around 27 meV), in a time of round 10 ns that is ten time smaller than the positronium lifetime.

In figure 3 are reproduced the results obtained by assuming that the remoderator is made by xenon atoms. The other quantities are the same as in figure 2. We see that the positronium dynamics is very similar to what obtained in the previous case, except that the time scale is

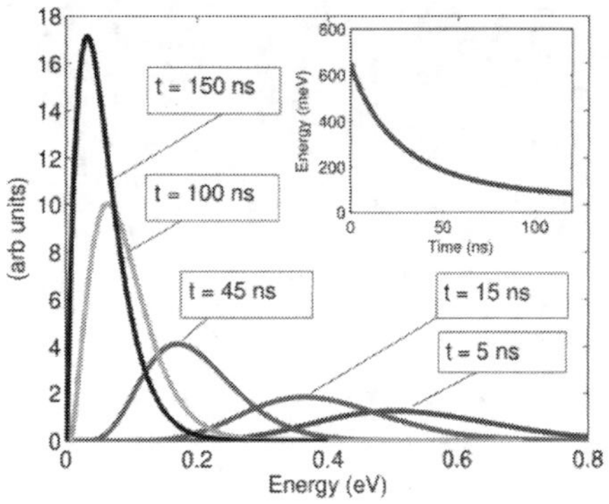

Figure 3. Dynamics of the positronium distribution function induced by collision with a gas made of Xe atoms. The Xe is under the same pressure and temperature conditions as in figure 3. The initial condition for the positronium gas is the same as in figure 2.

considerably increased. It is easy to understand the origin of the difference on the thermalization time when xenon or helium is used. According to the energy and momentum conservation laws, during the bi-atomic collision, we have that the energy transferred by positronium atom to the remoderator atom after one single collision is $\delta E = E_i m/M$ where E_i is the positronium energy before the collision. This consideration shows that the thermalization time scales as the inverse of the mass of the remoderator gas.

Our simulations revealed one important fact concerning the thermalization processes of the positronium induced by collisions with atoms (the same consideration applies also when the positronium collides with a solid or with some other molecules). In agreement with the experimental finding, our simulations show that the positronium energy equals the mean energy of the reservoir after few collision processes. As a conclusion, we can state that the positronium is thermalized before the annihilation. However, a finer analysis of the results reveals that, in contrast to the common assumption, the velocity distribution of the atoms is *not* described by a Bose-Einstein distribution function. A depletion of low velocity atoms can be clearly observed. The final positronium distribution (after 3.5 ns when the remoderator is helium and 150 ns when the remoderator is xenon) is more similar to a gaussian distribution peaked around the room energy (a MB distribution should show a peak when the velocity of the atoms goes to zero). Moreover, this conclusion is independent from the choice of the initial preparation of the positronium gas: even if the positronium is prepared according to the MB distribution with a temperature higher than the remoderator, after the interaction with the thermal bath, the positronium atoms show a non equilibrium configuration, where a depletion of very cold atom is present.

3. Conclusion

We have investigated the thermalization process of the positronium induced by the interaction with a gas (remoderator). We have considered two difference species of gas (Xe and He) characterized by a large mass difference. Our study shows that the positronium main energy becomes comparable with the gas energy in a time shorter than the positronium lifetime. However, the thermalization processe cannot be considered as completed. The numerical simulation show clearly that the positronium during all the thermalization process are not well described by a quasi-equilibrium Maxwell-Boltzmann distribution. In particular, the production of ultracold atoms results inefficient.

[1] Liszkay L et al. 2009 Appl. Phys. Lett. 95 124103
[2] Nagashima Y et al. 1995 Phys. Rev. A 52 258
[3] Nagashima Y Hyodoy T Fujiwarayx K and Ichimuraz A 1998 J. Phys. B: At. Mol. Opt. Phys. 31 329
[4] Shibuya K Kawamura Y and Saito H 2013 Phys. Rev. A 88 042517

[5] Saito H and Hyodo T 2006 Phys. Rev. Lett. 97 253402
[6] Murtagth D J Cooke D A and Laricchia G 2009 Phys. Rev. Lett. 102 133202
[7] Wada K et al. 2012 Eur. Phys. J. D 66 108
[8] Shibuya K et al. 2013 Phys. Rev. A 88 012411
[9] Saito H Nakayama T and Hyodo T 2009 J. of Phys. 194 012038
[10] Mitroy J and Novikov S A 2003 Phys. Rev. Lett. 90 183202
[11] Dixit A Hinschberger Y Zamanian J Manfredi G and Hervieux P-A 2013 Phys. Rev. A 88 032117
[12] Morandi O Hervieux P-A and Manfredi G 2014 J. Phys. B: At. Mol. Opt. Phys. 47 155202
[13] Mariazzi S Bettotti P and Brusa R S 2010 Phys. Rev. Lett. 104 243401
[14] Morandi O Hervieux P-A and Manfredi G 2013 Phys. Rev. A 88 23618
[15] Morandi O Hervieux P-A and Manfredi G 2014 Eur. Phys. J. D, 68 84
[16] Morandi O Hervieux P-A and Manfredi G 2014 Phys. Rev. A 89 033609
[17] Fedus K Karwasz G P and Idziaszek Z 2013 Phys. Rev. A 88 012704

Journal of Physics: Conference Series **618** (2015) 012012 doi:10.1088/1742-6596/618/1/012012

Positron annihilation spectroscopic studies of multiferroic $Bi_{1-x}Pr_xFeO_3$ nanocrystalline compounds

Jincemon Cyriac[1,2], M.T. Rahul[1], Nandakumar Kalarikkal[1,3] and P.M.G. Nambissan[4,§]

[1] School of Pure and Applied Physics, Mahatma Gandhi University, Kottayam, Kerala 686560, India
[2] Department of Physics, Devamatha College, Kuravilangad, Kottayam, Kerala 686633, India
[3] International and Inter University Centre for Nanoscience and Nanotechnology, Mahatma Gandhi University, Kottayam, Kerala 686560, India
[4]Applied Nuclear Physics Division, Saha Institute of Nuclear Physics, Kolkata 700064, India

[§] E-mail : pmg.nambissan@saha.ac.in

Abstract. Positron lifetime (PL) and coincidence Doppler broadening spectroscopic (CDBS) experiments were carried out on $BiFeO_3$ samples doped with Pr (in place of Bi) in concentrations (x) = 0, 0.05, ..., 0.35. In the initial stages of doping, the existing Bi^{3+} vacancies are occupied by Pr^{3+} ions and positron trapping is reduced. From $x = 0.15$ to 0.35, more number of positrons diffused to the surfaces of the nanocrystallites. The CDBS ratio curves showed enhancement of the peak at $p_L = 10.2\times10^{-3}$ m_0c due to positron annihilation with the $2p$ electrons of oxygen. It suggested increased trapping of positrons in newly created cationic vacancy-type defects due to the mismatch of ionic radii of Pr^{3+} substituting Bi^{3+} ions. The lattice parameters decreased (3.9661-3.9248Å) while the nanocrystallite sizes reduced and increased (28.6-34.2nm) during the doping.

1. Introduction

In the era of multiferroic nanocrystalline materials being rigorously looked into for their promising properties for physical and technological applications [1], $BiFeO_3$ has got the maximum attention because of the ability to tune its properties by adding or "substituting" trace amounts of lanthanide elements in place of a fraction of Bi so that the desired application can be performed without seriously compromising on its basic structure and/or ionic distribution. Still, defects are invariably bound to play a decisive role in controlling some of the useful properties down to even an atomic scale. The defects, especially of the vacancy-type, can result from either one or more out of a few factors such as charge imbalance, mismatch of ionic radii and lattice expansion or contraction. Positron annihilation spectroscopy is especially useful to investigate them and has been proved to be successful in a number of similar studies in the past [2, 3]. In this work too, we focus on the defects-related variations of the positron lifetimes and intensities and the coincidence Doppler broadened spectral line shapes in praseodymium (Pr) - doped $BiFeO_3$ and tried to understand the specific features responsible for such changes.

2. Experimental details

The synthesis of $Bi_{1-x}Pr_xFeO_3$ (x = 0.0, 0.05, … 0.35) samples was performed through the sol-gel route using hydrogenated bismuth nitrate ($Bi(NO_3)_3.5H_2O$), ferric nitrate ($Fe(NO_3)_3.9H_2O$) and praseodymium nitrate ($Pr(NO_3)_3.6H_2O$) as precursors and polyvinyl alcohol (PVA) as chelating agent. For the synthesis of the undoped bismuth ferrite ($BiFeO_3$), 0.1M solutions of $Bi(NO_3)_3$. $5H_2O$ and $Fe(NO_3)_3.9H_2O$ in 70%-concentric nitric acid and water respectively were mixed and sonicated for 20 minutes. PVA was added to the mixed solution with a weight equal to the weight of the metal ions (Bi^{3+}) and was stirred until it was completely dissolved. After the complete dissolution of PVA, the solution was evaporated at 333K till it gradually turned into a viscous sol and then to dark brown gel. The dried gels were calcined at 873K for 2 hours. Similar procedure was repeated for the synthesis of the other samples, the differences being the change in the molarity of the precursor solutions including $Pr(NO_3)_3.6H_2O$ in water and the weight of the PVA in order to keep the stoichiometry and nanocrystalline nature of the product (Table 1).

Table 1. Details of the amounts of the reagents used for the synthesis of the samples.

Samples	$Bi(NO_3)_3$. $5H_2O$ (g)	$Fe(NO_3)_3$. $9H_2O$ (g)	$Pr(NO_3)_3$. $6H_2O$ (g)	Bi (g)	Fe (g)	Pr (g)	PVA (g)
$BiFeO_3$	4.851	4.04	0	2.090	0.558	0	2.648
$Bi_{0.95}Pr_{0.05}FeO_3$	4.608	4.04	0.218	1.985	0.558	0.070	2.614
$Bi_{0.90}Pr_{0.1}FeO_3$	4.366	4.04	0.435	1.881	0.558	0.141	2.580
$Bi_{0.85}Pr_{0.15}FeO_3$	4.123	4.04	0.653	1.776	0.558	0.211	2.546
$Bi_{0.80}Pr_{0.20}FeO_3$	3.881	4.04	0.870	1.672	0.558	0.282	2.512
$Bi_{0.75}Pr_{0.25}FeO_3$	3.638	4.04	1.088	1.567	0.558	0.352	2.478
$Bi_{0.70}Pr_{0.30}FeO_3$	3.395	4.04	1.305	1.463	0.558	0.423	2.444
$Bi_{0.65}Pr_{0.35}FeO_3$	3.145	4.04	1.525	1.359	0.558	0.494	2.410

The samples were checked for their purity and characterized by x-ray diffraction. For positron annihilation measurements, a ^{22}Na source of strength ~ 400 kBq deposited on a thin (~ 2 mg.cm^{-2}) annealed (at 1273K for 2 hours in vacuum ~ 10^{-6} mbar) Ni foil is kept embedded in the sample (in powder form) taken in a glass tube. The glass tube is continuously evacuated to maintain the sample and source under moisture-free atmosphere throughout the experiments. For positron lifetime measurements, BaF_2 scintillators coupled with XP2020Q photomultiplier tubes were used as detectors and the slow-fast gamma-gamma coincidence spectrometer had a time resolution of 170 ps. The HPGe detectors used for the coincidence Doppler broadening (CBD) measurements had energy resolutions 1.27 and 1.33 keV at 511 keV. More than a million counts were collected under each positron lifetime spectrum and the data were analyzed using the PALSfit program [4]. About 20 million coincidence events were generated in each CDB spectrum and the analysis was done using the LAMPS software [5].

3. Results and discussion

The sizes of the nanoparticles were obtained from the x-ray diffraction patterns via the peak width analysis and using the well-known Scherrer equation [6]. Further, the lattice parameters were estimated from the Bragg diffraction formula and combining it with the lattice constant. These are shown in figures 1(a) and (b). The initial decrease of the particle sizes is attributed to the reduction of strain brought in by the occupancy of the Bi^{3+} vacancies by the Pr^{3+} ions and the subsequent increase to the strain developed by the substitution in place of Bi^{3+} ions by Pr^{3+} ions of slightly larger radii (1.013 Å in place of 0.96 Å).

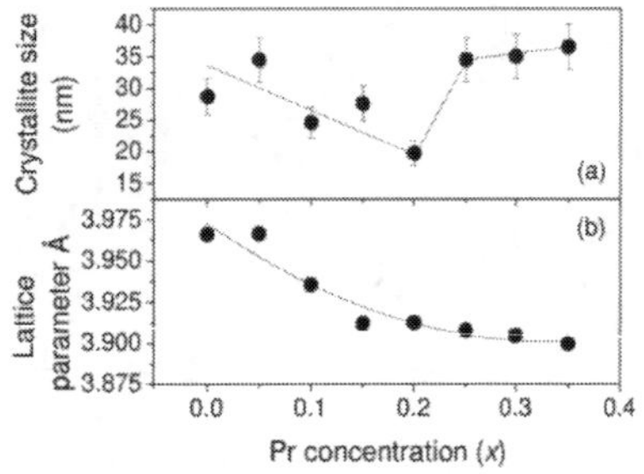

Figure 1. (a) The size of the nanocrystallites and (b) the lattice parameter versus the concentration of Pr in the samples.

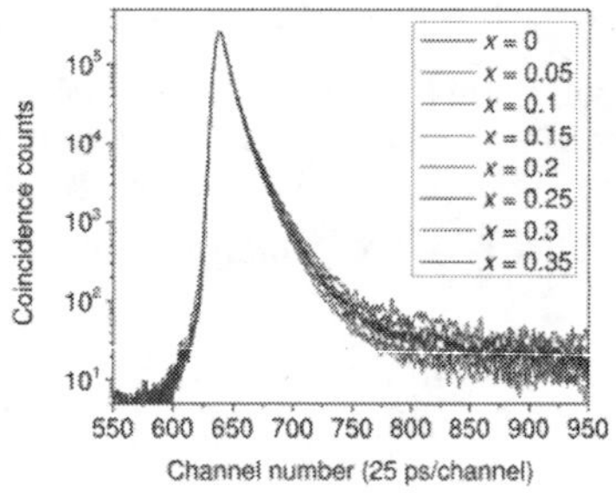

Figure 2. Peak-normalized positron lifetime spectra of all the samples.

There are a number of aspects of nanocrystalline material systems that can affect the values of the measured positron annihilation parameters. First and foremost is the ability of thermalized positrons to diffuse through the interstitial regions of the material over distances of the order of 10-100 nm to reach the crystallite surfaces during their lifetimes. If the crystallites are of sizes less than these thermal diffusion lengths and devoid of any positron trapping defects in their interior, the positrons will eventually diffuse out and annihilation will therefore take place on the crystallite surfaces or within the intercrystallite regions (as positronium in the latter case). It is therefore prudent to attribute positron lifetimes of the order of 400-600 ps to the diffused vacancy clusters on the crystallite surfaces. Often this gets admixed with the defect-specific positron lifetime and, as a result, lower values for τ_2 are obtained. The identity of the defects that trap positrons within the crystallites is often lost or difficult to retrieve in such cases.

The peak-normalized positron lifetime spectra of the samples shown in figure 2 are indicative of the sensitivity of this technique to identify the changing electronic environment within the samples due to doping and of the presence of extended vacancy-type defects of diverse nature and size in the samples. The positron lifetimes and their intensities are given in the figures 3(a) and 3(b). Lattice contraction results into the atoms coming closer and the sizes of the vacancy-type defects too can have a proportionate decrease. The resulting enhancement in the local electron density will therefore normally decrease the positron lifetime and the initial trend of variation of τ_2 is in agreement with this interpretation. An anomaly is observed as the subsequent rise in τ_2 which is remarkable and is indicative of an enhancement in the number of positrons reaching the nanocrystallite surfaces, thereby tilting the admixed lifetime τ_2 to higher values. The intensity I_2 decreases with doping concentration, which signifies the cancellation of a number of

Bi^{3+}-vacancies by the doped Pr^{3+} ions. Continued substitution results into strain being developed due to the larger ionic radii of the Pr^{3+} ions which replace the Bi^{3+} ions. The immediate consequence is the increase in size of the crystallites, as indicated in figure 1(a), and the probable increase in the number of positrons annihilating within the defect-free regions of the crystallites. In other words, the fall of I_2 can be well thought of as an independent and opposite change in I_1 and the initial fall and following near-constancy of τ_1 supports this argument. It should be mentioned that the shortest lifetime τ_1 incorporates in it the lifetime of positrons undergoing "free" annihilation and the Bloch state residence time of the trapped positrons. In nanocrystalline solids, the concept of "free" annihilation is a misnomer as the positrons would have by then thermally diffused to the crystallite surfaces and got trapped there. We however could not eliminate such a lifetime component in the data analysis and it should have a definite physical origin. We therefore attribute the term "free" to positrons getting annihilated in flight during the thermal diffusion to the crystallite surfaces. Although the orthopositronium lifetimes (τ_3) and intensities (I_3) do show similar changes, considering the negligibly small values of I_3, they are not further discussed.

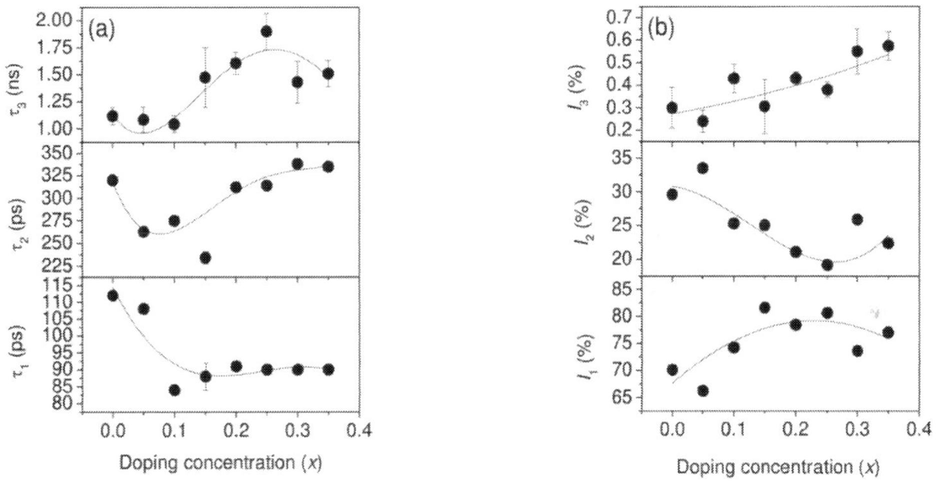

Figure 3. (a) Resolved positron lifetimes and (b) their relative intensities versus the concentration of Pr in the samples.

The results from coincidence Doppler broadened spectra support these explanations. The ratio curves generated with respect to well annealed (at 1273K for 2 hours in vacuum ~ 10^{-6} mbar) single crystalline Al (99.999% pure) are shown in figure 4. The peak observed at an electron momentum $p_L = 10.2 \times 10^{-3}$ m_0c results from the annihilations with the $2p$ electrons of oxygen ions and its amplitudes are shifted upward after an initial decrease till the doping concentration $x = 0.15$ (figure 5). For annihilation with the oxygen electrons, the positrons are to be trapped in cationic vacancies and the initial trend of variation are consistent with the cancellation of such vacancies by the doped Pr^{3+} ions. That the intensity of annihilation with the oxygen electrons thereafter enhances is an indication of new vacancy type defects generated due to the lack of effective substitution by Pr^{3+} ions which have got larger radii compared to the Bi^{3+} ions. Intriguingly, this feature was not reflected in the variation of the intensity component I_2.

We also calculated the lineshape parameters *S* and *W* from the counts under the usual energy intervals of the projected one-dimensional spectra and their variations too indicated changes taking place in the samples due to doping (figures 6(a) and (b)). For positrons to be sensitive to the changes in electronic environment due to the substitution of Bi^{3+} by Pr^{3+} ions, they should have got trapped and annihilated at the lattice sites vacated by the oxygen ions. Normally such a possibility is ruled out because of the positive charge associated with the oxygen vacancies. But

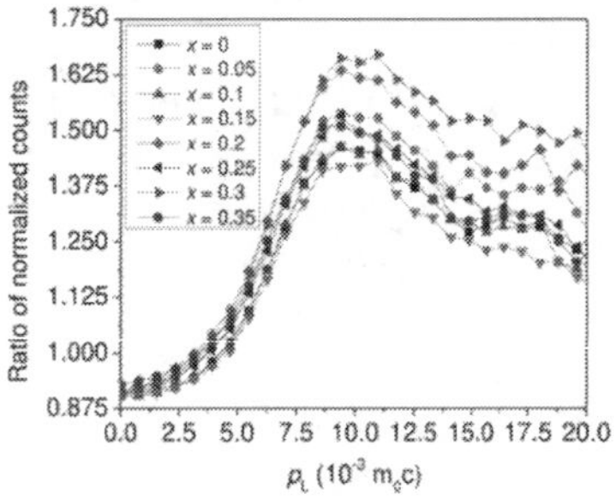

Figure 4. The ratio curves generated with respect to Al from the CDB spectra.

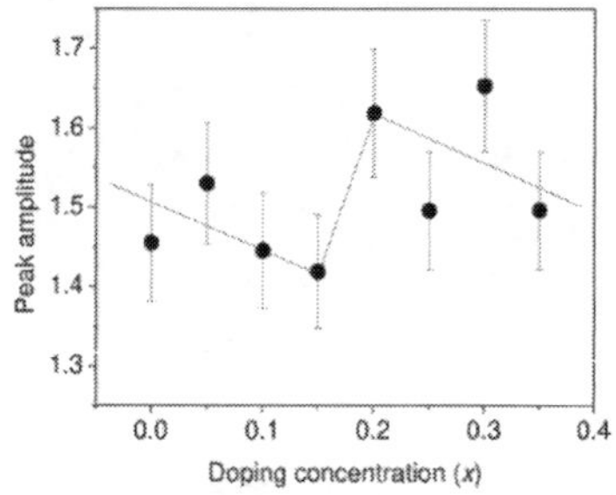

Figure 5. The amplitude of the peak at $p_L = 10.2 \times 10^{-3}$ m_0c of the ratio curves versus the Pr^{3+} doping concentration in the samples.

the situation is favorable for positron trapping when the defects are vacancy clusters composed of neighboring anionic and cationic monovacancies in numbers such that the resultant vacancy clusters are either neutral or negatively charged. Since, in this study, we have been not able to separately resolve the positron lifetime in the defects, it can only be postulated that the vacancy clusters are either divacancies of the form V_{Bi+O} (neutral) or trivacancies $V_{Bi+O+Bi/Fe}$ (negatively charged).

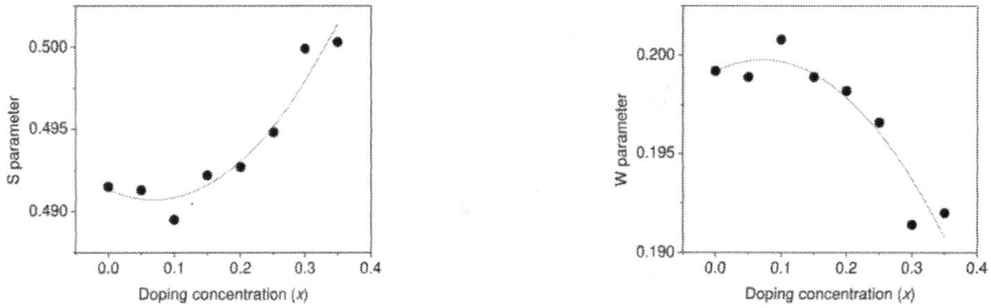

Figure 6. (a) *S* and (b) *W* versus Pr^{3+} doping concentration in the samples.

Bi^{3+} has the electronic configuration [Xe] $4f^{14}5d^{10}\,6s^2\,6p^0$ whereas Pr^{3+} has got partially filled 4*f* and 5*d* orbits in it due to the [Xe] $4f^2\,5d^0\,6s^0$ structure. As Pr^{3+} ions continue to substitute for the Bi^{3+} ions, the absence of the $6s^2$ electrons will be increasingly felt by the positrons and the *W* parameter will decrease. The *S* parameter consequently increases reflecting further as a linear fall with the *W* parameter in the *S*-*W* plot (figure 7). The empty 5*d* orbit in Pr^{3+} will further facilitate improved ferromagnetic properties for the multiferroic samples and can be a topic of independent studies in future.

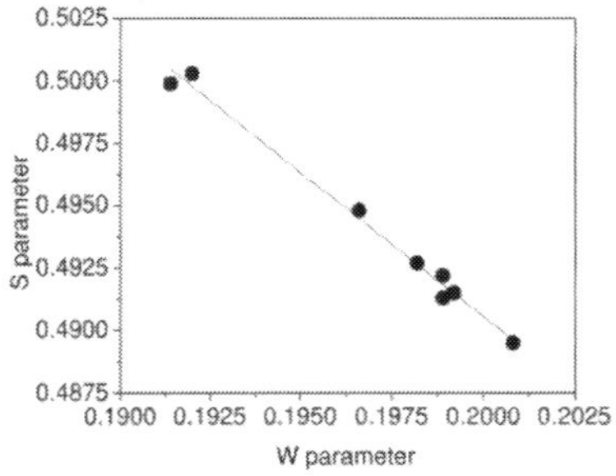

Figure 7. The *S*-*W* plot.

4. Summary and conclusions

Doping by Pr^{3+} ions in place of Bi^{3+} in multiferroic nanocrystalline $BiFeO_3$ resulted in a gradual decrease of the lattice parameter whereas the average crystallite size initially decreased and then increased. An interplay between the strain associated with vacancy type defects and the surface stress due to substitutional effects is the deciding factor on the sizes of the doped crystallites [7]. The role of the defects is reflected in the positron lifetimes and intensities. A number of Bi^{3+} vacancies, originally present due to the non-stoichiometry of the compound, got filled up by the Pr^{3+} ions. Further doping led to substitutional effects and resulted into strain and strain-induced defects whereby the defect-specific lifetime increased. CDBS results also supported the concept of fresh cationic vacancies being the dominant trapping centres for positrons. Positronium formation, although of small intensities, cannot be ignored totally as the lifetimes and intensities often exhibit systematic variation with grain size, doping concentration, temperature etc. These information can be highly useful in studies involving defect-related properties such as magnetism, heat treatment, cold-work, metallurgy etc of similar samples.

References

[1] Spaldin N A, Cheong S W and Ramesh R 2010 *Phys. Today* **63** 38
[2] Mukherjee A, Banerjee M, Basu S, Nambissan P M G and Pal M 2013 *J. Phys. D : Appl. Phys.* **46** 495309 (9pp)
[3] Nambissan P M G 2013 *J. Phys.: Conf. Ser.* **443** 012040
[4] Olsen J V, Kirkegaard P, Pedersen N J and Eldrup M 2007 *Phys. Stat. Sol. (c)* **4** 4004
[5] Chatterjee A, Ramachandran K, Kumar A and Behere A 2013 http://www.tifr.res.in/~pell/lamps.html
[6] Patterson A 1939 *Phys. Rev.* **56** 978
[7] Manuel Diehm P, Ágoston Péter and Albe Karsten 2012 *ChemPhysChem* **13** 2443

Positron probing of phosphorus-vacancy complexes in silicon irradiated with 15 MeV protons

N Arutyunov[1,2,4], V Emtsev[2], R Krause-Rehberg[1], M Elsayed[1,3], C Kessler[1], V Kozlovski[5] and G Oganesyan[2]

[1] Martin Luther University Halle, Department of Physics, 06120 Halle, Germany
[2] Ioffe Physico-Technical Institute, St. Petersburg 194021, Russia
[3] Minia Univrsity, Faculty of Science, Physics Department, Minia 61519, Egypt
[4] Inst. Ion-Plasma&Laser Technol. (Inst. of Electronics), Tashkent 700187, Uzbekistan
[5] St. Petersburg State Polytechnic University, St. Petersburg 195251, Russia

E-mail: n_arutyunov@yahoo.com

Abstract. Defects in phosphorus-doped silicon samples of floating-zone material, *n*–FZ–Si(P), produced under irradiation with 15 MeV protons at room temperature are studied by positron annihilation lifetime spectroscopy over the temperature range of ~ 30 K – 300 K and by low-temperature Hall effect measurements. After annealing of *E*–centers and divacancies, we detected for the first time high concentrations of positron traps which had not been observed earlier. These defects are isochronally annealed over the temperature interval of ~ 320 °C – 700 °C; they manifest themselves as electrically *neutral* deep donor centers in the material of *n*-type. A long-lived component of the positron lifetime, $\tau_2(I_2 \leq 60\%)$ ~ 280 ps, attributed to these centers, suggests a relaxed configuration involving two vacancies. The enthalpy and entropy of annealing of these centers are E_a ~ 1.05(0.21) eV and $\Delta S_m \approx 3.1(0.6)k_B$, respectively. It is argued that the microstructure of the defect consists of two vacancies, VV, and one atom of phosphorus, P. The *split* configuration of the VPV complex is shortly discussed.

1. Introduction

It is widely believed that accumulation of defects in *n*–FZ–Si(P) material takes place similarly under irradiation with fast electrons and protons of MeV energies, and the phosphorus-vacancy pairs (*E*–centers), together with divacancies, are believed to be dominant acceptor-type defects responsible for degrading *n*–type conductivity of silicon wafers at low doses of irradiation. However, this similarity vanishes under proton irradiation, and the generally adopted model of dominating formation of *E*–centers appears to be inadequate for interpreting experimental data [1]. It has been revealed that electrically neutral defects trapping positrons are formed in *n*–FZ–Si(P) material as a result of proton irradiation [2]. The microstructure of these defects is still an unresolved problem. Below we present some *new* data on their positron probing and analyze their possible configurations.

2. Experimental

Samples cut from Si crystal grown by the floating-zone technique were studied. They were subjected to irradiation with 15 MeV protons at room temperature. Two doses of proton irradiation, $4 \cdot 10^{13}$ and $8 \cdot 10^{13}$ protons/cm^2, were used. According to data of IR spectroscopy, the concentration of oxygen in

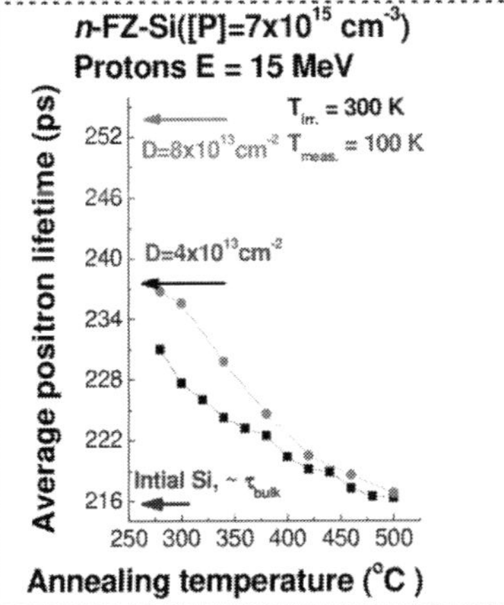

Figure 1. The average positron lifetime (τ_{av}) vs annealing temperature for the phosphorus-multivacancy centers having deep donor states at $E_{DD} > E_c - 0.24$ eV in silicon (see text). Arrows – τ_{av} after irradiation of material.

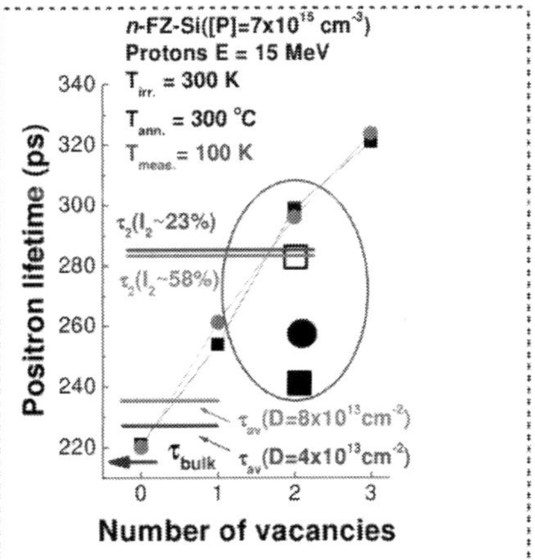

Figure 2. Positron lifetime vs number of vacancies in the vacancy cluster: ■ – Ref. [3]; ● – [4]; ■ – [5]; □ – [6]; ● – [7]. *This work*: upper lines – the long-lived component, $\tau_2(I_2)$; lower lines – τ_{av} (T_{anneal}=300 °C).

this carbon-lean material was less than ~ $2 \cdot 10^{16}$ cm^{-3}. The thickness of samples was ≤ 900 μm, which is much less than the stopping range of the protons (~1500 μm). We have conducted Hall effect measurements as a complementary method to that of positron annihilation lifetime; in a similar way this material was studied in [2]. Data on the average positron lifetime τ_{av} ($\Delta T \approx$ 30–300 K) were used for obtaining the shorter (τ_1) and longer (τ_2) values of lifetime; $\tau_{av} = \tau_1 I_1 + \tau_2 I_2$; the intensities are $I_1 + I_2 = 1$ (LT9 and LT10 programs were applied for data processing). The ratio I_2/I_1 was used for estimating the positron trapping rate k_i (see equation (1)). The irradiated samples were isochronally annealed over the temperature range of 80°C –700°C. The rate of annealing was $\alpha \approx h \cdot t_h^{-1}$ (t_h =10 min.; h = 20°). *For convenience*, we use the temperature units K and °C while discussing the data of measurements and isochronal annealing, respectively.

3. Thermally stable electrically neutral positron traps: phosphorus-multivacancy complexes

Isochronal annealing leads to decreasing the average positron lifetime (i) *proportionally* to the dose of irradiation: the ratios of τ_{av} (100K)$|_{irr.}$ values obtained before annealing (shown by arrows in figure 1) and after annealing at 300 °C τ_{av} (100K)$|_{irr.}${anneal. at 300 °C } are 1.039 and 1.078 at doses $4 \cdot 10^{13}$ and $8 \cdot 10^{13}$ protons/cm^2, respectively. This proportionality suggests (ii) that the thermally stable defects of a vacancy type are formed as a result of irradiation rather than in the course of isochronal annealing.

A clearly seen correlation between a decrease in the average positron lifetime τ_{av} and an efficiency of restoration of electrical activity of phosphorus atoms shown in [2] points out the involvement of phosphorus atoms in the microstructure of these defects: their concentration is estimated to be (iii) N_{DD} ~ 10^{15}cm^{-3} and ~ $5 \cdot 10^{14}$ cm^{-3} for the higher and lower doses of proton-irradiation, respectively.

The conclusions (i)-(iii) indicate that the defects survive at earlier stages of annealing when *E*–centers and divacancies are annealed in the temperature intervals of ~ 100 °C – 160 °C and ~ 180 °C – 280 °C, respectively. As we should expect for the carbon-lean *n*–FZ–Si(P) material [1, 2], any oxygen-related centers have *not* manifested themselves in neither electrical nor positron annihilation data.

Being deep donors with their levels $E_{DD} > E_c$–0.24 eV, the centers turned out to be electrically neutral (or even, perhaps, electrically inactive) in the investigated material of *n*–type [1]. However, owing to an open volume related to the presence of vacancies in their microstructure, the deep donors are *effective* positron traps in the irradiated material [2]: the positron trapping cross-section for these defects is estimated to change from ~$5.7 \cdot 10^{-14}$ cm^2 (230 K) to ~$3 \cdot 10^{-12}$ cm^2 (66 K). According to the data of Hall effect measurements, the isochronal annealing of these defects is finished at ~ 650 °C – 700 °C [1]; by means of PAL spectroscopy they have been detected up to ~ 500 °C (figure 1).

4. Open volume of thermally stable phosphorus-vacancy complex: a split VPV configuration

If trapped, positrons live longer in the states related to structural defects and this fact indicates the presence of vacancies in their configurations. The positron lifetime calculated *ab intio* for the open volume of a vacant site is longer than the one for the regular crystal lattice of silicon; at a divacancy the positron lives still longer [3, 4, 5]: its lifetime is getting somewhat shorter owing to the relaxation effects [6, 7]. The lifetime of the long-lived component, $\tau_2\,(100\text{K})|_{\text{irr.}}\{\text{ann. } 300\,°\text{C}\}$, proves to be ~ 280 ps at the both doses of proton irradiation (see figure 2): the value of τ_2 depends on both the temperature of annealing and measurement and it covers a range of ~ 275 – 305 ps, i.e. the one close to that being characteristic of a divacancy. As seen from figure 2, the calculated positron lifetime associated with the divacancies and the lifetime τ_2 ~ 280 ps, related to the phosphorus complexes of a vacancy type, are in the range of ~ 242 – 300 ps, thus inviting us to conclude that the phosphorus-related complex under discussion should also contain two vacancies. There is no information on *ab initio* calculations of the positron lifetime for such complexes in the literature so far. A possible atomic configuration of the complex as a VPV combination [1] is depicted in figure 4. The position of the central atom (P) may be *most likely* distorted because of a relaxation of the surrounding atoms. Earlier, a similar atomic configuration has been considered for the *E*–centre containing one full vacancy split into two semi-vacancies, i.e. 0.5V–P–0.5V [8]. We think that the thermally stable phosphorus-vacancy complex having *two* full vacancies may be configurated similarly, with a distortion of VPV configuration, which may provide suitable positron lifetime, ~ 250 ps $\leq \tau \leq$ ~280 ps, which corresponds to a pair of vacancies involved in the defect (see figure 2). *Ab initio* calculations of the structural parameters in both configurations, VVP and VPV, might shed light on this point. In this connection it will be noted a model of the tin-vacancy pair in which tin atom resides in a position halfway between two normal silicon atom sites (i.e. tin atom is in the center of a "divacancy" [14]).

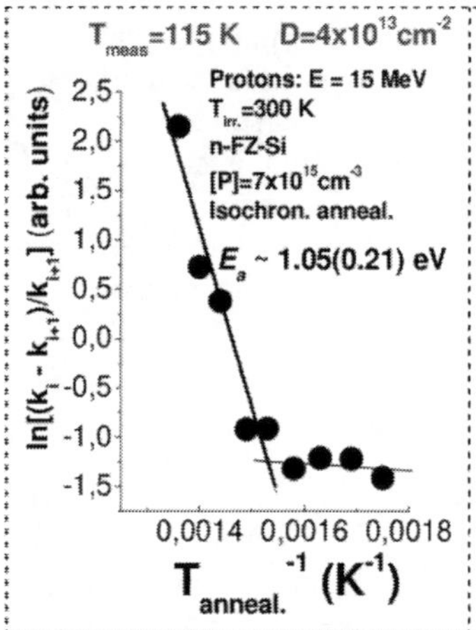

Figure 3. Arrhenius plot of equation (1) for obtaining enthalpy (E_a) of annealing of the phosphorus-multivacancy complexes. For E_a value the error of fitting is shown.

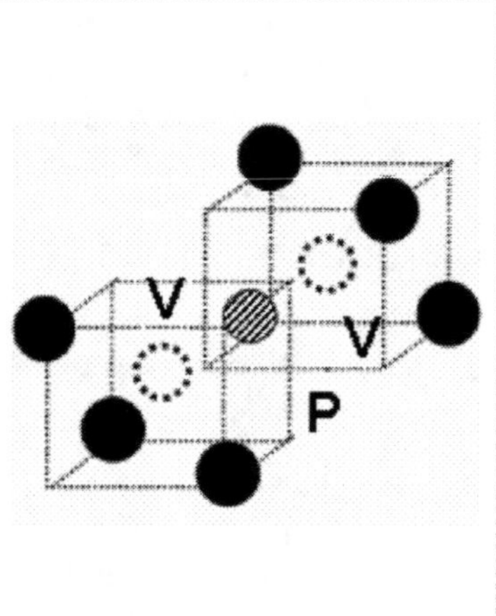

Figure 4. VPV configuration of the complex in crystal of silicon: atom of phosphorus (P) is the shaded circle between two vacancies (V). *Relaxations are not shown* (see text).

5. Enthalpy and entropy of annealing of thermally stable phosphorus-multivacancy complexes

Below we consider just *a part* of the data obtained at $T_{\text{meas.}} = 115$ K for *a low* dose of irradiation ($4 \cdot 10^{13}$ protons/cm^2), when the positron trapping is still effective (see figure 3): isochronal annealing of the defects under study is described by the equation of quasi-chemical reaction of the 1st order:

$$\ln \frac{k_i - k_{i+1}}{k_{i+1}} \cong A - \frac{E_a}{k_B T_{a(i+1)}}\,, \qquad (1)$$

where the positron trapping rate $k_i\,(T_{\text{meas.}})$ is derived at the annealing temperature T_i from the average positron lifetime [9, 11]; E_a is the enthalpy of annealing; $A \approx K_0 \cdot h \cdot \alpha^{-1}$. Factor K_0 was estimated making use of a generalized approach [10]: $K_0 \approx 4\pi r_0 \cdot N_{DD} \cdot \beta \cdot (a/2)^2 \cdot v_0 \cdot exp\,(\Delta S_m/k_B)$, where $r_0 \approx 10^{-8}$ cm; $\beta \approx$

1, $a \approx 2.35 \cdot 10^{-8}$ cm (the interatomic distance in the crystal lattice of silicon); $\nu_0 \leq 10^{13}$ s^{-1} is the characteristic frequency of atomic vibrations; $\Delta S_m/k_B$ is the change of entropy due to dissociation of the vacancy-impurity complex (k_B is the Boltzmann constant). The positron trapping rate was estimated by well-known formula [9]: $k_i \approx \lambda_b \cdot I_2/I_1 \approx \mu \cdot N_{DD}$ (μ is the positron trapping coefficient; see also, e.g., [11] for details). Using equation (1) we have evaluated the enthalpy of annealing and the entropy as equal to E_a ($\Delta T_{\text{ann.}}$= 360 – 460 °C) $\geq$ 1.05(0.21) eV and $\Delta S_m \approx 3.1(0.6) \cdot k_B$, respectively. This E_a value for activation of annealing ($\geq$ 1 eV) is close to the binding enthalpy of divacancy (~ 0.94 eV at T=0 K [12]; according to EPR data [13], a lower limit on the binding energy of two vacancies is $\geq$ 1.6 eV, and the activation energy of dissociation of divacancy is equal to ~ 1.3 eV).

The configurational entropy ~ $1 < \Delta S_m/k_B < 4$ is close to the one discussed for a short-range distortion around *point* defects in the cubic crystal lattice of diamond type. This conclusion is also supported by the positron lifetime value $\tau_2 \leq 280$ ps which we attribute to the thermally stable divacancy phosphorus-impurity complex, i.e. to the *point* defect (see also sections 3 and 4). The entropy estimated is known to characterize the effective frequency of oscillations of the atoms in the defect towards a saddle point where the defect dissociation is most probable (for more detail see, e.g., [10]). Values of ΔS_m ~ 2–3k_B include implicitly a contribution related to the distortion of the crystal lattice around the defect; an assessment of this contribution is a challenge, so knowledge of these ΔS_m values is indispensable for elucidating a role of entropy in thermal decomposition of the complexes: which configuration, a *non-split* PVV or a *distorted* VPV, is preferable, will be discussed elsewhere.

6. Conclusion

Thermally stable vacancy complexes including an impurity atom of phosphorus have been revealed in *n*–FZ–Si(P) material irradiated by 15 MeV protons. These defects are believed to be deep donors at $E_{DD} > E_c - 0.24$ eV. Isochronal annealing of these complexes in the interval of ~ 300 °C – 650 °C leads to the restoration of electrical activity of phosphorus impurity atoms as shallow donors. The enthalpy of annealing (~ 1 eV) and the configurational entropy (2–3k_B) estimated on the basis of data of the positron lifetime measurements suggest a dissociation process of the complexes consisting of one phosphorus atom and two vacancies, most likely, in a split (distorted) VPV configuration.

7. Acknowledgments

We are indebted to German Academic Exchange Service (DAAD) for supporting this work.

References

[1] Emtsev V V, Ivanov A M, Kozlovski V V, Lebedev A A, Oganesyan G A, Strokan N B and Wagner G 2012 *Fiz. Tekh. Poluprov.* **46** 473
[2] Arutyunov N Yu, Elsayed M, Krause-Rehberg R, Kessler C, Emtsev V V, Oganesyan G and Kozlovski V V 2014 *Sol. St. Phenom.* **205–206** 317
[3] Hakala M, Puska M J and Nieminen R M 1998 *Phys. Rev.* B **57** 7621
[4] Staab T E M, Haugk M, Sieck A, Frauenheim T, and Leipner H S 1999 *Physica* B **273–274** 501
[5] Staab T E M, Sieck A, Haugk M, Puska M J, Frauenheim Th and Leipner H S 2002 *Phys. Rev.* B **65** 115210
[6] Kuriplach J, Morales A L, Dauwe C, Segers D, and Sob M 1998 *Phys. Rev.* B **58** 10475
[7] Makhov D V and Lewis L J 2004 *Phys. Rev. Lett.* **92** 255504
[8] Telezhkin V A and Tolpygo K B 1973 Sov. Phys. Solid. State **15** 740
[9] Arutyunov N Yu, Elsayed M, Krause-Rehberg R, Emtsev V V, Oganesyan G and Kozlovski V V 2013 *Mat. Sci. Forum* **733** 224
[10] Vineyard G J 1957 *J. Phys. Chem. Solids* **3** 121
[11] Brandt W 1974 *Appl. Phys.* **5** 1
[12] Van Vechten J A 1986 *Phys. Rev.* B **3** 2674
[13] Watkins G D and Corbett J.W 1965 *Phys. Rev.* **138** A543
[14] Watkins G D 1975 *Phys. Rev. B* **12** 4383

Journal of Physics: Conference Series **618** (2015) 012014 doi:10.1088/1742-6596/618/1/012014

Investigation of microstructural changes of ODS steel EP450 after Helium implantation

I Bartošová[1,2], M Dománková[3], F A Selim[2] and V Slugeň[1]

[1] Institute of Nuclear and Physical Engineering, Slovak University of Technology in Bratislava, Ilkovičova 3, 812 19 Bratislava, Slovak Republic
[2] Department of Physics, Bowling Green State University, 104 Overman Hall, OH 43403, USA
[3] Department of Materials Science, Faculty of Materials Science and Technology in Trnava, Slovak University of Technology in Bratislava, Paulínska 16, 917 24 Trnava, Slovak Republic

E-mail: iveta.bartosova@stuba.sk

Abstract. Oxide dispersion strengthened steels (ODS) are good candidates for the first wall of the actual proposed nuclear fusion reactors. This work focuses on the investigation of EP450 ODS by Positron annihilation lifetime spectroscopy (PALS) and Transmission electron microscopy (TEM) in "as received" state and after Helium implantation. Helium implantation produces radiation damage which in many respects is similar to the one related to neutron irradiation. Additionally, the selected area electron diffraction (SAED) together with TEM revealed the nature of Y_2O_3 particles and identified chromium-rich carbides $M_{23}C_6$ in the base material. Positron measurements indicate a high concentration of small-scale defects distributed in the sample. TEM was used for identification of microstructural changes after Helium implantation.

1. Introduction

Oxide dispersion strengthened steels are promising materials for advanced nuclear systems. In fission reactors as well as fusion technologies the envisioned environment is much more severe. Therefore these steels will have to endure higher temperatures, doses and -for example- high corrosion in the Super-critical pressurized water reactor [1, 2]. The main advantage of ODS steels is their excellent high temperature strength. At present time, steel EP450 is used for wrapper tubes of reactors BN-600 and BOR-60 [3].The microstructure of ODS is enhanced by dispersed particles of Y_2O_3 which influence dislocation movement. In positron measurements these particles act as trapping sites for positrons due to high positron affinity of Yttrium as was shown in previous work [4]. Therefore PALS can specifically probe the microstructure around Y_2O_3 particles.

In this work we study the behaviour of yttria oxide particles in as-received and helium-implanted samples. Helium implantation often serves as a certain simulation of irradiation damage [5] as neutron irradiation inside a nuclear reactor is not always available. This method allows one to avoid difficulty associated with high level of radioactivity that makes sample handling very difficult.

2. Experimental

The chemical composition of ODS ferritic steel used in this study is given in table 1. Special care has been paid to prepare samples for TEM measurements. Since the implantation depth is only about 1μm from the surface we had to modify the sample preparation. The kinetic energy of ^{2+}He ions reached 500keV, achieving irradiation induced damage roughly 10 dpa, which is relative to $0.1 C/cm^2$. Temperature during implantation of ^{2+}He ions into the samples was about 52-60°C.

Table 1. Chemical composition of EP450 ODS steel [3].

Element	Fe	Cr	V	Ti	C	Si	Mn	Mo	Ni	Nb	Y_2O_3
Wt.%	83,24	13.1	0.18	0.25	0.13	0.23	0.44	1.61	0.12	0.35	0.35

2.1 Positron annihilation lifetime spectroscopy measurements

A standard positron source ^{22}Na deposited on a Kapton foil was used. The estimated positron source activity is about 1MBq. We applied a sandwich setup in which two identical samples cover the source to achieve higher statistics. Each spectrum contained at least $1x10^6$ counts and was evaluated with PATFIT'88 program [6]. Decay components due to annihilation in NaCl (≈430ps) and kapton foil (≈382ps) were subtracted in the procedure.

2.2 Transmission electron microscopy and selected area electron diffraction

In the case of as-received sample, a thin steel foil with thickness of 0.1mm was prepared by mechanical grinding. Afterwards targets with 3mm diameter were further thickened in Tenupol 5 device in a solution of: 300 ml HNO_3 + 700 ml CH_3OH, voltage 15V and electrolyte temperature 0°C. After cleansing in ethanol and drying, the sample was prepared for TEM.
As mentioned above, the preparation procedure of the thin foil of the implanted sample was modified. For the final thinning process ion bombardment of accelerated Argon particles was used. The acceleration voltage was 5kV, maximum current 2.5A and preparation time 15 hours.
Investigations by TEM were carried out on JEOL 200CX electron microscope with accelerating voltage 200 keV. SAED for identification of lattice parameters supplemented TEM analyses.

3. Results

3.1 TEM results

Figures 1 and 2 show TEM images for the sample before Helium implantation. Figure 1 shows non-uniform distribution of quasi-globular particles in ferritic matrix. Size and pattern of segregated globular particles is strongly heterogenic. The particles size and distribution were evaluated from TEM images. The size varied from 5 to 450 nm. Almost 60% of all particles were in the 20-40nm range. These particles were identified by SAED as particles of Y_2O_3. Aside from Yttria particles also particles of irregular shape were observed. These particles precipitate mainly on ferritic grain boundaries and were identified as $M_{23}C_6$ carbides as can be seen in figure 2. Presence of these carbides is a result of the chemical composition; however they do not have a beneficial effect on the microstructure.

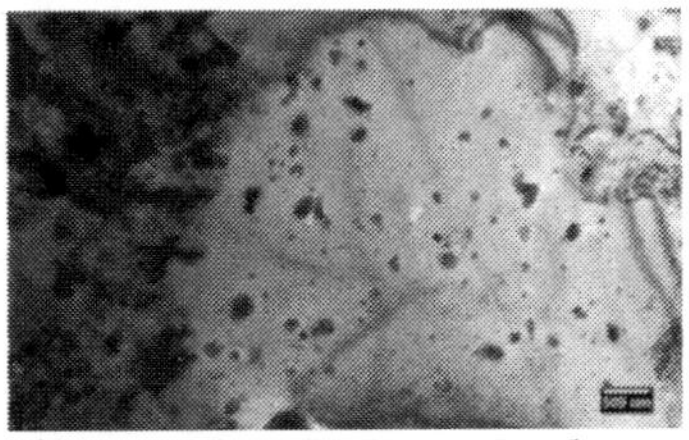

Figure 1. *Before implantation: Dispersed particles (Y_2O_3 oxides) of quasi-globular shape in ferritic matrix.*

Figure 2. *Before implantation: Detail on $M_{23}C_6$ particles of irregular shape precipitating on ferritic grain boundaries.*

Figures 3-6 represent the sub-structure of the Helium implanted sample. The matrix is composed of ferrite, which retained its polyedric morphology. In comparison to the as-received state we are able to observe increased density of dislocations in the surface layer. Also a decrease in overall Yttria particle density was observed in comparison to as-received state. The particles are still arranged heterogeneously and their size ranged from 4-200 nm, however the dominant size is still 20-40 nm.

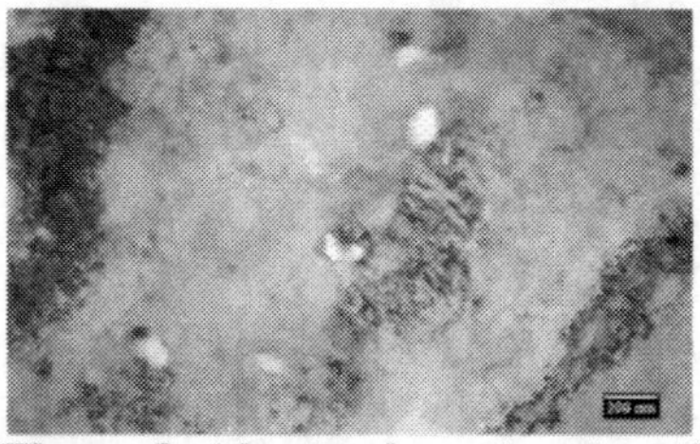

Figure 3. *After implantation: Detail on sporadic occurrence of fine Y_2O_3 particles in ferritic matrix.*

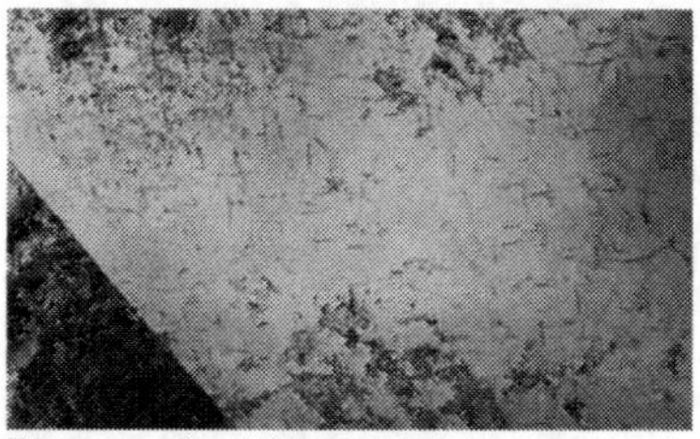

Figure 4. *After implantation: Ferritic grain with needle shaped particles.*

Fine needle shaped particles can be also observed in the ferritic matrix (figures 4, 5). From the nature of these particles behaviour, we can assume the formation of coherent aggregations of alloying elements, which probably conjoined by the influence of the implantation. These aggregations segregate in the ferritic matrix in a very homogenic manner and they also preserve their preferential orientation. From analyses of the diffraction spectrum, we conclude that the nature of segregation is related to the {211} type habitual plane of ferrite. In figure 6 we can also observe interactions of dislocations with fine globular particles of Yttria.

Figure 5. *After implantation: Detail on needle shaped particle in ferritic matrix.*

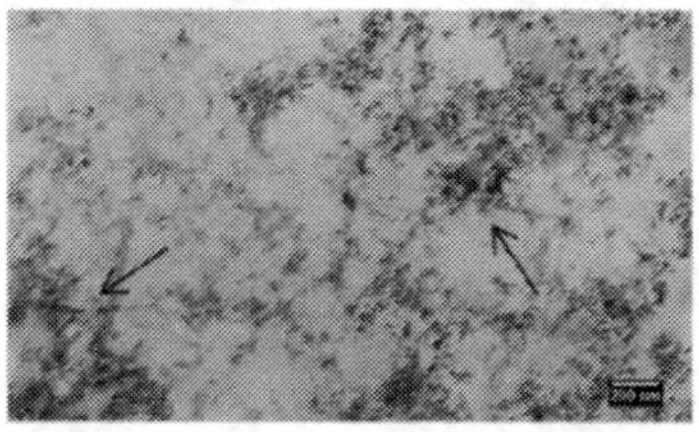

Figure 6. *After implantation: Interaction of dislocations with fine yttria particles.*

3.2 PALS results

PALS is based on the principle that electron–positron annihilation is time-dependent on the electron density in matter. If the metal contains defects such as vacancies, vacancy clusters and dislocations i.e., regions of less than average atomic density, positrons may become trapped at these defects [7, 8, 9]. The trapping of positrons into defects takes place in competition with the annihilation of the positrons in the bulk. Lifetimes τ_1, τ_2 and intensities are shown in figure 7 and 8. The shorter component τ_1 represents the reduced lifetime, which is the bulk lifetime reduced by an amount that depends on positron trapping. The longer component τ_2 represents contribution of the positrons trapped at one or more type of defects. The second component τ_2 in as-received state is about 174ps. This lifetime corresponds to dislocations and/or mono-vacancies. Figure 8 suggests that 60% of positrons get trapped and annihilate at these sites. The average lifetime value was calculated using equation (1).

$$\tau_{avg} = \tau_1 I_1 + \tau_2 I_2 \tag{1}$$

The average lifetime τ_{avg} in "as-received state" is about 133ps. After Helium implantation both values of τ_2 and τ_{avg} rose. Lifetime of positrons in defects increased by ≈70ps and the average lifetime by

Journal of Physics: Conference Series **618** (2015) 012014 doi:10.1088/1742-6596/618/1/012014

25ps. This refers to formation of larger defects, i.e. vacancy clusters. The τ_2 value of 240ps probably corresponds to clusters of 3-vacancies in average, however also larger defects may be present in the samples due to the fact that Helium can occupy vacancies and therefore the defect appears smaller. The intensity of this long-lived component is about 40%. In fact the decrease of I_2 after He implantation (see Fig. 8) indicates that yttria oxide particles have filled some of the vacancies and open volume defects and prevented positron trapping. In figure 8 we can observe interactions of dislocations with Yttria particles. It is probable that dislocations also form around the needle shape particles and together with Y_2O_3 particles create attractive sites for positron trapping and annihilation.

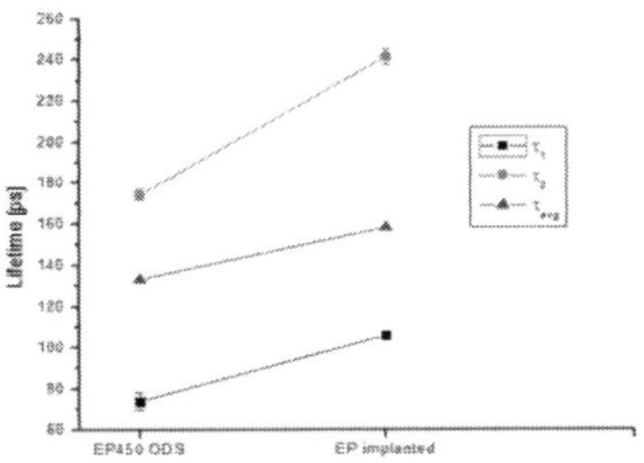

Figure 7. Reduced lifetime in bulk (τ_1) and defect (τ_2). Average lifetime (τ_{avg}) was calculated.

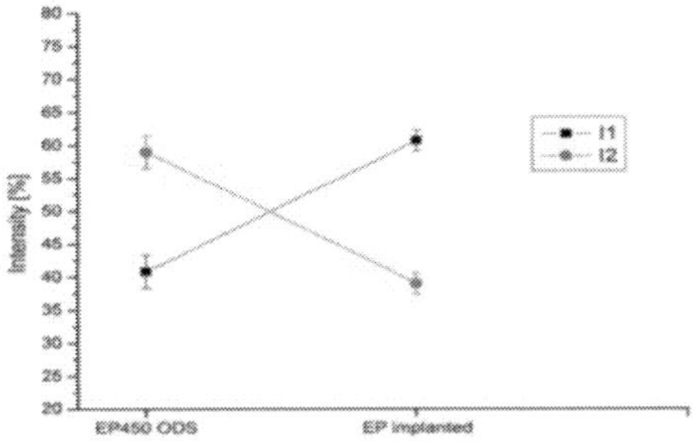

Figure 8. Intensities of lifetime components. Intensity I_1, I_2 corresponds to τ_1 and τ_2 respectively.

4. Conclusion

EP450 ODS was studied by PALS and TEM in "as-received" state and after Helium implantation. After implantation we observed formation of needle shaped particles. These are probably clumps of alloying elements which formed due to the implantation effect. The re-localization of these elements produced larger vacancy clusters and dislocations. After Helium implantation Y_2O_3 particles became more sporadic and not visible by TEM technique. Therefore implantation also influences the state of Y_2O_3 particles. In this case we observed that Yttria particles became finer and their amount decreased in comparison to "as-received" state. PALS revealed clusters of 3-vacancies, however also larger defects may be also present in the samples.

References

[1] Novotny R, Janíka P, Penttilä S, Hähner P, Macák J, Siegl J, Hausil P 2013 *J. Of Supercritical Fluids* **81** 147
[2] Jian Li, Zheng W, Penttilä S, Liu P, Woo O T, Guzonas D 2014 *J.Nucl.Mat* **454** 7
[3] Nikitina A A, Ageev V S, Chukanov A P, Tsvelev V V, Porezanov N P, Kruglov O A 2012 *J.Nucl.Mat.* **428** 117
[4] Bartošová I, Čížek J Lukáč F, Slugeň V 2014 *Acta Phys. Pol. A* **125** 702
[5] Troeva T, Popova E, Staikova P, Nankova N,Yoshiieb T2012 *J Phys: Conference series* **207** 012033
[6] Kirkegaard P, Pedersen N J, Eldrup M, Risø National Laboratory, DK–4000 Roskilde, Denmark, 1989
[7] Puska M J, Lanki P, Nieminen R M 1989 *J.Phys.:Condens. Matter* **1** 6081
[8] Krause-Rehberg R, Leipner H: Positron Annihilation in Semiconductors, Springer-Verlag, 1999.
[9] Hautojarvi P: Positrons in Solids, Springer-Verlag, Heidelberg, 1979.

Measurement of the positron diffusion constants in polycrystalline molybdenum by the observation of positronium negative ions

Takuji Suzuki, Simpei Iida, Takashi Yamashita and Yasuyuki Nagashima

Department of Physics, Tokyo University of Science, 1-3 Kagurazaka, Shinjuku, Tokyo, 162-8601, Japan

E-mail: 1213705@ed.tus.ac.jp and ynaga@rs.kagu.tus.ac.jp

Abstract. We have measured the positron diffusion constants in polycrystalline molybdenum by the observation of positronium negative ions (Ps^-). The Ps^- ions emitted from the sample surface coated with Na were accelerated. The γ-rays from the accelerated Ps^- ions were Doppler-shifted and thus the signals of self-annihilation of the Ps^- ions were isolated from those of self-annihilation of para-positronium (p-Ps) or pair-annihilation of positrons in the bulk. Clear and reliable values of the diffusion constants have been obtained.

1. Introduction

Recently, we developed a method to investigate positron diffusion in metals by the observation of positronium negative ions (Ps^-) [1]. This method has been enabled by the recent studies of efficient emission of Ps^- ions from alkali-metal coated surfaces [2–5]. The γ-rays emitted from self-annihilation of accelerated Ps^- ions are Doppler-shifted [2] and the Ps^- component can be isolated easily in the γ-ray energy spectrum. Thus reliable studies of positron diffusion are available. In the present work, we have applied this method for the measurement of the positron diffusion constants in polycrystalline molybdenum.

2. Experiment

The experimental system used was the same as that in the previous work [1]. The base pressure of the sample chamber was 4×10^{-8} Pa.

The sample was a polycrystalline molybdenum foil of 25 μm thickness. Its purity was 99.95 %. It was annealed *in situ* at 1500 K or 1800 K for 30 min by the passage of an electric current. After the sample was cooled down to room temperature, the surface was coated with Na to increase Ps^- emission efficiency. The deposition source was an alkali-metal dispenser purchased from SAES Getters S.p.A. The thickness measured by the quartz crystal oscillation deposition monitor was 9×10^{14} atoms cm^{-2}.

The sample was biased at -1 kV. Slow positrons were incident onto the sample after passing through a grounded grid located 1 mm in front of the sample. The energy of the positrons incident onto the sample, E_{e^+}, was varied from 1 keV to 30 keV. The Ps^- ions emitted from the surface of the sample were accelerated by an electric field between the sample and the grid.

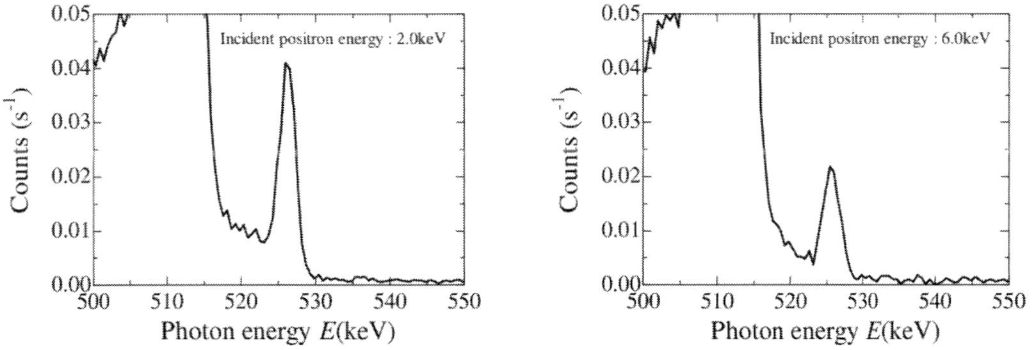

Figure 1. γ-ray energy spectra at incident positron energies of 2 keV and 6 keV after annealing of the molybdenum sample at 1800 K.

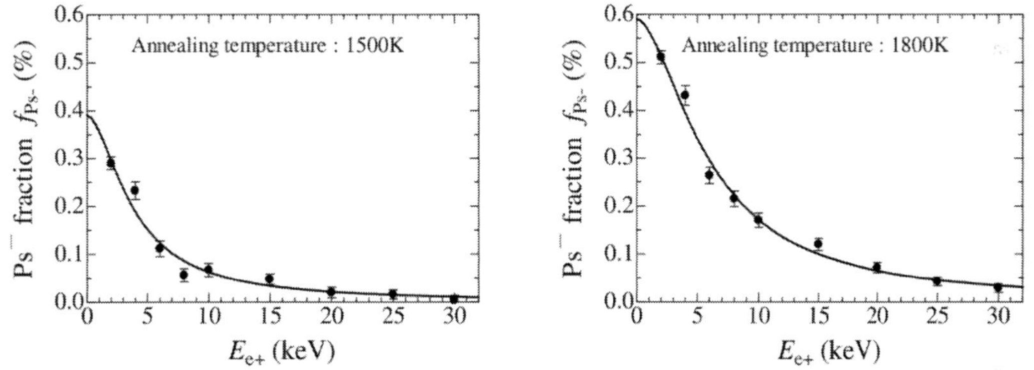

Figure 2. Ps$^-$ fractions f_{Ps^-} plotted against incident positron energy E_{e^+}.

The γ-rays from self-annihilation of the accelerated Ps$^-$ ions were detected by a Ge detector in coincidence with signals from a NaI(Tl) scintillation detector placed behind the sample to reduce the background.

At each value of E_{e^+} the data were accumulated for short time intervals (150 s) and the measurements were repeated 16 times.

3. Result and discussion

The annihilation γ-ray energy spectra for $E_{e^+} = 2$ keV and 6 keV are shown in figure 1. These spectra have been normalized to the measurement time. The blue-shifted peaks due to the self-annihilation of the Ps$^-$ ions are clearly isolated from those of positron pair-annihilation in the bulk or self-annihilation of p-Ps.

The yields of Ps$^-$ ions were obtained by fitting the annihilation γ-ray spectra with the sum of a Gaussian function and a background. The yields were normalized to the number of incident positrons at each E_{e^+} obtained from the count of the 511 keV-peak for the unannealed sample in order to calculate the Ps$^-$ emission efficiency, f_{Ps^-}.

Figure 2 shows the experimental values of f_{Ps^-} as a function of E_{e^+}. The Ps$^-$ emission

Table 1. Positron diffusion constants and diffusion lengths for the molybdenum sample annealed at 1500 K and 1800 K.

T_a (K)	E_0 (keV)	D_+ (cm^2·s^{-1})	L_+ (nm)
1500	3.8±0.5	0.13±0.09	40±14
1800	6.0±0.5	0.6 ±0.4	87±29

efficiency is expressed as follows:

$$f_{\mathrm{Ps}^-}(E_{e^+}) = \frac{f_{0\mathrm{Ps}^-}}{1+(E_{e^+}/E_0)^n} \ , \tag{1}$$

where E_0 is the incident energy at which half of the positrons return to the surface via diffusion motion, n is the power dependence coefficient of the mean positron stopping depth and $f_{0\mathrm{Ps}^-}$ is the Ps^- emission fraction. The relationship between the positron diffusion length $L_+ = \sqrt{D_+\tau}$ and E_0 is expressed as follows:

$$L_+ = A{E_0}^n \ . \tag{2}$$

Here, D_+ is the positron diffusion constant, τ is the positron lifetime in the sample (121 ps) [6] and A is a constant value which depends on the mass density of the sample, ρ, as follows [7–11]:

$$A = \frac{40\,\mathrm{nm}\cdot\mathrm{keV}^{-n}\cdot\mathrm{g}\cdot\mathrm{cm}^{-3}}{\rho} \ . \tag{3}$$

We fitted equation (1) to the experimental values for the annealing temperature, $T_a = 1500$ K and 1800 K, simultaneously. The optimized value of n was 1.7±0.1. The values of E_0, D_+ and L_+ are listed in table 1. The dependence of D_+ and L_+ on the annealing temperature indicates that more vacancies were removed by the annealing at higher temperature. The diffusion constant for molybdenum single crystal was measured by Huomo *et. al.* from the energy dependence of the positronium fraction emitted from the surface [12]. However the present values can not be compared with it because the annealing temperatures were low and the sample was polycrystalline.

In the future, we are planning to measure the positron diffusion constants in single crystals of tungsten, molybdenum and other metals where efficient emission of Ps^- ions has been confirmed [13, 14].

4. Acknowledgement

This work was supported by a Grant-in-Aid for Scientific Research (No.24221006) from the Ministry of Education, Science and Culture of Japan. We thank Luca Chiari for revising the spelling in this paper.

References

[1] Suzuki T, Terabe H, Iida S, Yamashita T and Nagashima Y 2014 *Nucl. Inst. Meth.* B **334** 40
[2] Nagashima Y and Sakai T 2006 *New. J. Phys.* **8** 319
[3] Nagashima Y, Hakodate T, Miyamoto A and Michishio K 2008 *New. J. Phys.* **10** 123029
[4] Terabe H, Mishichio K, Tachibana T and Nagashima Y 2012 *New. J. Phys.* **14** 015003
[5] Nagashima Y 2014 *Phys. Rep.* **545** 95

[6] Mackenzie I K 1983 *Positron Solid-State Physics: Proc. Int. School Phys. 'Enrico Fermi', course 83* ed W Brandt and A Depasquier (Amsterdam: North-Holland) p 196
[7] Schultz P J and Lynn K G 1988 *Rev. Mod. Phys.* A **60** 701
[8] Lynn K G and Lutz H 1980 *Phys. Rev.* B **22** 4143
[9] Mills A P Jr 1978 *Phys. Rev. Lett.* **41** 1828
[10] Wilson R J and Mills A P Jr 1983 *Phys. Rev.* B **27** 3949
[11] Nielsen B, Lynn K G and Vehanen A 1985 *Phys. Rev.* B **32** 2296
[12] Huomo H, Vehanen A, Bentzon M D and Hautojärvi P 1987 *Phys. Rev.* B **35** 8252
[13] Nagashima Y, Hakodate T, Miyamoto A, Michishio K and Terabe H 2009 *J. Phys. Conf. Sci.* **194** 012039
[14] Nagashima Y, Hakodate T, Miyamoto A, Michishio K and Terabe H 2009 *Phys. Status. Solidi.* C **6** 2291

Journal of Physics: Conference Series **618** (2015) 012016 doi:10.1088/1742-6596/618/1/012016

Positron annihilation studies of Eu and Dy doped α'-Sr_2SiO_4

S K Gupta, K Sudarshan*, S K Sharma, P K Pujari and V Natarajan
Radiochemistry Division, Bhabha Atomic Research Centre,
Mumbai 400085, India

*E-mail: kathis@barc.gov.in

Abstract: Sr_2SiO_4 is an important inorganic host for lanthanide doped white light emitting diodes (LEDs). Strontium silicate (Sr_2SiO_4) samples doped with 1.0 mol % of Eu^{3+} and Dy^{3+} content were prepared via sol-gel route and characterized by X-ray diffraction (XRD), Raman spectroscopy and positron annihilation spectroscopy (PAS). The concentration of the dopant ion and the temperature of annealing were optimized for maximum luminescence intensity. The positron annihilation lifetime and coincidence Doppler broadening (CDB) measurements indicated that the local environment around the positron annihilation site is different in Eu^{+3} doped and Dy^{+3} doped samples. The results could be explained based on the different local site occupancy of Eu^{+3} and Dy^{+3} in the matrix.

1.0 Introduction

Oxide-based hosts have received considerable attention for use in flat-panel displays due to their luminescent characteristics, stability in high vacuum, and the absence of corrosive gas emission under electron bombardment, as compared to currently used sulfide-based phosphors [1]. Therefore, oxide-based phosphors are likely to emerge as the choice for field emission diodes (FED) green or red phosphors. Among these, strontium silicate is an excellent matrix due its stable crystal structure, good mechanical strength and high thermal stability provided by the tetrahedral silicate $(SiO_4)^{2-}$ group [2]. Sr_2SiO_4 has attracted current interest due to its special structural features and potential application in developing white LEDs. GaN (400 nm chip) coated with Sr_2SiO_4 : Eu^{2+} was shown to exhibit better luminous efficiency than that of industrially available products such as InGaN (460 nm chip) coated with YAG:Ce [3]. The optical band gap of alkaline earth silicate is 4.7eV and therefore these materials are characterized by good transmission properties in the visible part of the electromagnetic spectrum. Strontium silicate exists in monoclinic (β-Sr_2SiO_4) phase at low temperatures and in orthorhombic (α'-Sr_2SiO_4) phase at high temperatures with a transition temperature of ~85 ^{0}C [4]. It has been reported that the crystal structures of α'-Sr_2SiO_4 (orthorhombic), and β-Sr_2SiO_4 (monoclinic), are very similar. In β→α' transition a small rotation in SiO_4 tetrahedra cause appearance of (100) plane, which is absent in β form [4, 5]. In both the phases there are two different Sr sites: one having coordination number 9 and the other with coordination number 10. The ten coordinated Sr(1) sites form linear three-membered rows of (Si–O–Sr(1)–O–Sr(2)) whereas the nine coordinated Sr(2) sites form zig-zag chains of (Sr(1)–O–Sr(2)–O–Sr(1)) along the b-axis [3]. Pictorial representation of the structure is given in figure 1.

It is known that the luminescence properties are strongly influenced by the presence of lattice defects [6] like mono vacancy, vacancy cluster, voids or pores which are formed naturally during synthesis. The positron annihilation spectroscopy is a powerful and sensitive tool to identify the defects in the solids, distinguish the type of defects and provide relevant information about the size and environment around the defects [7].

In the present work, $Sr_2SiO_4:Eu^{3+}$ and $Sr_2SiO_4:Dy^{3+}$ phosphors are synthesized by sol–gel method using tetraethyl orthosilicate as precursor. The samples were characterized by XRD for phase purity. The positron annihilation lifetimes and coincidence Doppler broadening of annihilation radiation has been measured and the results are discussed.

2.0 Experimental

2.1. Sample preparation:

All the chemicals used in the sample preparation were of Analytical reagent (AR) grade and procured from Sigma Aldrich. The alkaline earth silicate samples were prepared via a sol-gel route using tetraethyl orthosilicate (TEOS) and strontium nitrate. For preparation of lanthanide doped sample (1.0 mol %), appropriate quantities of $Eu(NO_3)_3$, and $Dy(NO_3)_3$ were added to the hydrolyzed TEOS solution. The details of synthesis are mentioned in our earlier work [8, 9].

2.2. Instrumentation:

The phase purity of the prepared phosphors was confirmed by X-ray diffraction (XRD). The measurements were carried out on a STOE X-ray diffractometer equipped with Ni filter, scintillation counter and graphite monochromator. The diffraction patterns were obtained using monochromatic Cu-K_α radiation (λ= 1.5406Å) keeping the scan rate at 1sec/step in the scattering angle range (2θ) of 10° to 60°. The $K_{\alpha 2}$ reflections were removed by a stripping procedure to obtain accurate lattice constants.

Positron annihilation life time spectroscopy (PALS) measurements were carried out by using a fast-fast coincidence system with plastic scintillation detectors. The time resolution of lifetime spectrometer was 260 ps (^{60}Co) with time calibration of 25 ps/channel. ^{22}Na, (10 μCi) deposited and dried between two kapton films of thickness 8 micron was used as positron source. The positron source was completely immersed in powder samples to stop all positrons in the sample. For each of the three spectra ~10^6 counts were recorded. The computer program, PATFIT-88 was used for the analysis of the spectra. Two HPGe detectors each having resolution of 1.1 keV at 514 keV of ^{85}Sr, located at 180° relative to each other were used to record two annihilation photons *E*1 and *E*2 emitting from a single event.The sum and difference in the energy of the two gamma-rays is plotted. The details of the coincidence Doppler broadening system and data analysis are reported in ref [10].

3.0. Results and discussion

3.1. Structural analysis: Powder X-ray Diffraction (PXRD)

The purity and crystallinity of the as-prepared Sr_2SiO_4 and $Sr_2SiO_4:Ln^{3+}$ (Ln = Eu, Dy, Sm) samples were characterized using powder X-ray diffraction (XRD). As seen from the XRD pattern in figure 2, the diffraction peaks of both Sr_2SiO_4 and $Sr_2SiO_4:Ln^{3+}$ (Ln = Eu and Dy) samples could be indexed to the orthorhombic phase of α'- Sr_2SiO_4 (JCPDS card No. 39-1256). Incorporation of lanthanide ion has not changed the XRD pattern and no impurity peaks were observed confirming that doping has not distorted the structure of strontium silicate and dopants are efficiently dissolved in the Sr_2SiO_4 host lattice by replacing the Sr^{2+} because of the similar ionic radius and very small level of doping (1.0 mol %).

3.3. Positron annihilation spectroscopic studies

From the coincidence Doppler broadening measurements, the momentum distribution ratio curves of Sr_2SiO_4, and Eu^{3+}, Dy^{3+} doped Sr_2SiO_4 with respect to silicon as reference are shown in figure 3. It is clearly seen from the figure 3 that the shape of the momentum ratio curves in all cases are similar but the heights of the peaks differs considerably in the case of Dy^{3+} doped Sr_2SiO_4. The similar shape of the curves indicates that the elements around the site of positron annihilation in each case are similar. The peak around 10×10^{-3} m_0c in ratio curves of different oxide samples with respect to silicon is attributed to the positron annihilations with 2p electrons of oxygen [11].

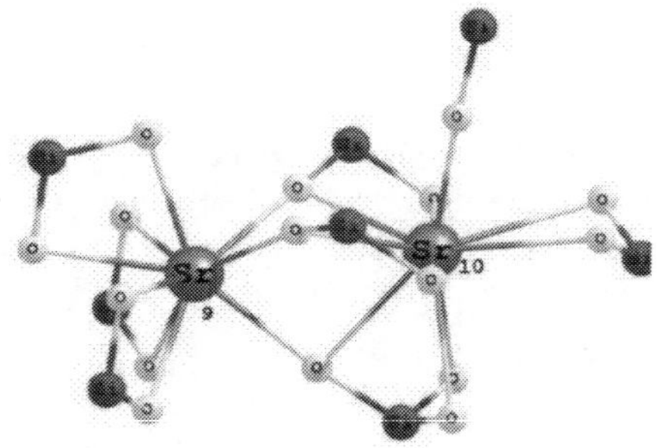

Figure 1: Pictorial representation of 9- and 10-coordinated Sr^{2+} in strontium silicate

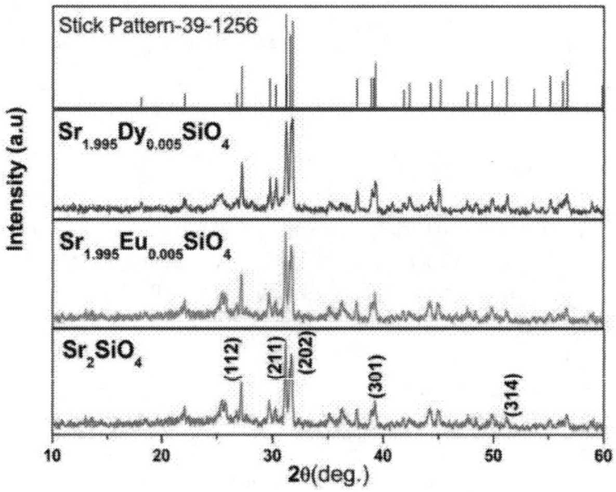

Figure 2: XRD pattern Sr_2SiO_4 and Sr_2SiO_4:Ln^{3+} (Ln = Eu and Dy)

The positron annihilation lifetime spectra could be fitted to three components. The lifetime of the long lived component was 1-1.2ns with ~1% intensity. The origin of this component is not clear. The two other major components observed are given in table 1 along with the respective intensities. As seen from Table 1, τ_1 in Sr_2SiO_4 increased from 184 ps to 212 on Eu doping and to 236 ps on Dy doping. The increase in τ_1 is higher than seen due to oxygen vacancies. In general, life time changes are in the range of 10 ps for oxygen vacancies, 10-20 ps in the case of oxygen clusters and 30-50 ps for cation vacancies [12]. The lifetime changes suggest that the changes are due to cation vacancies. However, the change in lifetime upon Dy doping is more drastic than Eu doping. Similar results are observed in the case of coincidence Doppler broadening measurements too. The second lifetime component is similar both in Eu and Dy doped samples but is higher than that in the undoped sample. This lifetime could be due to vacancy clusters with contributions expected from the positron annihilations from surface of the particles.

Table 3: Positron life time values with their intensity

Sample	τ_1 *(ps)*	τ_2 *(ps)*	I_1 *(%)*
Sr_2SiO_4	*183.6 ±2.2*	*343.8 ±3.8*	*48.3±1.0*
Eu- doped	*212.5 ±1.9*	*395.0 ±7.1*	*67.0±1.0*
Dy- doped	*235.9 ±1.9*	*399.6 ±8.6*	*70.9 ±1.1*

It has been reported from time resolved fluorescence spectroscopy (TRFS) studies that Eu(III) ions occupy both Sr(1) (10 coordinated) and Sr(2) sites (9 coordinated), whereas Dy(III) ions occupy only Sr(2) (9 coordinated) sites [8, 9]. Though lanthanide doping is expected to increase the cation vacancies there by increasing the positron annihilation lifetimes, the nature of vacancies could be different due to preferential occupation of Dy in only the 9-coorinated sites. This would leave more cation vacancies at Sr(1) site with 10 oxygens around in the Dy sample. This might be the reason for enhanced peak in the coincidence Doppler broadening momentum ratio curve as well as the large change in the τ_1.

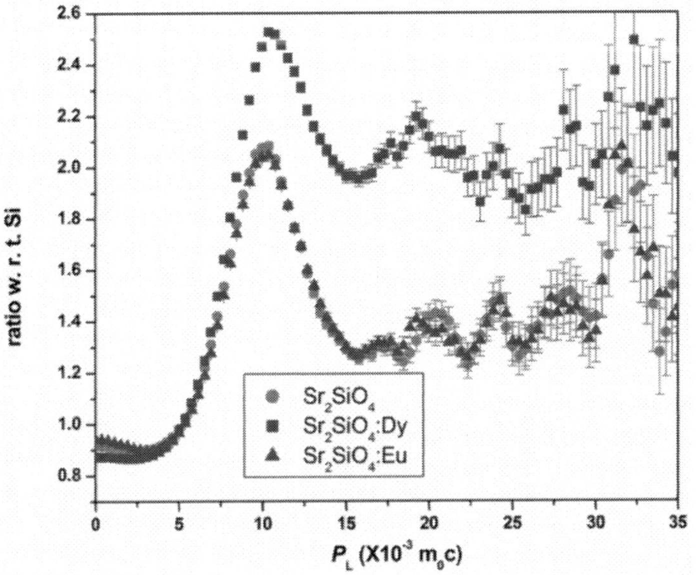

Figure 3: Ratio curves with respect to Si single crystal from coincidence Doppler broadening measurements on different Sr_2SiO_4 samples

4. Conclusions

The positron annihilation studies in Eu^{3+} and Dy^{3+} doped Sr_2SiO_4 showed that though positrons are annihilating from the cation vacancies in these samples, the local environment around the annihilation site is different in both the cases. The results are consistent with the reported time resolved fluorescence results where it has been shown Dy preferentially occupies 9-coordinated site of the Sr_2SiO_4 matrix. This results in more cation vacancies in 10-oxygen coordinated site in the lattice in Dy doped sample and it manifests as larger peak in the coincidence Doppler broadening momentum ratio curve as well as large positron lifetime.

References:

[1] Gupta S K, Sahu M, Krishnan K, Saxena M K, Natarajan V and Godbole S V 2013 *J. Mater., Chem. C* **1** *7054*

[2] Denault K A, Brgoch J, Gaultois M W, Mikhailovski A, Petry R, Winker H, Denbaars S P and Seshadri R 2014 *Chem. Mater.* **26** 2275

[3] Nag A and Kutty T R N 2004 *J. Mater. Chem.* **14** 1598

[4] Catti M, Gazzoni G and Ivaldi G 1983 *Acta Crystallogr. Sect. C.* **39** 29.

[5] Hyde B G, Seller J R and Stenberg L 1986 *Acta Crystallogr. Sect. B.* **42** 423.

[6] Lin Y -H, Li M, Nan C -W and Zhang Z 2007 *J. Am. Ceram. Soc.* **90** 2991.

[7] Kuramoto E, Tsutsumi T, Ueno K, Ohmura M and Kamimura Y 1999 *Comput. Mater. Sci.* **14** 28.

[8] Gupta S K, Mohapatra M, Kaity S, Natarajan V and Godbole S V 2012 *J. Lumin.* **132** 1329

[9] Gupta S K, Kumar M, Natarajan V and Godbole S V 2013 *Opt. Mater.* **35** 2320

[10] Sharma S K, Sudarshan K, Maheshwari P, Dutta D, Pujari P K, Shah C P, Kumar M and Bajaj P 2011 *Eur. Phy. J. B* **82** 335.

[11] Sharma S K, Pujari P K, Sudarshan K, Dutta D, Mahapatra M, Godbole S V, Jayakumar O D and Tyagi A K 2009 *Solid State Communications* **149** 550.

[12] Vance E R, Hanna J V and Hadley J H 2012 *Adv. Appl. Ceram.* **111** 94.

Positron annihilation studies in Li-implanted alumina

Prashant V Gaikwad[1,2*], S K Sharma[2], S Mukherjee[2], K Sudarshan[2], P Maheshwari[2], P K Pujari[2] and A Kshirsagar[1]

[1] Department of Physics, Savitribai Phule Pune University, Pune, India 411007.
[2] Radiochemistry Division, Bhabha Atomic Research Centre, Trombay, Mumbai, India 400085.

E-mail: prashant@physics.unipune.ac.in*, pujari@barc.gov.in

Abstract. Depth dependent Doppler broadening of annihilation radiation (DBAR) measurements are carried out for a sample of Li ion implanted in alumina. The effect of Li ion implantation and the subsequent isochronal annealing at the temperatures up to 1100 °C on the Doppler broadening annihilation parameters (S-parameter) are studied. The S-parameter around the Li implantation depth ($\sim$191 nm) increased with annealing temperature up to 700 °C and reduced beyond. The results suggest possible Li cluster formation in annealed sample.

1. Introduction

Metal or semiconductor nano-clusters are well known for their interest in optical properties as well as their scope in the fields of quantum dots [1, 2]. Ion implantation followed by annealing is one of the popular techniques used for synthesizing embedded nano-clusters in a host matrix [3]. It has been used to synthesize clusters of Si in SiO_2; Se, Cd and CdS in Al_2O_3; Li, Zn, Ag and Au in MgO;etc. [3-5]. Superiority of ion beam implantation technique is due to its ability to implant insoluble ions near the surface of host matrix. If the impurity is insoluble in host matrix then on thermal annealing they form precipitates followed by nano-clusters or nanocrystals [3].

Positron annihilation spectroscopy with the tunable energy slow positron beams is a very useful technique to probe specific depths. When heavy ions are implanted in a matrix, their implantation is localised around their range. Beam based positron annihilation studies are used to understand the radiation damage/defects created due to ion bombardment [4, 6]. Positron annihilation spectroscopy is also a very useful technique in studying the embedded systems. In many favourable cases, the differential positron affinity between the host matrix and embedded particles enables the selective study of the embedded systems. More positive the difference in positron affinities of the cluster and host, better is the confinement of positron in clusters. The positron annihilation studies on the effect of radiation and heavy ion implantation in semiconductors are well documented [7-9]. Besides the radiation effects, the Li clusters formed in high band gap materials like MgO were studied using this differential positron affinity between MgO and Li clusters to the advantage. The difference in the positron affinities causes most of the (92 %) positron localisation and subsequent annihilations inside the Li clusters, despite low percentage of metal doping (Li/MgO $\sim$ 1.3 %)[4]. Similar studies were reported on metal nano precipitates growth in alloys using coincidence Doppler broadening technique. About 90 % positrons were shown to annihilate in Cu nano precipitates in Fe even at very low concentration of Cu doped in Fe (Fe:Cu 1 % Cu)[10]. Although previous studies confirm formation of metal or semiconductor clusters inside the insulators in certain combinations, mechanism of cluster formation, dependency of growth rate of clusters on annealing temperature, defect formation due to irradiated ion beam and their annealing behaviour etc. are not fully understood. In present studies DBAR measurements have been carried out on Li ion implanted alumina for better understanding of these phenomenons.

2. Experimental Technique

Li ions of 50 keV from Low Energy Accelerator Facility (LEAF) at Bhabha Atomic Research Centre (BARC) have been implanted in single crystal α-Al_2O_3. The samples are of about 2.54$\times$ 2.54 cm^2 in area and 0.1 cm in thickness. The beam current was 200 nA and irradiation was done for 22 hours on about 1 $\times$ 1 cm ^{2} area of sample. Hence, the total fluence of implantation was 1×10^{16} ions/cm^2. The temperature change in the sample due to implantation of ions was not monitored. The ion implanted sample was isochronally annealed for 30 minutes at various temperatures in air. DBAR measurements are carried out using the slow positron beam at the Radiochemistry Division of BARC. The energy of the positrons is varied by floating the target to requisite voltage. A 30 % relative efficiency HPGe with resolution of 1.2 keV for 514 keV of ^{85}Sr is used for Doppler broadening measurements. The S-parameter is evaluated as fractional peak area in the region of 511$\pm$0.68 keV.

3. Results and Discussion

The range of 50 keV Li ions in the alumina matrix calculated using SRIM-2013 is about 191 nm. The S-parameter is evaluated from the DBAR spectra and the S-parameter profiles in the as received alumina and Li implanted alumina in the positron energy range up to 20 keV are shown in figure 1. The mean implantation depth of the positrons at various energies is calculated in nanometer (x) using equation 1, where ρ is the density of alumina in g/cm^3 (3.95 g/cm^3), and E is the energy of positrons in keV.

$$x(nm) \quad = \quad \frac{40}{\rho(g/cm^3)} \times E^{1.6}(keV) \tag{1}$$

The mean implantation depth of positron are also shown in figure 1. The calculated mean implantation depth of positrons of energy 6.3 keV corresponds to the range of 50 keV Li ions in the alumina matrix. The range of Li ions in the matrix is also indicated in figure 1.

As seen from figure 1 the S-parameter values in Li implanted sample are lower than that of as received sample till the depth corresponding to the positron energy value of 12 keV and there is no difference between the two samples beyond 12 keV. The lowering in S-parameter in Li implanted sample is maximum around the region of Li implantation, where the linear energy transfer of Li ions is also maximum. This indicates that the lowering in the S-parameter might be either due to the creation of vacancies by the Li ion or presence of Li impurity. The positron trapping in Al vacancies would enhance positron annihilation with comparatively higher energy valence electrons of oxygen, thereby reducing the S-parameter. The sample was annealed for 30 minutes in air at $500, 700, 900$ and 1100 °C temperatures and characterised by DBAR technique between each temperature change. The corresponding S-parameter profiles at these annealing temperatures are shown in figure 2 along with representative error bars. The relative variation of normalised S-parameter at 191 nm depth with annealing temperature is given in inset of figure 2.

It is seen from the figure that average S-parameter around Li implantation depth increases with annealing temperature and is maximum in the sample annealed at 700 °C. On further increase in the annealing temperature (1100 °C), the average S-parameter starts to decrease and the value falls even below the un-annealed sample. The increase in the S-parameter upon annealing at 500 °C or 700 °C could be attributed to either the agglomeration of the defects

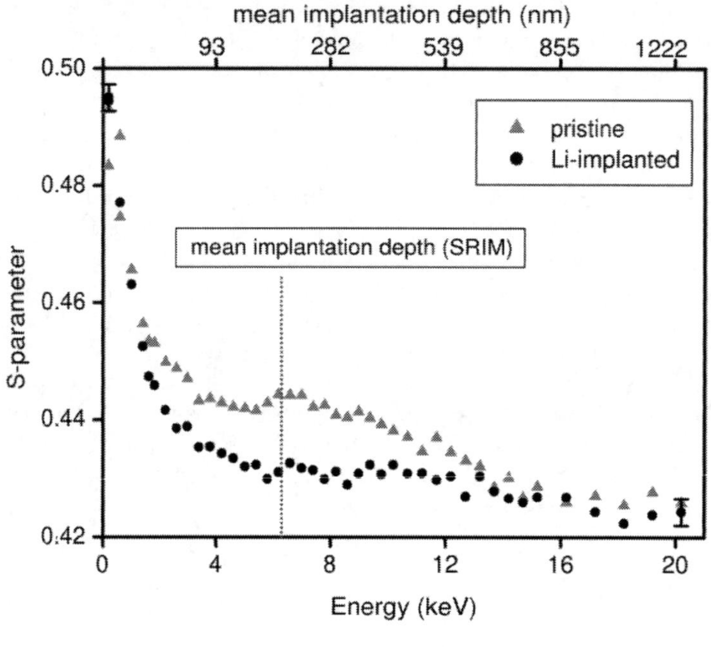

Figure 1. Depth dependent S-parameter profile with incident positron energy for the pristine sample and irradiated with Li-ion beam of 1×10^{16} fluence. Representative error bars are shown.

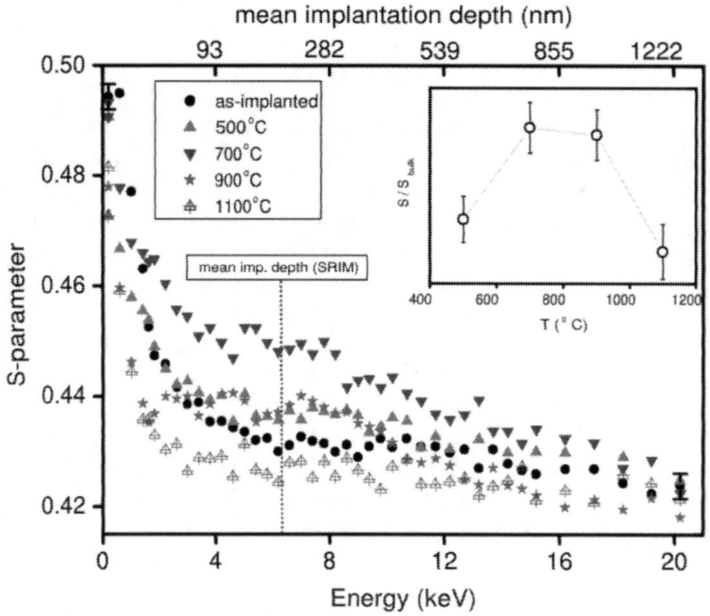

Figure 2. S-parameter profiles in Li implanted (1×10^{16}) alumina annealed at different temperatures. Inset shows variation in S-parameter at the implantation depth of Li-ions.

or positron trapping in the Li clusters formed. Similar behaviour of S-parameter was observed previously in the case of Li-ion implanted MgO on isochronal annealing and was attributed to the confinement of positron in the Li-clusters [4]. Similar phenomenon is also expected in the present case. On annealing at still higher temperatures (900 °C) the S-parameter has decreased. Have the changes in S-parameter with annealing temperature been due to agglomeration or annealing out of defects, the S-parameter should have been equal to the as received sample at highest annealing temperature. Lower S-parameter than even the as irradiated alumina in the case of the sample annealed at 1100 °C suggests that Li clusters might have migrated or dissolved leaving behind vacancies. Present experimental results suggest that the formation of Li clusters upon annealing up to 700 °C, and their breaking and migration beyond this temperature is more possible. However theoretical and experimental studies at higher Li fluences may be necessary to confirm this conclusion.

4. Conclusion

The Li ion implantation in alumina reduced the S-parameter around the Li implantation depth due to creation of large number of vacancies. Upon isochronal annealing, the S-parameter increases till 700 °C and then reduces beyond that. The increase in the S-parameter can be explained either as annealing out of the defects or positron trapping in Li-cluster formed upon annealing. However, the reduction in the S-parameter at higher annealing temperatures can not be explained based on defects annealing. Though Li-cluster formation and migration of these cluster at higher temperatures is the likely reason for the observed positron systematics, further theoretical studies are required to support the conclusion.

Acknowledgments

The authors thank the staff of LEAF, BARC for ion irradiation facility, Dr. Jyotiprakash for high temperature furnace facilities. P. V. Gaikwad thanks BARC for financial support and Dr. Dhanadeep Dutta for continuous valuable discussions.

References

[1] Fukumi K, Chayahara A, Kodano K, Sakaguchi T, Horino Y, Miya M, Jujii K, Hayakawa J and Satou M 1994 *J. Appl. Phys.* **75** 3075.
[2] Chakraborty P 1998 *J. Mater. Sci.* **33** 2235.
[3] White C W et. al. 1998 *Nuclear Instruments and Methods in Phys. Res. B.* **141** 228.
[4] Van-Huis M A, Van-Veen A, Schut H, Falub C V, Eijt S W H, Mijnarends P E and Kuriplach J 2002 *Phys. Rev. B* **65** 085416.
[5] Van-Huis M A, Van-Veen A, Schut H, Eijt S W H, Kool B J, De Hosson J Th M and Hibma T 2003 *Rev. Adv. Mater. Sci.* **4** 60.
[6] Zhi-Quan C, Maekawa M and Kawasuso A 2006 *Chin. Phys. Lett.* **23** 675.
[7] Schaefer H and Forster M 1989 *Mater. Sc. and Eng.* **A109** 161.
[8] Zinkle S J and Kinoshita C 1997 *J. Nuc. Mater.* **251** 200.
[9] Karwasz G P, Zecca A, Brusa R S and Pliszka D 2004 *J. Alloys and Compounds* **382** 244.
[10] Onitsuka T, Takenaka M, Kuramoto E, Nagai Y and Hasegawa M 2001 *Phys. Rev. B.* **65** 012204.

Positron Annihilation Study of Zr-2.5 wt.% Nb alloy Irradiated by Ar^{9+} heavy ions

Aruna Devi[a,*], Ranjini Menon[b], Priya Maheshwari[c], S. Neogy[a], P. Mukherjee[b] P.Y. Nabhiraj[b], P.K. Pujari[c], D. Srivastava[a] and G.K. Dey[a]

[a] Materials Science Division, Bhabha Atomic Research Centre, Trombay, Mumbai 400085, India.
[b] Variable Energy Cyclotron Centre, 1/AF Bidhannagar, Kolkata 700064, India.
[c] Radiochemistry Division, Bhabha Atomic Research Centre, Trombay, Mumbai 400085, India.

dsrivastavabarc@yahoo.co.in

Abstract. Zr-2.5 Nb alloy is used as a pressure tube material in pressurized heavy water reactor (PHWR). It is one of the most critical component which decides the lifespan of the reactor. The in-reactor degrading phenomenon of prime concern is dimensional changes caused by irradiation induced creep and growth processes. The present study aims to understand the mechanism of irradiation damage by irradiating the alloy with heavy ion. Such type of irradiation study would facilitate larger damage of material in a shorter time. Zr-2.5Nb alloy samples were irradiated using 315 keV Ar^{9+} ion for different durations. The irradiation doses were varied in the range of $3.1X10^{15}$ to $4.17X10^{16}$ Ar^{9+}/cm^2. SRIM calculation was carried out to evaluate damage profile in the irradiated samples. Beam based Positron Annihilation Spectroscopy (PAS) technique was used for depth profiling to characterize defect distribution in the alloys. The no. of defects generated is seen to increase with the increase in the fluence.

1. Introduction

Zirconium based alloys are used as structural materials in pressurized heavy water reactors (PHWRs) and replaced from Zircaloy-2 due to their low neutron absorption cross-section, good mechanical strength at operating temperature, low irradiation creep rate, high corrosion resistance, irradiation creep resistance and lower hydrogen absorption rates [1-3].The pressure tube carries the nuclear fuel and the high temperature, high pressure (typically 573 K and 10 MPa, respectively) water coolant, and it is subjected to fast neutron irradiation (typical neutron flux of 3.0 X 10^{17} n/m^2s, E > 1 MeV neutrons) [4,5]. Under this critical operating environment, the pressure tube material undergoes irradiation enhanced deformation and change in microstructure properties over a period of time. The nature of irradiation damage in Zr-2.5Nb is affected by the type of ion, irradiation dose and impurity content [6]. In case of heavy charged particle irradiation, having moderate energy, a significant amount of line defects in form of dislocations are also formed along with considerable amount of point defects [7]. Ion irradiation is not only an interesting method to qualify reactor materials but also to simulate the in-reactor damage by neutron irradiation [8]. There exist similarities and correlation between the damage or defects in materials generated by ion irradiation and neutron irradiation. Irradiation by heavy ion provides many advantages such as: (1) higher displacement rate which leads to shorter irradiation time, (2) lesser nuclear reactions, (3) controlled bombardment conditions and (4) lower irradiation cost compare to neutron irradiation experiments [9,10]. Heavily charged Ar^{9+} ions have been used in present study as it leads to dense cascade similar to those observed in neutron beam damage [11].

Positron annihilation spectroscopy (PAS) is a powerful technique for characterizing the vacancy-type defects such as micro-voids, open volume defects in solids materials [12]. The slow positron technique, Doppler broadening spectroscopy (DBS) using mono-energetic or variable energy positron allow to measure the defect profiles along the depth in solids, and it is most suitable for analyzing the surface and near-surface defects in ion-irradiated materials [13,14].

The focus of this work has been to characterize the irradiation damage of Zr-2.5Nb alloy following irradiation by 315 keV energy Ar^{9+} ions in the fluence range of 10^{15}–10^{16} Ar^{9+}/cm^2. The formation of vacancy-like defects due to irradiation was investigated by PAS technique. Analytical calculations using the SRIM-2013 Monte Carlo Code were carried out for ascertaining the dpa (displacement per atom) level at different fluences of ion irradiation. The findings of the slow positron DBS measurements are correlated with SRIM-2013 [15] simulation results.

2. Experimental methods

2.1. Material and sample preparation

Zr-2.5Nb pressure tube material was received from Nuclear Fuel Complex, Hyderabad, India. Samples of approximately 1 cm X 1 cm X 0.5 cm size were machined out and the surfaces were mechanically prepared following standard metallographic practice. The sample surfaces were finally electrochemically polished using an electrolyte consisting of 20% v/v Perchloric acid in Methanol at a temperature of -48 °C and voltage of 20 V to remove any stresses introduced during sample preparation.

2.2. Argon ion irradiation

A set of five samples were irradiated at room by 315 keV Ar^{9+} ion with the fluence of $3.1X10^{15}$, $6.25X10^{15}$, $9.3X10^{15}$, $2.08X10^{16}$, and $4.17X10^{16}$ Ar^{9+} ion/cm^2 at Variable Energy Cyclotron Centre, Kolkata, India. All the irradiations were carried out at room temperature. SRIM-2013 simulation software has been used to evaluate the depth profile of Ar^{9+} ions and damage distribution in terms of displacement per atom (dpa). The dpa was calculated from Norgett-Robinson-Torrens (NRT) formula [16];

$$dpa = \frac{0.8}{2E_d} \times \left(\frac{dE}{dX}\right)_n \times 10^8 \times \frac{\phi}{N}$$

where E_d = 40 eV [17] is the threshold displacement energy, Φ is the fluence in Ar^{9+}/cm^2, N is the atomic density (4.316 X 10^{22} atom/cm^3) and dE/dX in eV/Å is the linear energy transfer to the target material in nuclear process. dE/dX was calculated from SRIM-2013, which is equal to the sum of the phonon and binding energy distribution profile.

2.3. Doppler broadening spectroscopy(DBS) measurement

The DBS experiments were carried out at room temperature using slow positron beam with implantation energy ranging from 0.2-20 keV. Positron from a 50 mCi ^{22}Na source was moderated by a 1 μm tungsten foil and guided to the target chamber with the help of electric and magnetic field. A high purity germanium detector (HPGe) with a resolution of 2.0 keV at 1332 keV photopeak of ^{60}Co is used for Doppler broadening measurements. Doppler broadened annihilation γ-radiation characterized by the line shape *S*-parameter which is defined as the ratio of integral counts within ~ 2.0 keV energy window centered at 511 keV to the total photo peak area is evaluated as a function of positron implantation energy. The implantation depth of positron is related to incident energy by the relation, $z_0 = AE^n/\rho$, where, z_0 is expressed in nm, E is positron energy in keV, ρ is density of the

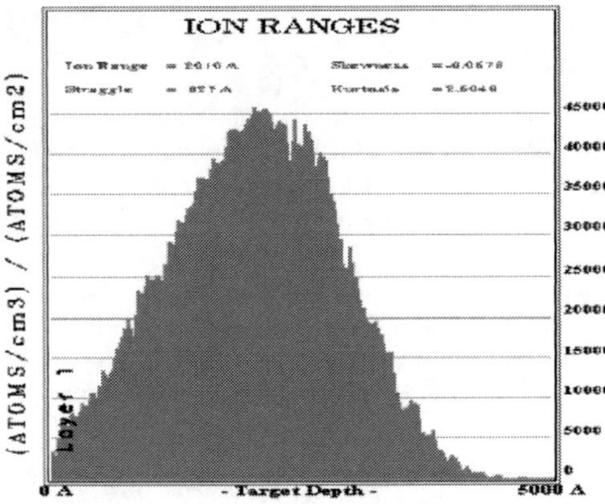

Figure 1. Ar^{9+} ion distribution profile in Zr-2.5Nb alloy with energy 315 keV.

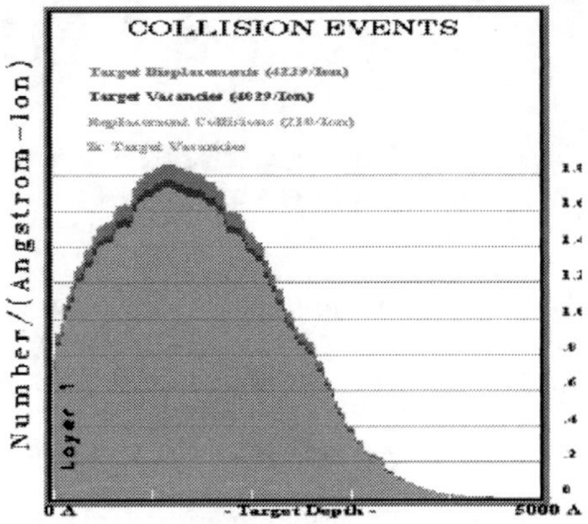

Figure 2. Damage profile of Zr-2.5Nb irradiated by 315 keV Ar^{9+} ion.

medium in g/cc, n ≈ 1.6 for positron incident on most of the materials and *A* is a material dependent constant.

3. Results and discussion

In this section the SRIM simulation and PAS results of the irradiated Zr-2.5Nb alloy samples are presented and discussed. The purpose of this irradiation study was to understand the irradiation degradation mechanism for the assessment of pressure tube behavior during in-service operation. Figure 1 shows the distribution of 315 keV Ar^{9+} ions in Zr-2.5Nb binary alloy. The mean implantation depth and maximum range of penetration of 315 keV Ar^{9+} ions in this alloy was found to be around 100-200 nm and 450 nm, respectively. The energy of Ar^{9+} ion decides the extent of damage in the form of number of displacement of atoms. Figure 2 shows the damage profile of Zr-2.5Nb alloy indicated by total number of target atoms displaced in the collision events. The peak damage was found to occur at a depth of around 100-200 nm (Figure 2) as indicated by the maximum number of dpa at this depth. This indicates an inhomogeneous distribution of the ion energy along the path traversed by the Ar^{9+} ions, which is inherent characteristic of damage caused by heavy ions in a medium and results into a distribution of defects generated across the depth of the samples. Figure 3 represents the damage depth profile for the samples irradiated with different fluences indicating the number of dpa as a function of depth in the sample. It is seen that number of dpa increases rapidly with the increase in the fluence. The increase is seen to be highest at the mean implantation depth of the ions in the alloy indicating maximum number of defects generated in this region. The inset in Figure 3 shows the number of dpa as a function of Ar^{9+} ion fluence indicating the increase in the defect concentration with the increase in the fluence.

The defect depth profiling using Doppler broadening has been carried out to corroborate the damage profile obtained through simulation. It is seen that S-parameter for irradiated samples is higher as compared to unirradiated one upto ~ 15 keV positron implantation energy (depth ~ 450 nm). The high S-parameter indicates presence of defects in the alloys which are generated due to ion irradiation. At higher positron implantation energies (E> 15 keV), the S-parameter for irradiated samples approaches to the value corresponding to bulk of the unirradiated sample. This shows that irradiation by 315 keV Ar^{9+} ion results in the generation of defects from the surface to ~ 450 nm depth of the sample. The S-parameter shows maximum at the depth ~ 100-200 nm indicating the region with maximum number of defects. This defect distribution is consistent with that obtained from SRIM simulations. Heavy ion irradiation leads to generation of different vacancy type defects viz. monovacancies, divacancies, vacancy clusters, vacancy loops etc. The trapping of positron in these defects results in increase in S-parameter compared to its value in defect free materials [18]. In addition to high value of S-parameter in irradiated samples, the S-parameter is seen to increase with the increase in fluence of the ions. The S-parameter profile at the lowest and highest fluencies is shown in Figure 4. The S-parameter profiles for other fluences are seen to lie between these two extremes and are not shown here. The large increase in S-parameter for the sample irradiated at the lowest fluence compared to unirradiated one indicates the generation of large number of defects even at this fluence. On further increasing the fluence the increase in defect concentration is not so pronounced.

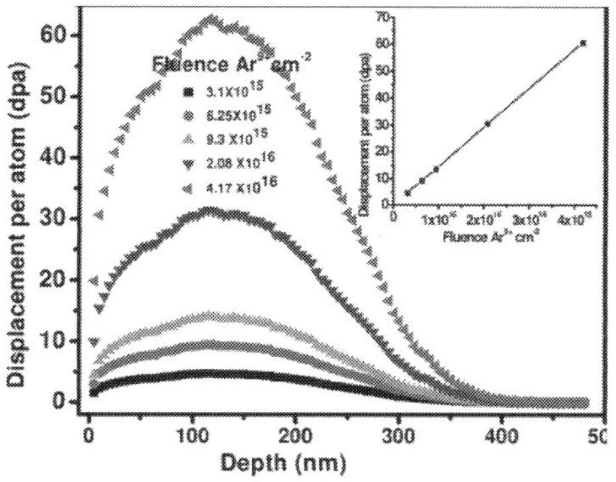

Figure 3. dpa Vs Depth profile and in inset dpa Vs fluence profile at 145 nm depth of Zr-2.5Nb alloy. irradiated by 315 keV Ar^{9+} ion.

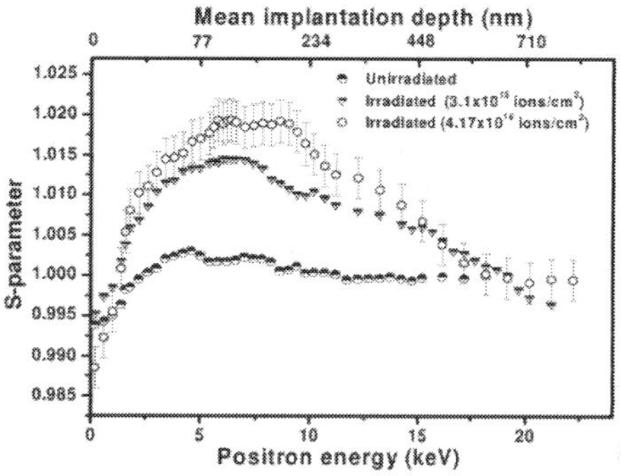

Figure 4.S-parameter as a function of incident positron energy (mean implantation depth) in unirradiated and irradiated Zr-2.5Nb alloy.

4. Conclusions

The defect depth profiling has been carried out in Zr-2.5Nb alloys irradiated by 315 keV Ar^{9+} ions at different fluences using slow positron beam. The S-parameter profiles indicated the generation of defects from the surface to ~ 450 nm depth of the alloy. This region corresponds to the implantation depth of 315 keV Ar^{9+} ions as calculated using SRIM simulation. The study has shown the increase in the number of defects with the increase in the fluence. The maximum damage due to irradiation is seen to be at a depth of 100-200 nm as shown by the maximum in S-parameter profile and consistent with the mean implantation of 315 keV Ar ions calculated using SRIM. The S-parameter profiles indicated the presence of large number of defects even in the sample irradiated at the lowest fluence.

References

[1] Mihalache M, Ionescu V, Meleg T and Pavelescu M, 2011 *Rom. J. Phys.* **56** 952-962.
[2] Chowdhury P S, Mukherjee P, Gayathri N, Bhattacharya M, Chatterajee A, Bhrat P and Nambissan P M G, 2011 *Indian Academy of Sci.* **34** 507–513.
[3] Mihalache M, Radu V, Ohai D and Palelescu, 2010 U.P.B. *Sci. Bull., Series* B **72**.
[4] Nanker P P, Mangsulikar M D, Cleveland J and Shah B K, 2006 *Asia-Pacific Conference on NDT, Auckland, New Zealand.*
[5] Sinha R K, Sinha S K and Madhusoodanan K, 2008 *J. Nucl. Mater* **383** 14-21.
[6] Jin H J and Kim T K, 2014 *Annals of Nuclear Energy,* **75** 309–315.
[7] Li B S, Du Y Y, Wang Z G, Shen T L, Li Y F, Yao C F, and Sun R J, 2014 *Nucl. Instrum. Meth in Physics Research* B **337** 21–26.
[8] Peng D Q, Bai X D and Pan F, 2007 *Physica* B **391** 72-78.
[9] Peng D Q, Bai X D and Pan F, 2006 *Vacuum* **81** 507-516.
[10] Gordo P M, Liszkay L , Kajcso Z, Havancsa'k K, Skuratov V A, Ko"gel G, Sperr P, Egger W,

Journal of Physics: Conference Series **618** (2015) 012018 doi:10.1088/1742-6596/618/1/012018

Lima A P and Marques M F F, 2008 *App Surface Sci.* **255** 254-256.
[11] Was G S, 2007 *Fundamentals of Radiation Materials Science: Metal and Alloy.,* Springer, Verlag Berlin Heidelberg.
[12] Liu X, Wang R, Ren A, Wu Y, Jiang J, Zhang C and Wang X, 2012 *Radiation Physics and Chemistry* **81** 1586-1552.
[13] Qiu J, Ju X, Xin Y, Liu X, and Wang B Y, 2011 *J. Nucl. Mater* **411** 20–24.
[14] Qiu J, Xin Y, Liu S, Wang Y L, Wang H B and Tang D, 2009 *Nucl. Instrum. Meth. Phy.Res.* B **267** 3162–3165.
[15] Ziegler J F and Biersack J P, 2013 *Nucl. Instr. And Meth,* **B 268**, 1818-1823.
[16] Wan Q, Shu G, Wang R, Ding H, Peng X, Zhang Q and Lei J, 2012 *Nucl. Instrum. Meth Phys.* Research B **287** 148–152.
[17] Krause-Rehberg R and Leipner H S, 1998 *Positron Annihilation in Semiconductors: Defect Studie*, Spinger-Verlag Berlin.
[18] Hautojarvi P and Vehanen A, 1979 *Analytical Characterization of Aluminum, Steel, and Superalloys* Spinger-Verlag Berlin.

Positron annihilation spectroscopy of Uranium Molybdenum alloys

S. Mukherjee[1], S. K. Sharma[1], V. Sinha[2], P. K. Pujari[1,*]
[1]Radiochemistry Division, [2]Metallic Fuels Division,
Bhabha Atomic Research Center, Mumbai, India, 400085.

E-mail: pujari@barc.gov.in

Abstract. Positron annihilation spectroscopy has been carried out on as-cast and heat treated U-Mo alloys with Mo concentration varying from 8-10 wt. %. The positron lifetime data suggest that, upon heat treatment at 900 °C, the bulk lifetime decreases. The chemical surrounding of the positron annihilation sites is similar for the as-cast and heat treated samples suggesting a uniform distribution of Mo around defect sites.

1. Introduction

Currently there is an intense drive for low power research reactor to transition from highly enriched uranium (HEU) based fuels to low enriched uranium (LEU) based ones (^{235}U < 20%) [1]. In that regard, U-Mo alloys have provided a nice alternative with comparable density as HEU. One of the main problems with the uranium metal based fuel is the presence of α phase (the room temperature phase of U) which has a very low solubility for most of the alloying elements (Zr, Mo etc). Other main problem is that under irradiation, α-U tends to preferentially grow in one direction which causes serious structural problem in the reactor. This is due to the fact that this phase has highly anisotropic thermal expansion with thermal expansion coefficient in (001) direction being an order more than that along (100) or (010) directions [2]. The γ phase, on the other hand, has a body centred structure and exhibits isotropic thermal expansion along with better mechanical properties under irradiation [3]. It is well known that γ-U is stable at high temperatures (> 774.8 °C) and phase diagram of uranium suggests that upon cooling to room temperature the γ phase will transform into α-U and MoU_2. Recently Sinha et al. [4] have shown that for Mo wt .% more than 8, it is possible to make alloys with 100% γ phase at room temperature. In this paper, we have carried out positron annihilation spectroscopy to probe the various allotropes of U and the effect of heat treatment on the microstructure of the alloy.

2. Experimental

Alloys with different Mo concentration were synthesized in induction furnace [4]. Chemically pure (≥ 99.8%) pellets of U and Mo were used as starting materials. The U pellets were cleaned electrochemically while the Mo pellets were ultrasonically cleaned. Yttria lined graphite crucibles were used to melt the pellets at 1300 °C under vacuum of 10^{-5} mbar. The molten alloys were extracted after being furnace cooled to room temperature. As-cast samples were hot rolled at ~ 650 °C which corresponds to the γ phase. The hot rolled samples were annealed at 900 °C for 24 hours. This set of samples is referred to as heat-treated samples. Positron annihilation lifetime spectroscopy (PALS) was carried out using two identical BaF_2 detectors with a timing resolution of ~240 ps. ^{22}Na source of 10μCi strength was sandwiched between two Kapton films of 8 μm thickness. The contribution in the PALS spectra from positrons annihilated in the Kapton foils (21%) was subtracted before analyzing the data [5]. Coincidence Doppler broadening (CDB) spectroscopy was carried out using two identical high purity Ge detectors with resolution of ~2.1 keV at 1332 keV (^{60}Co). In order to magnify the changes and trends, the CDB spectra of all the metals and alloys were normalized to that of pure Si. Prior to PALS or CDB measurement, the alloys samples were electrochemically cleaned in a H_3PO_4 solution (5% by vol) with a DC supply of 5V. PALS measurements were carried out in a vacuum of ~50 x 10^{-3} mbar.

3. Result and discussion

In earlier study, The XRD pattern of the as-cast alloys prepared using same method shows that they are predominantly in γ phase [4]. The positron lifetime data for both as-cast and heat treated samples, show predominantly one component. The lifetime component for the as-cast samples are higher than the heat treated samples indicating that some defects (vacancy) are present in the as-cast samples as a result of hot rolling which are annealed out to an extent after heat treatment of the alloys (Fig. 1). The positron lifetime in the heat treated samples are very close to the positron lifetime (157 ps) in γ-U [6] which is supposed to be lower than positron lifetime in α-U mainly due to difference in their respective lattice structures. It has been observed that the lattice parameter of γ-U-Mo alloy reduces with Mo concentration [7] which can be the cause of reduction in positron lifetime in as-cast samples with Mo concentration.

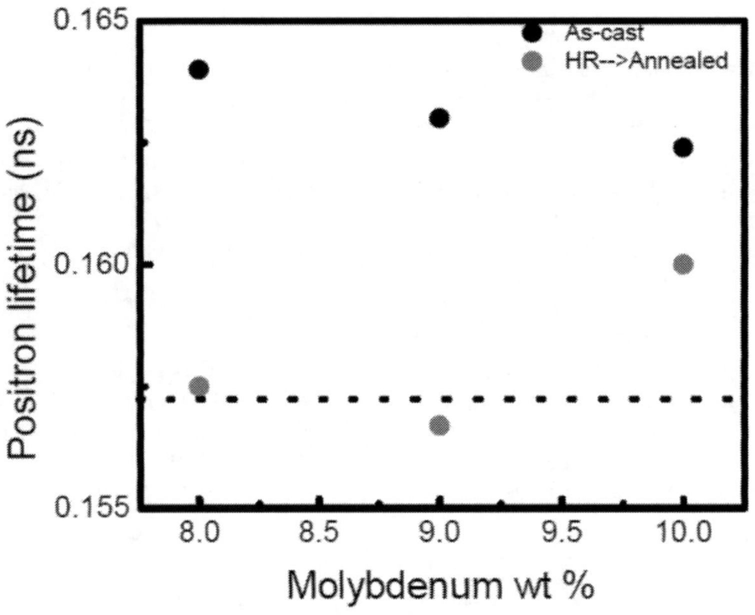

Figure1. Variation of positron lifetime in as-cast and heat treated samples as a function of Mo wt %. All the alloys show one lifetime except as-cast U-8%Mo which has a 450 ps component of ~ 5% intensity. Errors on the data points are ≤ 1.0%. The reduced χ^2 in all the cases is better than 1.2. The dashed line corresponds to bulk positron lifetime (157 ps) in γ-U.

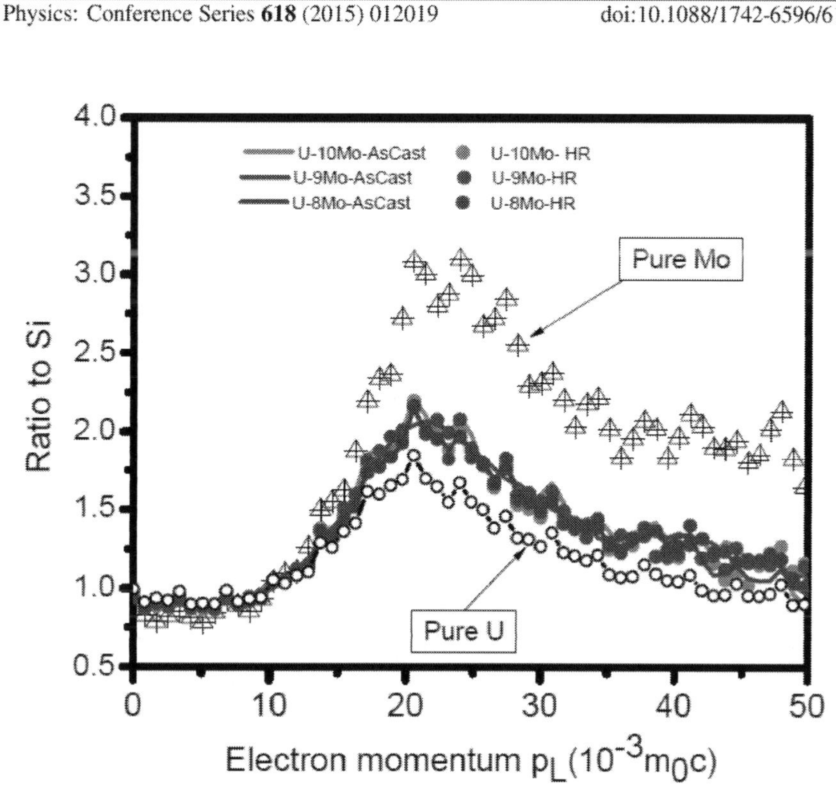

Figure 2. Coincidence Doppler broadening spectra of U-Mo alloys. All the spectra have been normalized to Si. CDB spectra of pure U and Mo are also shown for comparison.

Figure 2 shows the CDB ratio curves for U-Mo alloys samples along with pure U and Mo metals. The spectra for as-cast samples show higher area compared to pure U in the electron momentum range ($P_L > 20 \times 10^{-3}\ m_0c$). This slight deviation can be explained by the presence of ~ 20 atomic% Mo in U matrix. Based on XRD data [4], it has been suggested that for as-cast alloys the Mo distribution may not be uniform in the metastable γ phase. Our data suggests that the chemical environment seen by the positrons annihilating from the γ phase is similar in all the as-cast alloys. On heat treatment, CDB spectra of the alloys do not differ from the as-cast samples indicating the Mo distribution at positron annihilation sites remain unchanged. This also shows that no phase segregation or Mo clusters are present in the heat treated samples. In recent future, we have plan to do various heat treatment to these samples, wherein the γ phase is decomposed to different phases as observed from XRD. The positron annihilation measurement in these samples will be carried out to study the changes in the microstructure of these alloys as a result of different phase formation.

4. Conclusion

Positron annihilation spectroscopic measurement has been carried out on dilute U alloys containing 8, 9 and10 wt % Mo. PALS data supplemented by XRD results suggest that the as-cast alloys are predominantly in γ phase while annealing reduces the defects in the as-cast samples. The CDB measurements suggest that the Mo distribution around the defect sites, where positron predominantly annihilates, remains unchanged on heat treatment. The CDB measurements on heat treated samples also rule out any possibility of Mo segregation in the samples. Further measurements are in process to evaluate the microstructure of these alloys as a result of γ phase decomposition.

Journal of Physics: Conference Series **618** (2015) 012019 doi:10.1088/1742-6596/618/1/012019

5. References

[1] Snelgrove J L, Hofman G L, Meyer M K, Trybus C L and Wiencek T C 1997 *Nucl. Eng. Des.* **187** 119.

[2] Lloyd L T 1961 *J. Nucl. Mater.* **3** 67.

[3] Bleiberg M L, Eichenberg J D, Fillnow R H and Jones L J (Eds.), *Radiation stability of uranium-base alloys-Part 4*, Bettis Plant 1957 7.

[4] Sinha V P, Hegde P V, Prasad G J, Dey G K and Kamath H S 2010 *J. of Alloys and Comp.* **506** 253.

[5] Kirkegaard P and Eldrup M 1991 *Comput. Phys. Commun.* **17** 401.

[6] Dryzek J, Kuriplach J, Dryzek E, Quarles C A and Šob M 1999 *Sol. State Commun.* **111** 465.

[7] Sinha V P 2012 *Studies on the preparation and characterization of U3Si2 intermetallics and U-MO alloys* PhD Thesis, Homi Bhabha National Institute, Mumbai, India.

Journal of Physics: Conference Series **618** (2015) 012020 doi:10.1088/1742-6596/618/1/012020

Dispersion and interaction of graphene oxide in amorphous and semi-crystalline nano-composites: a PALS study

Frans H J Maurer and Carlos R Arza

Department of Chemistry, Polymer & Materials Chemistry,
Lund University,
P.O. Box 124, SE -22100 Lund, Sweden

E-mail: frans.maurer@polymat.lth.se

Abstract. The influence of dispersion and interaction of Graphene Oxide (GO) in semi-crystalline Polyhydroxy butyrate (PHB) and glassy amorphous Poly(*t*BP-oda) is explored by Positron Annihilation Lifetime Spectroscopy (PALS). The ortho-Positronium lifetimes which represent the main free volume hole size of both polymers are mainly affected by the large differences in internal stresses built up by the shrinkage of the polymers during their preparation, restricted by the platelet structure of GO. The ortho-Positronium intensities, which represent the ortho-Positronium formation probabilities, suggest a strong dependency of on the dispersion of the nano-particles and their aspect ratio.

1. Introduction

Two different types of polymers are mixed with Graphene Oxide (GO) to modify their physical properties. The first type, PHB, is a naturally occurring biodegradable semi-crystalline polyester which is produced by bacteria. The second type, glassy amorphous poly(*t*BP-oda), belongs to the group of Polybenzoxazines which are formed after self-activated polymerization of benzoxazine monomers at high temperatures. Physical properties of both polymers and their nano-composites as a function of GO content are presented elsewhere [1, 2]. Common feature of both systems is the forming of a network structure by the GO particles, characterized by an equilibrium modulus in the melt state for PHB and monomer liquid state for poly(*t*BP-oda). Two different onsets of percolation volume fractions result from the mixtures of GO in the two polymers.

GO exhibit excellent properties as a nano-filler. High aspect ratio (lateral size on the order of 1 μm and thickness in the order of 1 nm), a Young's modulus in the range of 200 GPa, and a versatile surface chemistry, possesses a high potential to effectively modify the physical properties of polymer nano-composites. The purpose of this contribution is to search for correlations between structural characteristics of the nano-composites and parameters resulting from Positron Annihilation Lifetime Spectroscopy (PALS) such as ortho-Positronium lifetimes $\tau_{o\text{-}Ps}$ and ortho-Positronium intensities I_3. The thickness of the GO platelets are in the nm range, suggesting an interaction of GO with radiation chemistry effects in the terminal spur of the positron thermalization track in the nano-composites at low volume fractions.

2. Materials and experimental

GO was prepared according to the procedures of the modified Hummers method from graphite flakes, exfoliated for the PHB nano-composites and dried in an oven at 130 ^{0}C [1]. Samples for PALS measurements of PHB nano-composites were compression moulded at 180 ^{0}C for 2 min. The detailed description of the preparation of the (*t*BP-oda) samples is reported elsewhere [2]. The mixtures (150 mg) were pressed into discs with a diameter of 15 mm using a support of a Teflon disc of 2 mm thick and the same diameter. The crosslinking reaction was completed by placing the discs in an oven while heating them in sequences at 100, 120, 140, 160, 180, and 200 ^{0}C for 2h each. After cooling to room temperature, the samples were separated from the Teflon support. Standard PALS spectra were collected with about 1.5 million counts per spectrum at 22°C. Five Spectra per sample were recorded and average values of three lifetime components were established with PALSfit. Only averaged ortho-Positronium lifetimes and intensities are reported here.

3. Results and discussion

Figure 1 shows the influence of GO content on the ortho-Positronium lifetime of PHB and poly(*t*BP-oda). The τ values of the unfilled polymers are 2.33 ± 0.02 for PHB and 1.96 ±0.02 for poly(*t*BP-oda) respectively. The glass transition temperatures (T_g) of the unfilled glassy amorphous polymer is 150 ^{0}C and the T_g of the semi-crystalline PHB is equal to 4 °C, which explains the large difference in lifetimes and corresponding free volume hole sizes when using the Tao-Eldrup relation correlating lifetime and free volume hole size [3, 4]. Remarkable differences can also be observed for the sensitivity of including minor concentrations of GO. The lifetime of *t*BP-oda is hardly affected by the incorporation of GO, while the lifetime of PHB is strongly increased. Reasons for this behaviour may be found in the large difference in shrinkage of the two nano-composite systems during their preparation.

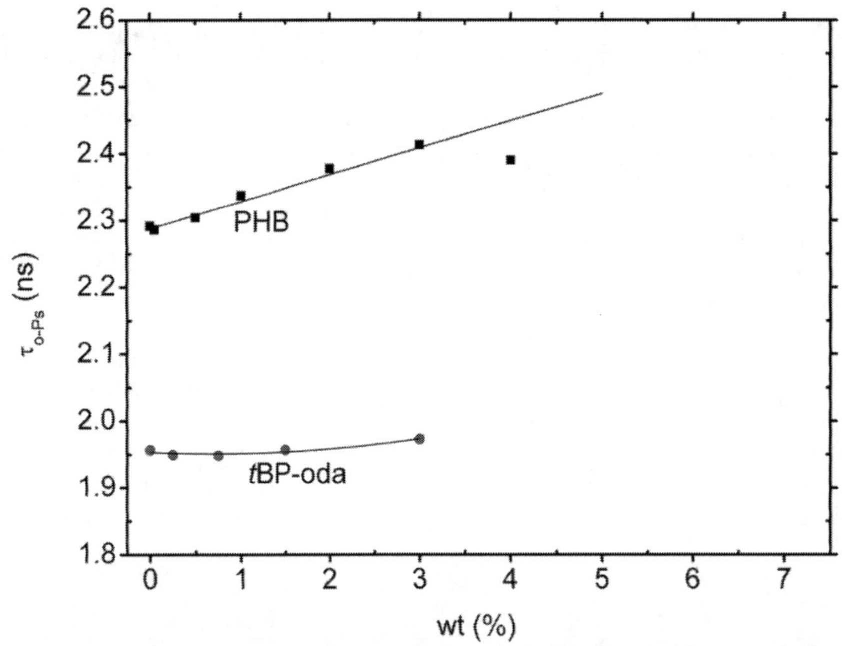

Figure 1. Ortho-Positronium lifetimes of PHB and poly(*t*BP-oda) as a function of wt (%) GO.

Polybenzoxazines are well known for their 'near-zero shrinkage' behaviour during polymerization [5]. In contrast, PHB crystallizes on cooling from the melt, resulting in a heat of fusion of 91.3 J/g, corresponding to a crystallinity of 63%. An appreciable shrinkage may occur when considering that the PHB amorphous density is 1.178 g/cm^3 and crystalline density 1.260 g/cm^3 [6]. Platelet fillers, like GO, hinder the shrinkage and reduce the thermal expansion, resulting in hydrostatic tensile stresses on the amorphous phase in semi-crystalline polymers, and tensile stresses on the tie-molecules which bridge the crystal lamellae. Hydrostatic tensile stresses will consequently increase the free volume cavity size.

From rheological dynamic measurements experiments, reported elsewhere, follows that GO shows network formation in both polymeric systems at very low GO concentrations [1, 2]. In rheology this effect is visible by a dynamic equilibrium shear modulus in the terminal zone. In figure 2 these equilibrium moduli are shown for PHB processed by using two different solvents and for the *t*BP-oda filled systems. For PHB the lowest volume fraction of GO where an equilibrium modulus could be established was 7.1 x 10^{-2} %. The corresponding equilibrium modulus was 11.7 Pa. The same equilibrium modulus is reached in the *t*BP-oda samples at a much higher volume concentration of 1.7 x 10^{-1} %. According to Ren et al. [7] an aspect ratio ζ can be established when knowing the perculation volume fraction ϕ_p, assuming randomly distributed disc-shape nano-fillers. The aspect ratio of the nano-filler ζ, the ratio of width to thickness, is:

$$\zeta = 3\ \phi_s/2\ \phi_p \qquad (1)$$

Where $\phi_s = 0.29$, the volume fraction at the onset of perculation of interpenetrating, randomly packed spheres. The aspect ratio of GO in PHB is determined 400 and the aspect ratio of GO in (*t*BP-oda) 165.

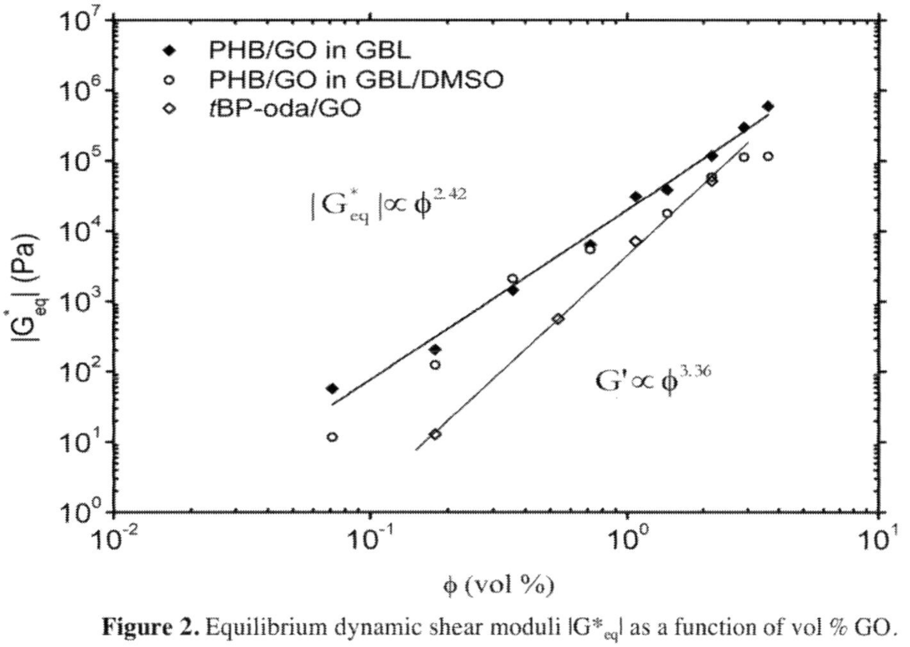

Figure 2. Equilibrium dynamic shear moduli $|G^*_{eq}|$ as a function of vol % GO.

In Figure 3 the relative ortho-Positronium intensities I_3/I_0 are presented as a function of GO concentration. The decrease of intensity with GO content for both systems is stronger than a weight average decrease assuming no ortho-Positronium formation in the filler particle. This is, in combination with the high aspect ratios and nm thickness of the fillers, a strong indication of inhibition reactions in the terminal spur by the nano-fillers. For the PHB/GO nano-composites a correction for the crystallinity is taken into account. Differences in heat of fusion as presented in Figure 4, and the assumption that only the amorphous phase contributes to the ortho-Positronium intensity, is taken into account for calculation of the PHB 'amorphous' presented data.
The influence of inhibition reactions on the ortho-Positronium intensity for nano-filled polymers at low concentrations may be represented by an equation similar to the one presented by Eldrup et al. [8]:

$$I_3 = I_0 \,[A + (1\text{-}A)/\,(1 + \sigma.D\,)] \qquad (2)$$

σ represents an inhibition constant, which is related to the efficiency of the scavengers in comparison to the positronium formation efficiency. Scavengers at the GO surface include electron-attracting moieties such as hydroxyl and epoxy groups. $D = \zeta$. wt (%) and $A.I_0$ is the saturation value of I_3.
Best fitting with A = 0.59 gives an efficiency factor for PHB of 0.34 and for poly(*t*BP-oda) of 0.10.
The better dispersed GO particles with larger aspect ratio are thus more efficient in the inhibition process for ortho-Positronium formation in the two nano-filled polymers studied so.

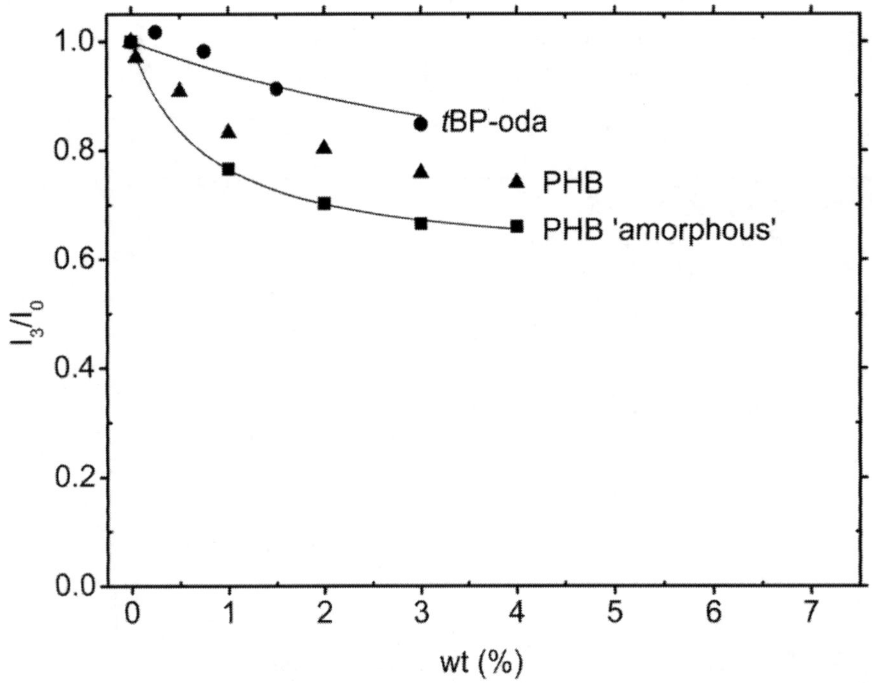

Figure 3. Relative ortho-Positronium intensities I_3/I_0 as a function of wt (%) GO.

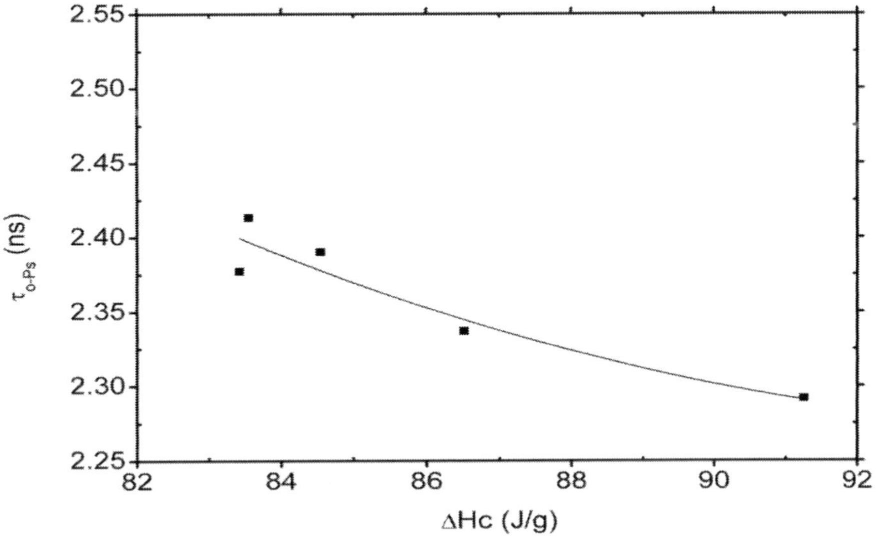

Figure 4. ortho-Positronium lifetimes as a function of heat of fusion for PHB.

3. Conclusion

PHB and poly(*t*BP-oda) nano-composites were prepared with low fractions of GO nano-fillers and studied with Positron Annihilation Lifetime Spectroscopy. Free volume hole sizes, represented by ortho-Positronium lifetimes $\tau_{o\text{-}Ps}$, increase with increasing GO content for semi-crystalline PHB samples but are merely independent of GO content for the glassy amorphous poly(*t*BP-oda). This remarkable difference in behavior is explained by the differences in volume shrinkage of the two polymers, leading to large hydrostatic internal stresses and dilatation of the free volume cavities for the PHB nano-composites.

The ortho-Positronium formation probability, represented by the ortho-Positronium intensities I_3, decreases appreciable for both polymers with increasing GO content, pointing to an inhibition of the ortho-Positronium yield by scavengers present at the surfaces of the GO nano-particles. The experimental results show that aspect ratio and dispersion of the GO nano-particles have an appreciable influence on the positronium yield. Earlier hypothesis and the hope that I_3 represents a free volume hole density can not be kept in the light of understanding of the mechanism of positronium formation [9]. The present work may contribute to this understanding.

References

[1] Arza C R, Jannasch P and Maurer F H J 2014 *European Polymer Journal* **59** 262
[2] Arza C R, Ishida H and Maurer F H J 2014 *Macromolecules* **47** 3685
[3] Tao S J 1972 *J. Chem. Phys.* **56** 5499
[4] Eldrup M, Lightbody D and Sherwood J N 1981 *Chem. Phys.* **63** 51
[5] Ishida H and Allen D J 1996 *J. Polym. Sci.:Part B: Polymer Physics* **34** 1019
[6] Barham P J, Keller A, Otun E L and Holmes P A 1984 *J. Mater. Sci.* **19** 2781
[7] Ren J, Silva A S and Krishnamoorti R 2000 *Macromolecules* **33** 3739
[8] Eldrup M, Shantarovich V P and Mogensen O E 1975 *Chem. Phys.* **11** 129
[9] Dlubek G 2010 in *Polymer Physics, Eds Utracki L and Jamieson A M* Wiley Singapore

When some elementary free volumes in polymers are not seen by positron annihilation experiments

V.P. Shantarovich[1], V.G. Bekeshev[1], A.V. Pastukhov[2], V.A. Davankov[2], O.K. Krasil'nikova[3], E.V. Belousova[1], I.B. Kevdina[1], M.K. Filimonov[1], V.W. Gustov[1]

[1] N.N. Semenov Institute of Chemical Physics Russian Academy of Sciences, Kosygin st. 4, 119991 Moscow, Russia, shant@center.chph.ras.ru
[2] A.N. Nesmeyanov Institute of Organoelement Compounds Russian Academy of Sciences, Vavilov st. 28, 119991 Moscow, Russia
[3]A.N. Frumkin Institute of Physical Chemistry and Electrochemistry Russian Academy of Sciences, Leninskii prosp.,31, 119071 Moscow Russia

Abstract Size distributions of elementary free volumes have been studied in mesoporous micro-heterogeneous polymer sorbents. Positron annihilation lifetime spectroscopy (PALS), low temperature gas adsorption (BET) and thermo-stimulated luminescence (TSL) measurements are employed as complementary instruments for the study. It is shown that small admixtures of rubbers are very effective for variations of the pore size distribution. While BET technique was very informative for measurements of mesopores (2-50 nm), positron annihilation was sensitive to micropores (<2 nm), but not for mesopores. The last specificity is explained by the limited positronium diffusion length in a polymer and also by inhomogeneous distribution of mesopores in heterogeneous systems. TSL measurements gave information on sizes of rubber inclusions in compositions.

1.Introduction:

Positron annihilation lifetime spectroscopy PALS is used traditionally for the studies of free volumes in polymers [1-4]. We continue this direction on examples of membrane materials and sorbents. The specific feature of this study is the including modified materials. They are modified in a way to obtain some new unique properties, for example, the specific distribution of elementary free volumes for sorbents. It was shown previously that, for membranes, preparation of nanocompositions consisting of polymers and highly dispersed non-organic admixtures (mixed membrane matrices MMM) [5-7] is very promising. Another way infers saturation of polymers with gases in supercritical state with the further heating [8]. In this study we follow the first way and apply this approach to sorbents.

Another special feature of this study is a combination of the three supplementary experimental methods: PALS, thermostimulated luminescence (TSL) and low temperature gas adsorption (BET) [3, 9]. The effect of TSL is emission of light by a substance, irradiated preliminary at low temperature, on heating. As it was stressed recently, PALS and TSL methods are generally interconnected by the same radiation-chemical nature of the processes (spur process) [10-17]. At low temperatures, intensity of positronium formation [10-13] and integral intensity of the TSL [14, 15] are dependent on the storage of the spur electrons (weakly bound electrons) and influenced by bleaching. This is another proof of the spur mechanism of the positronium formation [16, 17]. Papers [18-21] demonstrated effectiveness of PAL-TSL combination for the free volume studies in some new membrane materials and sorbents.

It came out that local rigidity (stiffness) of a polymer structure [21, 22] (estimated in the TSL experiments as intensity α of high temperature component of TSL curve [18]) correlates well with the effective sizes of the free volumes holes (FVHs) and, correspondingly, with *o*-positronium annihilation lifetimes. Fine correlation was found between the pore size, parameter α and oxygen permeation $P(O_2)$ for a number of polymeric and composition materials [3, 6, 22]. The obtained results confirmed the idea [23-25] of interconnection between the permeation, free volume and local rigidity (local tensions, steric hindrances) in a polymer matrix. These results make it possible to conclude sometimes about membrane and sorbent properties (permeation, specific surface) even without corresponding measurements, only on the basis of PAL and/or TSL data [18-20]. However, for PAL studies of composition materials, there is a peculiarity, which we would like to discuss on example of sorbents made of mesoporous polydivinylbenzene.

2. Experimental:

2.1 Measurements

Specific surface measurements, where incremental volumes of pores of the given size ([cm^3 g^{-1} nm^{-1}] units, Fig.1) were measured using the data on low temperature gas (N_2) sorption. We used High vacuum system ASAP-2020 MP Micromeritics (USA) [3, 9] in the interval of relative pressures from 10^{-6} to 0.99. Adsorption-desorption isotherms obtained had a pattern typical for mesoporous adsorbents with a hysteresis loop. Size distributions of FVHs (see Fig.1, for example) are found from BJH (Barrett-Joyner-Halenda) analysis.

In the PALS measurements we used spectrometer EG&G ORTEC (USA) with time resolution (FWHM of the ^{60}Co prompt coincidence curve) 300 ps, ^{44}Ti radioactive source and integral statistic more than 10^6 coincidence events for each measurement. We applied well known PATFIT program for separating the *pick-off o*-Ps components τ_3, I_3 and τ_4, I_4.

We used TSL-4 (Russia) device for measurements of integral (on the wave length of light) yield of TSL on heating in the temperature range 80-370 K. Scanning time was 30 min. The intensities of high temperature components of the curves α were calculated as described earlier [18, 19] and were taken as measure of local rigidity of the studied matrices.

2.2. Materials

The objects for investigation were prepared on the basis of mesoporous polydivinylbenzene (MPD), where additional porosity is created by some porogen solvents during polymerization of divinylbenzene. The commercial analogous system is called XAD-4 (Table 1). For modification, small amounts (7 vol. %) of linear polymers (rubbers with linear structure and not very high molecular weight) were added to the reaction mixture on the stage of polymerization of divinylbenzene. The resulting systems are supposed to be microheterogeneous and known as mesoporous nanocompositions. Changing the kind of the rubber (polybutadiene PB, polyisoprene PI, polyisobutylene PIB) and conditions of polymerization, polymers with different size distributions of free volumes holes (FVH) had been prepared.

3. Results and discussion:

Results of PALS, low temperature gas (N_2) adsorption BET (treated according to BJH [9]), and TSL data are shown in Figures 1, 2 and in Tables 1, 2. It is seen that densities of the samples (ρ) are very low (≈ 0.5 g cm^{-3}) and fraction of the free volume is turned out to be about 50%. At the same time, the longest *o*-Ps lifetime and corresponding intensity are smaller than those for the cross-linked polystyrenes CPS [3] which have about the same specific surface (≈ 1000 m^2 g^{-1}). However, densities of polystyrenes CPS are essentially higher (≈ 0.7 g cm^{-3}). This is an evidence that the studied samples are mesoporous, but not microporous, and the Ps atom is not able to localize in the largest, the most distant free volumes. Actually, in table 1, average diameters of EFV D_{av}^{BJH} are essentially different from those determined from the longest *o*-Ps lifetime τ_4 (46-67 ns correspond to D_4=2.9-3.6 nm). The curve in figure 1 shows size distribution of incremental volumes ([cm^3 g^{-1} nm^{-1}] units) of pores with diameter D from the intervals ΔD=1 ns. It seems reasonable to suppose that the free volumes with D_4^{PAL} from the interval 1 ns are those forming the longest lived component I_4. In this case, $N_4=6V_4^{BJH}/\pi D_4^3$. For MPD, for example, this gives 0.6×10^{19} g^{-1}, and this is obviously the lowest concentration of EFV accessible for Ps in this polymer.

Table 1. Polydivinylbenzene-based polymeric sorbents. Positron annihilation data, BJH (*Barrett-Joyner-Halenda*) adsorption results and parameter α from the TSL curves. The figures in bold present data that are most frequently referred to in the discussion.

Sorbent	τ_3, ns I_3,%	τ_4, ns I_4,%	I_3+I_4, %	D_4^{PAL}, nm	D_{av}^{BJH}, nm	S_{meso}^{BJH}, m^2 g^{-1}	$\alpha\times10^3$
XAD-4	1.93±0.02* 23.73±0.39	50.37±1.41 16.53±0.20	40.26			900	
MPD	2.10±0.02 26.19±0.39	46.61±1.13 11.5±0.11	37.69	**2.86**	**10.24**	700	**240**
MD/PI	1.91±0.02 30.55±0.44	64.19±3.92 9.53±0.29	40.08	**3.50**	**8.83**	400	**200**
MD/PIB	2.19±0.02 33.80±0.30	45.81±1.32 9.58±0.11	43.38	**2.84**	**3.88**	270	**90**
MD/PB	1.92±0.01 32.95±0.35	66.58±4.19 6.28±0.21	39.23	**3.58**	**13.75**	310	**180**

*Corresponding numbers D_3^{PAL} for the all substances are about 0.6 нм.

Table 2. Additional characteristics found from the PAL and BET (BJH) measurements.

Sorbent	V_Σ^{BJH}*, cm^3 g^{-1}	ρ, g cm^{-3}	*Fvf*, %	V_4^{BJH}, cm^3 g^{-1} nm^{-1}	$N_4\times10^{-19}$, g^{-1}	$N_3\times10^{-19}$, g^{-1}	S_{micro}, m^2 g^{-1}
MPD	1.20	0.48	57	**0.072**	0.60	6.51	72.3
MD/PI	0.84	0.58	48	**0.045**	0.22	4.10	42.3
MD/PIB	1.29	0.46	59	**0.055**	0.50	8.35	86.8
MD/PB	0.24	0.51	12	**0.044**	0.80	25.00	63.7

*The range of the pore diameters D was from 1.7 to 300 nm.

From this, positronium diffusion coefficient can be estimated as $d^{Ps}=L^2/6\tau$, where $L = N_4^{-3}$ stands for non-localized Ps diffusion length and τ is accepted to be about 0.3 ns (free positron annihilation life time). For MPD, for example, this gives $d^{Ps}=3\times10^{-4}$ cm^2/s, which is close to data, published in [26]. The advantage of the PAL method compared to the gas sorption, is a possibility to determine concentration of the micropores N_3 (corresponding to τ_3, I_3, Tables 1,2) undetectable by BET: $N_3=N_4I_3D_4/I_4D_3$ [24]. The results obtained correspond to mesoporous character of the studied samples [27] ($S_{meso}^{BJH}>>S_{micro}$).

Finally, we mention some of TSL results: parameters α in Table 1 and TSL curves for MPD, (MD/PB 7%), and curve for pure 1,4 PB taken with intensity 7% of that for MPD (Figure 2). *First*, the data illustrate an obvious correlation between TSL parameter α (local stiffness, rigidity [19]) and mesoporosity characteristics (specific surface S_{mes}^{BJH} and D_{av}^{BJH} (bold numbers in Table 1)). *Second*, TSL for pure rubber (PB), though being much more intensive than that for MD, does not reveal itself in the TSL of (MD/PB 7%): there is no peak at 170K, typical for PB, in the TSL of composition. Therefore we have to conclude: though composition materials of the given type are considered as heterogeneous, effective size of micro-heterogeneities (phase inhomogeneities) is not higher than

200-300Å. In opposite case, characteristic curves of the components (positions of the characteristic peaks) would be easily seen.

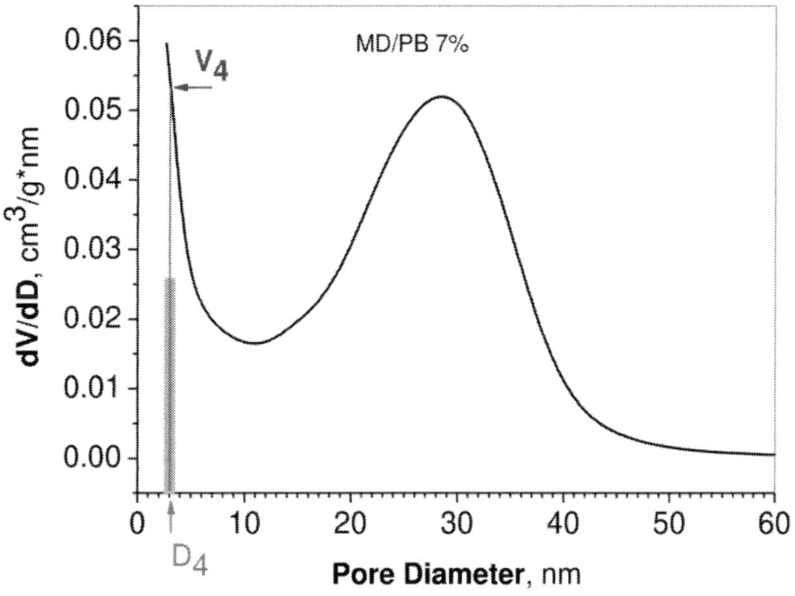

Figure 1. Comparison of the PAL and BET data on porosity of (MD/PB 7%) sample: positron annihilation (thick vertical line, presenting the area of about 1 ns width, centered in the vicinity of D_4) and low temperature gas sorption (dependence of incremental free volume on the pore diameter).

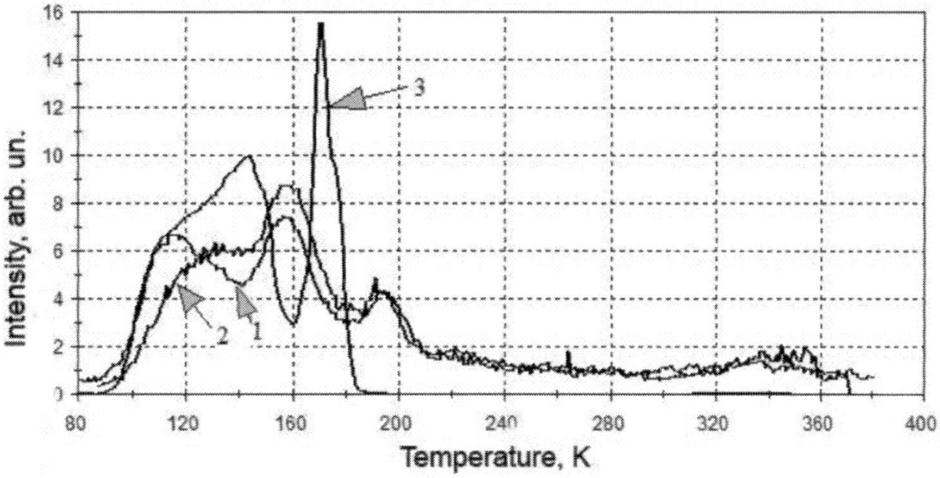

Figure 2. Thermostimulated luminescence curves for (1)- polymer composition (MD/PB 7%), (2)- basic polymer MPD and (3)- 1,4-PB component with intensity 7% relative to the curve for the basic polymer.

4. Conclusion

Some composition materials, novel polymer sorbents, consisting of mesoporous polydivinylbenzene and different rubbers of linear structure had been characterized by different shapes of mesopore size-distributions. The sorbents are studied using positron annihilation in combination with other methods: thermostimulated luminescence and low temperature gas adsorption. Due to combination of the methods, the limiting cases, when Ps atom loses its sensitivity to presence of elementary free volumes, still seen by other methods, are found. The results had been used for estimation of mobility of the positronium atom in the studied polymers.

References

[1] Rudel M., Krause J, Ratzke K, Faupel M, Yampolskii Yu, Shantarovich V and Dlubek G 2008 *Macromolecules* **41** 788

[2] Shantarovich V P, Suzuki T, Ito Y, Kondo K, Yu R S, Budd P M, Yampolskii Yu P, Berdonosov S S and Eliseev A A 2007 *Phys. Stat. Solidi* C **4** 3776.

[3] Shantarovich V P 2008 *J. Polym. Sci., Part B: Polym. Phys.* **46** 2485.

[4] Weber M H Lynn K G 2003 Positron Porosimetry *Principles and applications of Positron and Positronium Chemistry* ed. Y.C. Jean, P.E. Mallon P.E. and D.M. Schrader (New Jersey-London-Singapore-Hong Kong: World Scientific) pp 167-211.

[5] Winberg P, Desitter K, Datremont C, Mullens S, Vankelekom I F J and Maurer F H J 2005 *Macromolecules* **38** 3776.

[6] Bushell A F, Attfield M P, Mason C R, Budd P M, Yampolskii Yu, Starannikova L, Rebrov A, Bazarelli F, Bernardo P, Jason J C , Lanc M, Friess K, Shantarovich V, Gustov V and Isaeva V 2013 *J. Membr. Sci.* **427** 48.

[7] Shantarovich V P, Gustov V W, Medintseva T I, Polyakova A V, Belousova E V, Filimonov M K and Prut E V 2011 *Material Science Forum* **666** 81., 2011 *Doklady (Phys. Chem., Russ.)* **441** 771.

[8] Claes S, Vanderzande P, Mullens S, Van Bael M K and Maurer F H J 2011 *Macromolecules* **44** 2766.

[9] Gregg S G and Sing K S W 1967 *Adsorption Surface Area and Porosity* (London- New York: Academic Press).

[10] Ito Y, Hirade T, Hamada E, Suzuki T and Ito Y 1999 *Acta Physica Polonica* A **95** 433.

[11] Hirade T, Maurer F H J and Eldrup M E 2000 *Rad. Phys. Chem.***58** 465.

[12] Shantarovich V P, Hirade T, Kevdina I B, Gustov V W and Oleinik E F, 2001 *Acta Physica Polonika* A **99** 497.

[13] Shantarovich V P, Yu R S, Kino Ya, Hama Ya and Gustov V W 2011 *High Energy Chem. (Russ)* **45** 1.

[14] Nikolskii V G 1968 *High Energy Chem.(Russ)* **2** 271.

[15] Nikolskii V G 1982 *Pure and Appl. Chem.* **54** 493.

[16] Mogensen O E 1995 *Positron Annihilation in Chemistry* ed.: V.I. Goldanskii, F.P. Schaffer, (Berlin- Heidelberg- New York: Springer-Verlag) 66.

[17] Stepanov S V and Byakov V M 2003 Physical and Radiation Chemistry Applications of Positron and Positronium *Principles and Applications of Positron and Positronium Chemistry* ed Y C Jean, P E Mallon, D M Schrader (New Jersy-London-Singapore- New York: World Scientific) p 117.

[18] Shantarovich V, Gustov V, Polyakova A, Belousova E, Filimonov M and Yampolskii Yu 2009 *Phys. Status Solidi* C **6** 2387.

[19] Shantarovich V P, Gustov V W, Belousova E V, Polyakova A V, Bekeshev V G, Kevdina I B 2014 *Acta Physica Polonica* A **125** 806.

[20] Shantarovich V P, Gustov V W, Belousova E V, Polyakova A V, Bekeshev V G and Kevdina I

B 2014 *Rusian Gernal of Physical Chemistry B. Focus on Physics* **33** 559.
[21] Токarev A V, Bondarenko G N and Yampolskii Yu P 2007 *Polymer Sci.* A **49** 909
[22] Budd P M, Msaib K J, Tattershall C S, Ghanem B S, Raynolds K J, Mc Keown N B and Fritsch D 2005 *J. Membr. Sci.* **251** 263.
[23] Hofmann D, Heuchel M, Yampolskii Yu, Khotimskii V and Shantarovich V 2002 *Macromolecules* **35** 2129.
[24] Shantarovich V P, Suzuki T, He C, Davankov V A, Pastukhov A V, Tsurupa M P, Kondo K and Ito Y 2002 *Macromolecules* **35** 9723.
[25] Hofmann D, Entrialgo Costano M, Lerbert A, Heuchel M and Yampolskii Yu 2003 *Macromolecules* **36** 8528.
[26] Shantarovich V P, Novikov Yu A, Suptel Z K, Kevdina I B, Masuda T, Khotimskii V S and Yampolskii Yu P 2000 *Rad.Phys.Chem.* **58** 513.
[27] Gun'ko V M, Leboda R, Skubiszewska-Zieba J, Gawdzik B and Charmas B 2005 *Applied Surface Science* **252** 612.

Characterization of interfaces in Binary and Ternary Polymer Blends by Positron Lifetime Spectroscopy

C Ranganathaiah[1,2]

[1]Govt. Research Centre, Sahyadri College of Engineering & Management, Adyar, Mangalore – 575 007, India

[2]Department of Studies in Physics, University of Mysore, Manasagangotri, Mysore – 570 006, India

Email: cr@physics.uni-mysore.ac.in :crang1@rediffmail.com

Abstract. A miscible blend is a single-phase system with compact packing of the polymeric chains/segments due configuration/conformational changes upon blending. Differential Scanning Calorimetry (DSC) is the most employed method to ascertain whether the blend is miscible or immiscible. Positron Lifetime Spectroscopy (PLS) has been employed in recent times to study miscibility properties of polymer blends by monitoring the ortho-Positronium annihilation lifetimes as function of composition. However, just free volume monitoring and the DSC methods fail to provide the composition dependent miscibility of blends. To overcome this limitation, an alternative approach based on hydrodynamic interactions has been developed to derive this information using the same o-Ps lifetime measurements. This has led to the development of a new method of measuring composition dependent miscibility level in binary and ternary polymer blends. Further, the new method also provides interface characteristics for immiscible blends. The interactions between the blend components has a direct bearing on the strength of adhesion at the interface and hence the hydrodynamic interaction. Understanding the characteristic of interfaces which decides the miscibility level of the blend and their end applications is made easy by the present method. The efficacy of the present method is demonstrated for few binary and ternary blends.

1. Introduction

In the last few decades, it has been realized that many physical and mechanical properties of polymers can be significantly improved by the process of blending. The main objective of blending two or more polymers or copolymers is to produce a product with a set of desired properties for the end use of the material. Blending is a cost effective method to modify properties of polymers such as processability, heat distortion, impact strength etc. Polymer blends have been studied not only for their growing industrial applications but also for scientific and academic interest as well.

Depending on the structure and properties, polymer blends are generally classified into three types, *viz.* miscible polymer blends (M), immiscible polymer blends (IM) and partially miscible blends (PM). Thermodynamically, for miscible blends, the Gibbs free energy of mixing $\Delta G < 0$. Blends of immiscible polymers obtained by simple mixing show a strong separation tendency leading to a coarse structure and low interfacial adhesion. Complete miscibility is a rare phenomenon to occur when a mixture of two or more systems is made. This is particularly so in case of polymers due to their complex structure and relatively small entropy of mixing. Due to this reason, the miscibility level of a new polymer blend is often the first property to be investigated.

To characterize polymer blends, DSC is the most commonly used thermal analysis technique in which the glass transition temperature (T_g) is measured. Here, a single glass transition (T_g) is observed for a miscible blend, and evidence of phase separation is inferred from the observation of two or more T_gs. Microscopic techniques such as SEM, TEM have been also used to study the surface morphology of polymer blends. Scattering techniques such as SAXS and SANS provides information on polymer blend structures. Interactions in polymer blends are studied using the spectroscopic methods like FTIR and NMR. In this study, we have employed Positron Annihilation Lifetime Spectroscopy (PALS).

The interface in a blend is intimately connected to adhesion and interfacial tension, which quantify miscibility, and compatibilization capability. To understand better the nature of interfaces at the microscopic level, we used the novel approach proposed by Schnell and Wolf [1,2] for polymer/solvent systems in terms of the hydrodynamic interaction parameter (α) evaluated from viscosity measurements. This method was modified by Ranganathaiah et al [3] for polymer/polymer binary blends in solid phase for the characterization of a single interface by invoking the concept that free volume and viscosity are inversely related. The efficacy of this method has been verified for several binary blend systems under varying conditions [3,4] by determining hydrodynamic interaction parameter from the measured free volume data obtained from positron lifetime spectroscopy. The method is extended to ternary blends as well. However, the same approach is not directly applicable to ternary blends with three interfaces. So, the required mathematical formulations have been modified for the three distinct interfaces characterized three hydrodynamic interaction parameters $\alpha_{12,}$ α_{23} and α_{31} and this has been used to test the efficacy in few ternary blends. The results are compared with literature reports and found to be in good agreement.

2. Experiment

The blends of Polystrene-*co*-acrylonitrile/Polymethyl methacrylate (SAN/PMMA), Polymethyl methacrylate /Polyvinyl chloride (PMMA/PVC) and Polystyrene/ Polyvinyl chloride (PS/PVC) of different compositions were prepared by solution casting method using methyl ethyl ketone as the common solvent. We have also prepared the ternary blends of Polycaprolactone/Polyvinyl chloride /Polyvinyl acetate (PCL/PVC/PVAc) and Polycaprolactone/ Polystrene-co-acrylonitrile /Polyvinyl chloride (PCL/SAN/PVC) by the solvent casting method. The T_gs of pure polymers are: SAN:106°C, PVAc:-45°C PVC:74°C, PCL:-60°C. For ternary blends, some selected compositions were prepared keeping in mind results for these composition are available from literature so that our method's efficacy can be easily checked. Lifetime spectra for the blends and their respective individual polymers were acquired using positron annihilation lifetime spectrometer (PALS) with a time resolution of 220 ps. A 17 μCi ^{22}Na positron source on kapton foil was used. All the measurements were done at ambient room conditions. The lifetime spectra have been resolved into three lifetime components using computer program PATFIT-88 with proper source and background corrections. Further details of experimentation can be found from earlier works [4]. The *o*-Ps lifetime τ_3 and intensity I_3 are used to obtain the fractional free volume F_v of the blends as well as individual polymers. DSC studies of pure polymers and their respective blends were carried out as the first step of assessing a given blend as M, PM or IM based on the observation of single glass transition temperature (T_g), two T_{gs} respectively. The PM blends exhibited a broad single T_g smearing the two T_{gs} of the pure components. The inferences were inconclusive because no composition could be decided as the optimum to produce maximum miscibility for that blend.

3. Results and Discussion

For binary polymer blends, Kelly and Bueche [5] derived the composition dependent free volume additive rules in terms the free volume of blend components. According to them, at constant pressure and temperature, the fractional free volume F_v of the blend is given by,

$$F_v = w_1 F_{v1} + w_2 F_{v2} \qquad \ldots 1$$

here w_1, w_2, F_{v1} and F_{v2} are the weight fractions and fractional free volumes of component polymers 1and 2 respectively. However, it was later realized that free volume is not linearly additive unlike volume due to the complex nature of its formation. Hence, Jean et.al [6] proposed a relation, based on the work of Wu [7], by introducing a parameter called the inter-chain interaction parameter β to represent the interaction between dissimilar chains in a binary mixture and this is

$$F_v = F_{v1}\varphi_1 + F_{v2}\varphi_2 + \beta F_{v1}F_{v2}\varphi_1\varphi_2 \quad \ldots 2$$

Here, φ_1, φ_2, are the volume fractions of component polymers 1 and 2. It has been observed that free volume parameters of the blends exhibit negative and positive deviations from the linear additive rule (Equation 1) depending on whether the blend is a single phase or a two phase system. Accordingly, the inter-chain interaction parameter β acquires negative and positive values respectively (Equation 2). This implies that, for miscible blends negative values of β indicate favorable interactions between the component polymers and positive values of β indicate the absence of any interaction between the components and hence the blend is termed as immiscible [6,7].

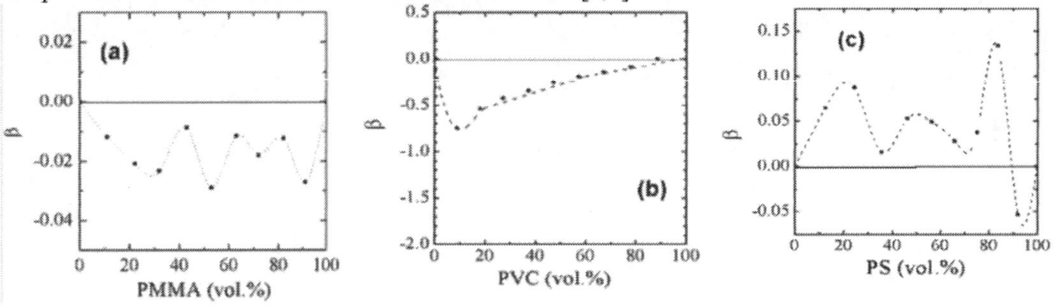

Figure 1 : Plot of β as a function of the volume percentage of the second polymer: (a) PMMA in SAN/PMMA, (b) PVC in PMMA/PVC, and (c) PS in PVC/PS.

In Figure 1, we notice that β *is* negative throughout the concentration range for SAN/PMMA blend and hence concluded as miscible. As miscibility varies with composition, figure1a fails to suggest which composition produces highest miscibility since it exhibits oscillatory behavior. Similarly, the PMMA/PVC blend is miscible. In case of PVC/PS, β is negative at 10% of PVC, but for other compositions it is positive hence immiscible. A comparison of the behavior of β in all three systems suggests that, although it could be taken as an indicator of miscibility, its behavior is not systematic (i.e., it exhibits complex behavior like oscillation) to infer a composition of highest miscibility [3].

3.1. Hydrodynamic Interactions parameter method: (Ranganathaiah's Method [3,10])

All materials exhibit visco-elastic properties in a given condition. These include amorphous polymers, semi-crystalline polymers, biopolymers etc. To be more specific, visco-elasticity is a molecular rearrangement. Polymers remain as solid materials even when these parts of their chains are rearranging in order to accompany the stress. As this occurs, it creates a back stress in the material. Hence, the adaptation of the theory developed to understand the polymer/solvent system in liquid phase to polymer/polymer blend in solid phase is thus justified. It is well known that free volume and viscosity are inversely related [8]. This concept was well adopted by Ranganathaiah et al. [3,10] to suitably modify the theory of hydrodynamic interactions for polymer/solvent systems formulated by Wolf et al. [1,2] based on the models developed to understand the behaviour of polymer solutions.

Ranganathaiah et al. [3,10] modified Wolf et al. theoretical equations [1,2] and deduced expressions for the hydrodynamic interaction parameter α suitable for polymer/polymer blend in solid phase. Here, the hydrodynamic interaction parameter is calculated by using the fractional free volumes of pure polymers and their respective blends which are experimentally measured from positron lifetime experiment. This approach was successful in revealing the composition dependent miscibility level and further discerns the nature of interfaces formed. The veracity of the method was also tested by bringing in phase modifications through irradiation with energetic electron beam and microwaves and found it clearly indicates the changes occurred at the interface/interfaces in binary and ternary blends [4,11].

The parameter α can be defined as the strength of hydrodynamic interaction at the interface. So also α can be viewed as a measure of the strength of adhesion at the interface. The final simple expression derived for the hydrodynamic interaction parameter α is shown to be

$$\Delta F_V = \left\{\delta[\gamma(1+\gamma\phi_2)^2 - \phi_1(1+\gamma)] + 2\alpha(1+\gamma)^2\phi_2 + e^{\left(\frac{1}{F_V}\right)}\rho\phi_1\right\}^{-1} \frac{(1+\gamma\phi_2)^3}{\phi_1\phi_2} \qquad ...3$$

here γ is the geometric parameter φ_1 and φ_2 are volume fractions of polymer 1 and 2, ρ is the density of the blend. Further ΔF_v is defined as

$$\Delta F_V = \left[\frac{1}{F_V} - \frac{\phi_1}{F_{V1}} - \frac{\phi_2}{F_{V2}}\right] \qquad ...4$$

i.e., difference in the reciprocal fractional free volumes of the constituent polymer 1 and 2 of the binary blends.

For Ternary polymer blend, it is characterized by three interfaces. We define φ_1, φ_2, φ_3 as the volume fractions of component polymers 1, 2 and 3 of the blend and α_{12}, α_{23}, α_{31} are the hydrodynamic interaction parameters corresponding to interfaces between polymer 1-2, 2-3 and 3-1 respectively, and their relationships are as follows.

$$\frac{1}{\Delta F_v} = \delta_1[\varphi_1 - \Omega_1(1+\Omega_1\Omega_2)] + \delta_2[\Omega_3(1-\Omega_2^2) - \varphi_3] - \delta_3\Omega_3^2\Omega_1$$

$$+ \sum_{\substack{i,j=1 \\ j=i+1}}^{3} \left[\frac{e^{1/F_{vB}}\rho\,\Omega_i^2\Omega_j}{1+\gamma_{ij}} + 2\alpha_{ij}(1-\Omega_i)\Omega_i\Omega_j\right] \qquad ...5$$

Here, the term on the left hand side is written as

$$\frac{1}{\Delta F_v} = \frac{1}{F_{vB}} - \frac{\varphi_1}{F_{v1}} - \frac{\varphi_2}{F_{v2}} - \frac{\varphi_3}{F_{v3}} \qquad ...6$$

where, F_{vB}, F_{v1}, F_{v2} and F_{v3}, are the measured fractional free volumes of the blend and its constituents 1,2 and 3 respectively. We define an effective hydrodynamic interaction parameter (α_{eff}) for the purpose of comparing the present results with published literature results and is expressed in terms of the individual α_{ij} and the respective volume fractions as [10,11]

$$\alpha_{eff} = (\varphi_1+\varphi_2)\alpha_{12} + (\varphi_2+\varphi_3)\alpha_{23} + (\varphi_3+\varphi_1)\alpha_{31} \qquad ...7$$

Details of derivation of mathematical equations can be found in our earlier works [3,10].

The friction associated with the visco-elastic flow results in energy dissipation at the interface and is influenced by the composition of the blend. In the case of miscible blends, the polymer chains are brought closer to each other so the friction at the interface increases. As energy dissipation increases the tension at the interface decreases. In Ranganathaiah's method, the negative sign of the hydrodynamic interaction parameter (α) is inferred as dissipation of energy at the interface. If miscibility is high at certain composition of the blend, α takes large negative values [3]. On other hand, for immiscible blends, α is close to zero or positive. Zero or positive value is interpreted as the absence of any favorable interactions between the constituent polymers and hence friction at the interface is very less.

For a miscible blend, when a chain moves, several other chains with which it has segmental level interactions pull it. This generates a good amount of friction in the system for which α attains large negative values. In Figure 2(a), we show a plot of α as a function of the PMMA percentage in the SAN/PMMA blend. The SAN/ PMMA blend is known to be miscible, not because of the intermolecular interactions but because of the intra-molecular repulsion between the SAN chains. It is important to note that several techniques have shown this blend to be miscible throughout the concentration range, but none of them have indicated the extent or degree of miscibility for different compositions. We can see from Figures 1(a) that this system exhibits miscibility from *β* parameter throughout the range of concentrations, but there is no indication of maximum miscibility at any composition. In contrast, a large negative value (-209) at 20% PMMA, observed in Figure 2(a), can be considered to indicate that the degree of miscibility is maximum because α shows only smaller values at a higher concentration of PMMA. SAN has high molecular weight and repulsion between the SAN chains make it easy for PMMA chains to slide in between. As the SAN content decreases, the repulsion also decreases, thus, PMMA with 20% content has the highest negative value of α.

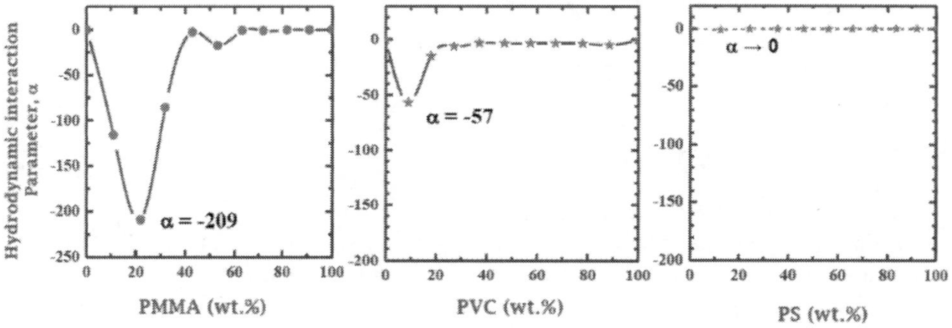

Figure 2 : Plot of α as a function of the volume percentage of the second polymer: (a) PMMA in SAN/PMMA, (b) PVC in PMMA/PVC, and (c) PS in PVC/PS.

In the case of PMMA/PVC blends, the interaction between the carbonyl group (C=O) of PMMA and α-hydrogen (H-C-Cl) of PVC drives the system to miscibility. However, this system is known to be partially miscible, and at some compositions, it becomes immiscible. α for this system in figure 2(b) is negative and maximum at 10% PVC (-57) and decreases as the PVC content increases. Beyond 30% PVC content, α attains smaller values closer to zero. The increase in the PVC concentration reduces interactions, and this results in smaller α values. The α values obtained for the PVC/PS blend system shown in Figure 2(c) are a clear indication that the α values are not exactly zero but are close to zero. This indicates that the system is immiscible throughout the concentration range suggesting friction at the interface is very negligible. In such immiscible blends with no specific interactions between polymer components, they form their own domains in the system, thereby reducing the possible contacts between the chains. In other words, the friction between the chains of the constituent polymers is negligible. Above method was verified by Xiaoyu Meng et.al [12] who employed this method to determine miscibility of binary polymer SBS/PS. Interestingly, the mechanical properties showed large improvement for highest miscibility composition.

For the ternary blend PCL/SAN/PVC, the α_{12}, α_{23}, α_{31} denote the hydrodynamic interaction parameters corresponding to respective interfaces between PCL/SAN, SAN/PVC and PVC/PCL, and evaluated according to equation 5. The interface α_{12} between PCL/SAN takes both positive and negative values depending on composition. Here, negative α indicates increased interfacial friction, observed when PCL is matrix and SAN is dispersed phase. This may be due to the interaction between the carbonyl group of PCL and tertiary hydrogen of SAN [13]. The second interface, α_{23} is positive at 34/33/33 composition while at other compositions it is negative. The negative values of α's is due to the repulsive force among the SAN chains paving way for PVC chains to slide in [14]. Interestingly the third interface in the blend that is between PVC/PCL, produces negative values for all compositions studied, indicating the formation of good interfaces attributable to the interaction between polar oxygen of PCL and proton of PVC [15]. Finally, the α_{eff}, plotted in figure 3, shows positive values at 20/40/40 and 34/33/33 compositions and thus we infer the blends are immiscible at these compositions. On the other hand, the negative values at 68/16/16 and 70/25/5 indicate the blends are miscible at these compositions. Importantly, 68/16/16 composition produces highest miscibility for this blend since α_{eff} has a high negative value. These results agree with literature data[11].

Journal of Physics: Conference Series **618** (2015) 012022 doi:10.1088/1742-6596/618/1/012022

In case of PCL/PVC/PVAc blend, three interfaces are designated as α_{12} for PCL/PVC, α_{23} for PVC/PVAc and α_{31} for PVAc/PCL. From figure 3, it can be seen that the interface between PCL/PVC is negative for all the four compositions studied. The second interface between PVC/PVAc shows positive values for three compositions suggesting the formation of weak interfaces due to the absence of favorable interactions between the components. The third interface is PVAc/PCL and is known to be miscible and it supported by negative α value for all compositions. The α_{eff}, also plotted in figure 3 shows that the blend is miscible at all compositions studied. Further, α_{eff} takes a high negative value of **-12.60** for the composition 80/10/10 inferring that this composition produces highest level of miscibility for this blend.

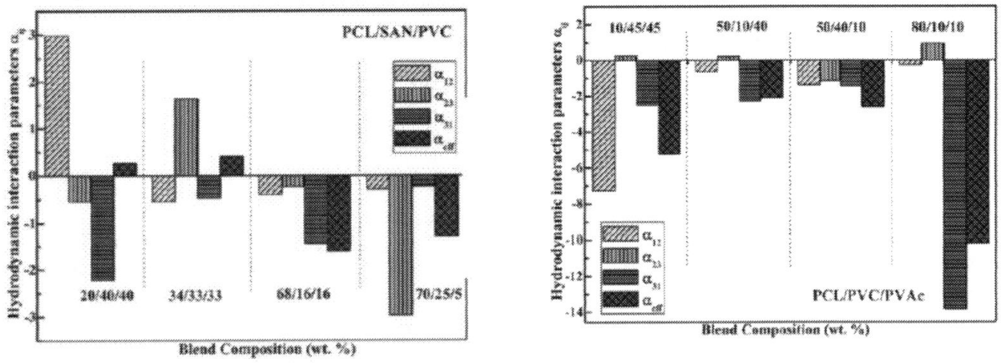

Figure 3: Plot of hydrodynamic interaction parameters α as a function of blend composition for the blend (a) PCL/SAN/PVC and (b) PCL/PVC/PVAc

4. Conclusion

A new method based on hydrodynamic interactions to determine the miscibility level of binary and ternary polymer blends is described. It has also been demonstrated that one can characterize the individual interfaces in ternaries by this method. The efficacy of the present method has been tested for few binary blends namely PS/PMMA, PMMA/PVC and PS/PVC and in ternary blends namely PCL/SAN/PVC and PCL/PVC/PVAc. The experimentally measured fractional free volume of pure polymers and their respective blends form the basis of evaluating the hydrodynamic interaction parameters.

Reference

[1] Schnell M and Wolf B A 2000 *J Rheol.* **44** 617
[2] Schnell M and Wolf B A 2001 *Polymer* **42** 8599
[3] Ranganathaiah C and Kumaraswamy G N 2009 *J Appl Polym Sci.* **111** 577
[4] Raj J M and Ranganathaiah C 2009 *J Polym Sci Part B: Polym Phys.* **47** 619
[5] Kelly F.N and Bueche 1961 *J. Polym. Sci.*, **50** 549
[6] Liu J, Jean Y.C and Yang H 1995 *Macromolecules* **28** 5774
[7] Wu S 1987 *J. Polym. Sci.: Part B Polym. Phys.* **25** 2511
[8] Williams M.L, Landel R.F and Ferry J.D, 1955 *J. Am. Chem. Soc.*, **77** 3701
[9] Jamieson A M, Olson B G and Nazarenko S 2010 *Polymer Physics: From Suspensions to Nanocomposites and Beyond* ed L A Utracki and A M Jamieson (John Wiley Hoboken) 491
[10] Meghala D and Ranganathaiah C, *Polymer*, 53 **(2012)** 842
[11] Meghala D and Ranganathaiah C, J. Appl. Polym. Sci. **130** (2013) 3335
[12] Xiaoyu Meng et.al,Polym. Engg.Sci. 54 **(2014)** 785
[13] Kwei T K, Frisch H L, Radigan W and Vogel S 1977 *Macromolecules*. **10** 157
[14] Hurang J C, Min K and White J L 1988 *Polym. Eng. Sci.* **28** 1085
[15] Kulshreshtha A K Singh B P and Sharma Y N 1988 *Eur. Polym. J.* **24** 29

Depth profiles and free volume in aircraft primer films

J D Van Horn[1], H Chen[1], Y C Jean[1], W Zhang[2] and M R Jaworowski[2]

[1] Department of Chemistry, University of Missouri-Kansas City, 5110 Rockhill Road, Kansas City, MO, 64110 USA

[2] Physical Sciences Department, United Technologies Research Center, 411 Silver Lane, East Hartford, CT, 06108 USA

E-mail: vanhornj@umkc.edu

Abstract. Positron annihilation lifetime spectroscopy (PALS) and associated techniques provide non-destructive methods to study the free volume inside polymeric materials, and to study material characteristics over a depth profile. Cast free films of organic- or aqueous-based, non-chromated aerospace primers, when cured for about one week, had very different water vapour transport (through-plane) behaviour. In addition, both types of primer films showed strong anisotropic behaviour in in-plane versus through-plane water vapour transport rates. We report the differences between the organic- and aqueous-based aircraft primer films samples and their surface depth profiles. In bulk PALS measurements, an aged, organic-based film exhibited typical lifetimes and intensities for a particulate-containing polymer film on both faces. In contrast, aqueous-based films exhibited face oriented-dependent differences. In all aqueous-based samples, the I_3 value of the back of the sample was smaller. The primer film samples were also evaluated with mono-energetic positron beam techniques to generate depth profile information. The heterogeneity in the samples was verified by Doppler broadening of energy spectroscopy (DBES). A model for the differences in the faces of the films, and their layered structure is discussed.

1. Introduction

Coatings used to inhibit corrosion of metal surfaces are complex mixtures comprised of polymer, molecular organic, pigments, inorganic particulate, and many other components [1]. Each component, along with any additives, plays a role in the physico-chemical structure, adhesion properties, and corrosion inhibition characteristics of a particular coating system. Typically, metal surfaces require a primer coat to enhance the adhesion performance of a subsequent or final coating system. Coatings may be generally classified into those with organic-based or aqueous-based carriers; another aspect is the addition of inhibitors to further mitigate corrosion at a coated surface.

It was previously found that a set of epoxy polyamide primers (one aqueous-based and one organic-based), with praseodymium inhibitor, exhibited differential water vapor transport rate behavior [2]. Water vapor transport is one critical aspect of corrosion formation under coatings. The aqueous-based primer (Deft-98) also showed an abnormal curing time dependent rate, while an organic-based primer (Deft-84) had almost no dependence on the curing time (after 7 days curing). The observation of dissimilar in-plane water vapor transport and curing time suggested further investigation using positron techniques. Positron annihilation spectroscopy (PAS) techniques provide a unique method for the structural and free volume evaluation of thin films, including heterogeneous coatings [3, 4].

2. Experimental

The details of the primer film samples' provenance, preparation, and characteristics were previously reported [2]. A special arrangement was made to have samples evaluated by PAS at 7- and 21-day curing times for comparison to 'aged' samples (> 1 month). General positron annihilation techniques and methods have been reviewed and reported from this lab [5, 6].

2.1. Bulk PALS.

PALS bulk measurements were performed by using a conventional fast-fast coincidence method at 25 °C. The time resolution of the spectrometer was <300 ps, measured using a ^{60}Co source in the ^{22}Na energy window settings [7]. Samples were prepared as before [2] and used as received under ambient or vacuum ($< 1 \times 10^{-4}$ torr) conditions at specified curing times. A $^{22}NaCl$ positron source (sealed between Kapton disks) was sandwiched between two identical, double-stacked film samples (30 mm × 30 mm × 0.35 mm). The stacked samples (~ 0.7 mm total per side) assure that all source positrons annihilate within the samples (d ~ 1.6 g/cc). Further, the film samples were oriented with respect to their front face (the side open to atmosphere during curing) or their back face (substrate side). Positron interaction with the source coating is taken into account in the data analysis.

All PAL spectra were resolved into three components by using the PATFIT program [8, 9]. The spectra were then fitted to three exponential lifetime components: the component with the longest lifetime (τ_3 = 2.0–2.2 ns) and its relative intensity (I_3) is assigned to pick-off annihilation of *o*-Ps in the void spaces of the film. Since the variations of τ_1 and τ_2 are much less than that of τ_3, the observed lifetime of *o*-Ps (the triplet state of Ps) may be directly correlated to the size of free-volume holes [10, 11]. Bulk phenomena and radius distribution were also evaluated using the MELT software [12].

2.2. Monoenergetic Positron Beam PALS and DBES Depth Profiling.

2.2.1. Monoenergetic Positron Lifetime Analysis. Conventional PALS [13] at varied positron implantation energy ($\propto$ depth) was used to measure the free volume properties in the coating film samples (~ 0.35 mm). PALS data contains quantitative information on the free-volume properties in polymeric systems from the surface, at interfaces, and into the bulk. For the slow positron beam at UMKC the lifetime resolution is ~500 ps for counting rates of 100-500 cps. Each PAL spectrum contained 1 million counts. The obtained PAL data were fit into lifetime components using the PATFIT program [8-9] and also into continuous lifetime distributions using MELT [12]. The smoothed lifetime distributions from MELT analysis are presented below.

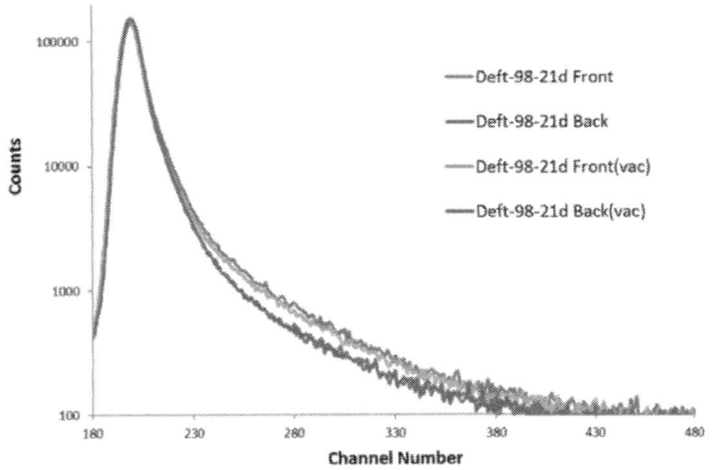

Figure 1. Representative positron annihilation lifetime spectra illustrating the differences between the front and back faces of the Deft-98, aqueous-base primer film, at 21-days. Lifetime spectra were taken under vacuum and ambient conditions.

Journal of Physics: Conference Series **618** (2015) 012023 doi:10.1088/1742-6596/618/1/012023

2.2.2. Depth Profiling Using DBES. The S-parameter of Doppler broadening energy spectra (DBES) was measured as a function of implantation energy using the slow positron beam (0-30 keV) at the Positron Science Laboratory, Department of Chemistry, University of Missouri-Kansas City [14, 15]. A vacuum in the sample chamber was maintained between 10^{-7} and 10^{-8} torr. DBES spectra were recorded at r.t. as a function of positron energy from about 0 to 25 keV. The DBES spectra were measured using an HPGe Detector (EG&G Ortec with 35% efficiency and energy resolution of 1.5 keV at 511 keV peak) at a counting rate of approximately 2,500 cps. The total number of counts for each DBES spectrum was 2 million. The obtained DBES spectra were characterized by the S-parameter, defined as a ratio of integrated counts between the energies of 510.3 and 511.7 keV (S width) to the total counts after the background is properly subtracted. Since the S parameter represents the relative value of the low momentum part of positron-electron annihilation radiation, it is sensitive to the change of the positron and Positronium (Ps) states due to microstructural changes. The S parameter has been widely used in detecting free volume depth profile in polymeric systems [16-18].

3. Results and Discussion

After an initial experiment, bulk and beam PALS and DBES measurements were carried out considering the orientation of the film to the positron source. Thus, films were evaluated in a face-to-face or a back-to-back fashion in bulk measurements, and also before and after exposure to beam conditions (high vacuum and incident irradiation). 'Front' here means the top, exposed face of a film laid down for curing; 'back' means the surface contacting the substrate upon which the film was cast.

3.1. Bulk PALS evaluation of primer films.

The Deft-84 organic-based primer film (an aged sample) exhibited typical *o*-Ps lifetimes (τ_3) and intensities (I_3) for a particulate-containing polymer film (data not shown). A comparison of the intensities (I_3) and estimated average pore sizes (R), for the organic-based films of either mixed faces or oriented faces indicated the relative homogeneity of these films. Measurements before and after exposure to high vacuum were done to address the possibility of residual or trapped solvent in the films over the curing time, and sample integrity under positron beam conditions. Data were the same in both conditions, indicating the aged samples were robust and experimental data not affected by residual solvent.

In contrast, the Deft-98 aqueous-base primer films exhibit a significant difference in their intensities (I_3), dependent on the orientation of the face (refer to Fig. 1 and Table 1). In all samples, 7-day cure, 21-day cure, and aged Deft-98, the I_3 value of the sample in the back-to-back orientation exhibited a lower value, and the associated free fractional volume decreased in value. In the average lifetime measurement, a 'homogeneous' material should exhibit the same lifetime, intensities, etc. regardless of orientation. In comparing the lifetime measurements here, the observed intensity difference is taken to mean that some heterogeneity exists in the internal structure of the Deft-98 primer films *vs.* Deft-84.

In addition, the average *o*-Ps lifetime (τ_3), and calculated pore size (R) decreased over the curing time of 7 and 21 day, to aged samples. Figure 2 illustrates the lifetime, pore size, and radius distribution differences for the Deft-98 primer films analyzed in this project. At 7 days the pore radius is approximately 3.0 Å and decreases to about 2.7 Å in the fully aged film (also compare lifetime data, Table 1). For comparison, the aged Deft-84 organic-base primer has a pore radius about 3.0 Å (data not shown, but compare *o*-Ps lifetime).

3.2. Monoenergetic DBES and PALS evaluation of primer films.

Depth profiling and lifetime measurements using the monoenergetic positron beam at UMKC corroborated the proposal of structural differences in the two faces of the Deft-98 primer film, as well as differences between the two types of primer films. The film layer differences may arise from the emulsion-to-film formation process, which is a key difference between the aqueous-based primers compared to organic-based systems.

Table 1. *Ortho*-positronium pickoff annihilation lifetimes in primer films. Fresh Deft-98 primer films (aqueous-base) exhibit significant heterogeneity in intensity when comparing the back and front faces of the film, and exhibit a decrease in lifetime over time.

Sample	Face	τ_3 (ps)	I_3 (%)	ΔI_3	FWHM(ps)
Deft-84 aged	Front	2190	6.5	0.1	253
Deft-84 aged	Back	2210	5.7	0.1	140
Deft-98 7 day	Front	2130	**9.9**	0.1	201
Deft-98 7 day	Back	2180	**5.0**	0.1	144
Deft-98 21 day	Front	2050	**9.6**	0.1	243
Deft-98 21 day	Back	2060	**6.5**	0.1	209
Deft-98 aged	Front	1900	**7.8**	0.1	221
Deft-98 aged	Back	1960	**5.6**	0.1	201

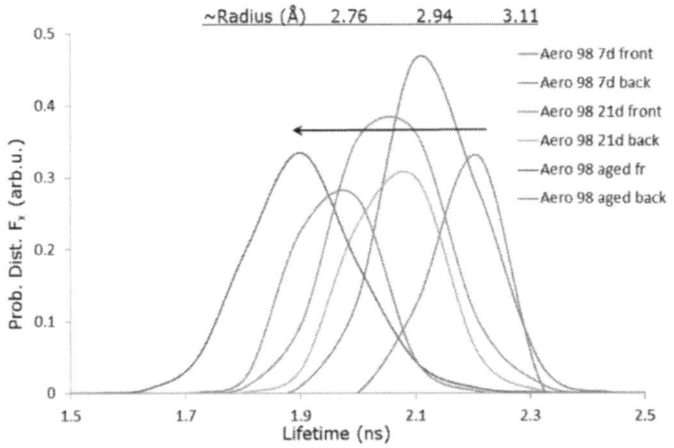

Figure 2. Mean lifetime, lifetime distribution, and approximate radius for front and back faces Deft-98 aqueous-base primer films over time. The arrow (←) indicates increased age of film.

The films' depth profiles are quite different comparing the front faces of the organic- or aqueous-based films (Figs. 3 & 4), or comparing the front and back faces of the Deft-98 film (Fig. 4). However, these differences may not have directly influenced the water transport phenomena previously reported.

The lifetime of *o*-Ps can be more correctly expressed as a distribution, and the correlation to hole radius (using a spherical approximation) can be expressed as a the free-volume probability-density function, PDF (sample, Fig. 5) [12]. The front faces of all the films, using lifetime measurements at varied positron implantation energies, demonstrate a decrease in the pore radius, roughly starting at a similar radius (3.7 Å) near the surface and having a bulk radius near 2.8 Å. For the facing sides the hole radius distribution is broader for Deft-98 at the deeper positron implantation energy (data not shown).

The front face of the Deft-98 sample exhibits a radius near the surface slightly smaller than that observed in the back faces (refer to τ_3 in Table 1 and estimated radius in Figure 2), and the distribution stays somewhat narrower at greater depth. This narrower distribution may be reflecting the thicker second layer in the front faces of the aqueous-base films, i.e. fewer inorganic particles leads to smaller void distribution. Further, at the largest implantation energy, the radius and distribution are similar between the front and back faces, suggesting that at the associated depth, the environment is now similar (e.g. a similar 'bulk' heterogeneous phase) when approached from the differing analytical directions

Journal of Physics: Conference Series **618** (2015) 012023 doi:10.1088/1742-6596/618/1/012023

(data not shown). In general, it appeared that the aqueous-based film exhibited broader void volume distributions than the organic-base primer film. An example of this is illustrated in Fig. 5, where the 7 and 21 day cured samples of the aqueous-base film are compared. The pore sizes in the Deft-98 may be trending to slightly smaller sizes, and in this and other comparisons the distribution of voids was characteristically broader.

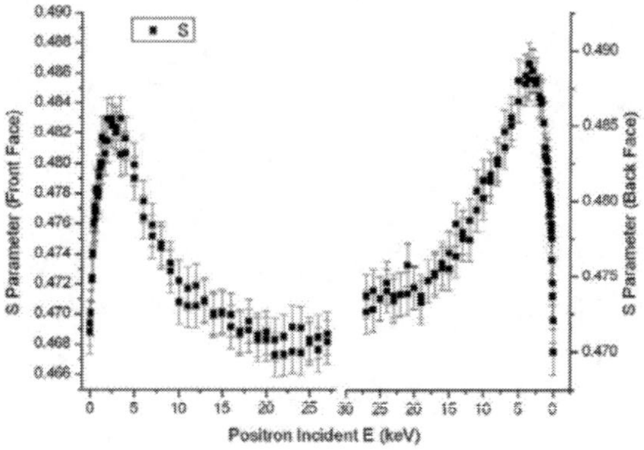

Figure 3. The S parameter, derived from Doppler-broadening energy spectroscopy, versus positron incident energy (or depth) in the front (left) and back (right) faces of the Deft-84 organic-base film. The interface layers are characteristically similar in nature.

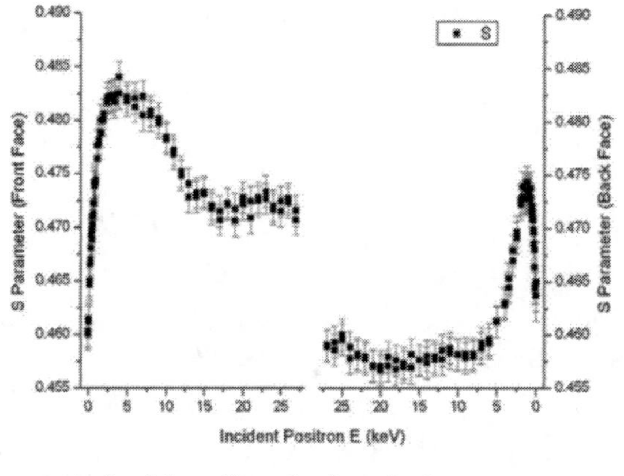

Figure 4. The S parameter, derived from Doppler-broadening energy spectroscopy, versus positron incident energy (or depth) in the front (left) and back (right) faces of the Deft-98 aqueous-base film. The S parameter fit to 2-layer models indicate a 500-600 nm layer on the front face compared to a <200 nm interface layer on the back (from VEPFIT [18]).

4. Conclusion

Positronium lifetime measurements of two different primer films indicated structural inhomogeneity in the aged, aqueous-based primer, as well as a smaller void radius in the film compared to the aged, organic-based primer film. Comparison of lifetime data at seven day curing time to longer curing times suggests that the film may still be undergoing structural development at the 7 day point. This observation may explain the differential curing and in-plane water transport phenomena observed previously [2]. Through-plane water transport phenomena may be impacted by the presence of the relatively thicker secondary layer of the front face of the aqueous-base primer film; this due to less pore interconnectivity from the lower distribution of inorganic particles in this part of the layer structure.

The depth profile of the aqueous-based primer film, observed via DBES, confirmed the structural differences in the faces of this film, and indicated a larger layer between the surface and bulk of the sample. Lifetime data indicate generally broader and changing void distributions through the aqueous film and curing process. The PAS and water transport results taken together point to the different curing and internal structural development and final forms of the two types of primer coating systems.

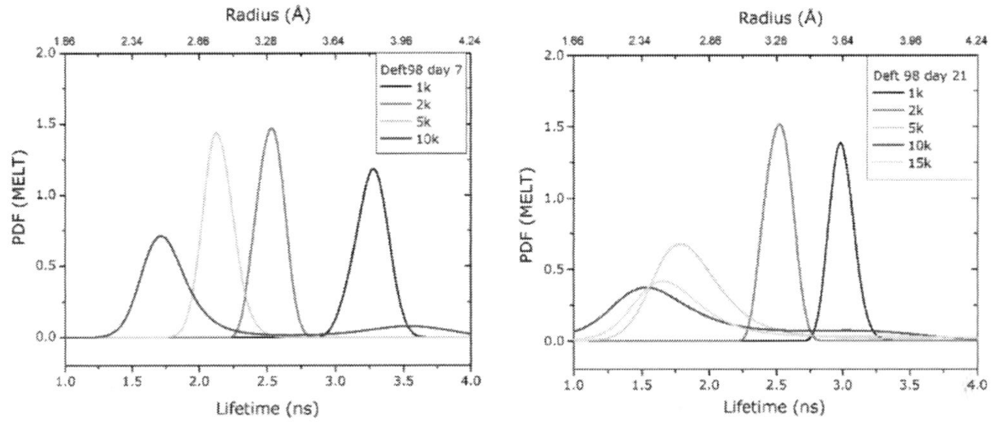

Figure 5. Free-volume radius distributions (PDF = probability density function, [12]) as a function of positron implantation energy (keV) for front faces of Deft-98 (7-day, left) and (21-day, right) samples.

5. Acknowledgements

This material is based on work supported by the U.S. Army Corps of Engineers, Humphreys Engineering Center Support Activity under Contract No. W912HQ-08-C-0011. We thank the U.S. Army Research Office for equipment funds (W911NF-10-1-0476). Views, opinions, and/or findings are of the authors and should not be construed as an official Department of Defense position or decision unless so designated by other official documentation.

References

[1] Harun M K H 2010 *Corrosion Protection by Paints and Organic Inhibitors* (Universiti Teknologi Mara, Shah Alam: University Publication Centre)
[2] Zhang W and Jaworowski M R 2011 *ECS Trans.* **35** 53
[3] Jean Y C *et al* 2005 *Prog. Org. Coat.* **52** 1
[4] Chen H *et al* 2007 *Macromolecules* **40** 7542
[5] Jean Y C *et al* 2013 *Macromolecules* **46** 7133
[6] Xia Z *et al* 2014 *Phys. Rev.* E **89** 022603
[7] Hong X *et al* 1996 *Macromolecules* **29** 7859
[8] Kirkegaard P *et al* 1981 *Comput. Phys. Commun.* **23** 307
[9] Kirkegaard P and Eldrup M 1972 *Comput. Phys. Commun.* **3** 240
[10] Tao S J 1972 *J. Chem. Phys.* **56** 5499
[11] Eldrup M *et al* 1981 *Chem. Phys.* **63** 51
[12] Shi E *et al* 2007 *Appl. Polym. Sci.* **104** 2468
[13] Zhang R *et al* 2000 *Rad. Phys. Chem.* **58** 639
[14] Chen H *et al* 2004 *Mater. Sci. Forum* **445-446** 274
[15] Kobayashi Y *et al* 1995 *Phys. Rev.* B **52** 823
[16] Jean Y C *et al* 1997 *Phys. Rev.* B **56** R8459
[17] Li Y *et al* 2003 *Biomacromolecules* **4** 1856
[18] Schut H and van Veen A 1995 *Appl. Surf. Sci.* **85** 225

Journal of Physics: Conference Series **618** (2015) 012024 doi:10.1088/1742-6596/618/1/012024

Characterization of the melting process of PTFE using positron annihilation spectroscopy

Y Honda[1] and S Nishijima[2]

[1] Institute of Scientific and Industrial Research, Osaka University, 8-1 Mihogaoka, Ibaraki, Osaka 567-0047, Japan

[2] Graduate School of Engineering, Osaka University, 2-1 Yamadaoka, Suita, Osaka 565-0871, Japan

E-mail: honda@sanken.osaka-u.ac.jp

Abstract. Poly(tetrafluoroethylene) (PTFE) is a semi-crystalline polymer and the lifetime of ortho-positronium (*o*-Ps) is known to be able to be separated into two components due to annihilation in the crystal region and in the amorphous region. The melting process of PTFE was investigated using positron annihilation spectroscopy and X-ray diffraction. The results indicated that volume expansion with an increase of temperature is dominantly due to the expansion of the amorphous region and a Ps bubble is formed at melting in both regions. The *o*-Ps relating to the crystal region definitely remains on the surface of crystal at the time of annihilation. The production of lower energy electrons at melting was deduced by the analysis of the Doppler broadened annihilation photopeak, and the increase in the number of such electrons was found to have great influence on the formation of the *o*-Ps and annihilation processes of positron and *o*-Ps.

1. Introduction

Whenever some properties related to the functional polymers are evaluated by using positron annihilation spectroscopy (PAS), special attention must be paid to the annihilation sites of positron and ortho-positronium (*o*-Ps), which is a triplet state of the electron-positron bound state. In many functional polymers, polarization due to functional groups or impurities would induce inhomogeneous formation of *o*-Ps and the annihilation of positron/*o*-Ps. We investigated the annihilation process of positron/*o*-Ps in perfluorinated polymer electrolyte membranes through the degradation process and found that the dominant part of *o*-Ps remains and annihilates near the hydrophilic site, i.e., just in the "skin" of inverse micelle that is formed by aggregation of the hydrophilic site [1], indicating that the lifetime of *o*-Ps is not significantly influenced by the condition of the hydrophilic site, but is sensitive to that of the micellar skeleton. Such localization of *o*-Ps would be caused by the presence of a local electric field due to hydrophilic groups. Because no such local electric field is found in poly(tetrafluoroethylene) (PTFE), which is a base polymer of the above-mentioned electrolyte membrane, annihilation of positron/*o*-Ps expects to occur homogeneously in PTFE. However, PTFE is known to be a semi-crystalline polymer, having a crystal region of hexagonal form and an amorphous region, and two types of lifetimes of *o*-Ps are known to be responsible for the space sizes of the crystal region and the amorphous region. PTFE is a well-known polymer that has several advantages, such as good chemical stability, high heat resistance and biocompatibility, and widespread use. The characterization of the melting process of PTFE is important not only from the academic perspective but also from an industrial application perspective. In such semi-crystalline polymers, special attention is on the individual behavior of the crystal region, the amorphous region and their interface during the melting process, and a recent theoretical model was also focused on the peeling of the crystal region during melting [2]. Many studies involving positrons for the characterization of PTFE have been reported so far. In particular, Dlubek et al. attempted to locate a precursor of deformation appearing on the surface of the crystal region at the phase transition in the range below the melting point [3]. In this work, the behavior of each region during melting is discussed.

2. Experimental setup

The measured sample was a homo PTFE plate brought from DAIKIN industries Ltd. In PAS, sodium-22 solution was dropped onto the sample with a thickness of 3 mm. The source was dried on the sample and then sandwiched with another homo PTFE plate, followed by wrapping with aluminum foil to prevent contamination. This source assembly was also sandwiched in between two ceramic heaters to suppress the temperature gradient and was heated to a temperature exceeding the melting point. The temperature was monitored using a thermocouple inserted between the PTFE plate and the aluminum foil. The intensity of the source was ca. 560 kBq. The temperature was controlled within a fluctuation of ±0.5 °C for the PAS and X-ray diffraction measurements mentioned below. The annihilated γ-rays were detected by the assembly of a BaF_2 scintillator and a photomultiplier tube together with a conventional time measuring system using a constant fraction differential discriminator and a time to pulse-height converter. Data analysis of the lifetime was mainly performed using the PATFIT program, in which a lifetime spectrum is represented by the formula of $N(t) = \sum_i \left(\frac{I_i}{\tau_i}\right) exp\left(-\frac{t}{\tau_i}\right)$, where the index i denotes each annihilation process of positron. The time resolution was approximately 290 ps throughout the measurements. Differential scanning calorimetry (DSC8270; RIGAKU) was used to evaluate the melting point. The temperature dependence of the structure was also investigated using X-ray diffraction (SmartLab; RIGAKU), where the applied temperature was monitored by the attached thermocouple to the sample.

3. Results and discussion

3.1. Differential Scanning Calorimetry (DSC)

DSC measurements were performed to evaluate the melting point. The rate of temperature increase was 10 °C per minute and endothermic peak appeared at 330 °C (Figure 1). The heat of fusion was evaluated to be 25 J/g based on the area of endothermic peak between 277.1 °C and 345.9 °C. As the heat of fusion for melted PTFE was reported to be 93 J/g elsewhere [4], the crystallinity of the sample was estimated to be 26.9 %. However, in a semi-crystalline polymer, it is better to describe the melting process based on a three-phase model consisting of a crystal region, an amorphous region and a rigid amorphous region rather than a two-phase model consisting of a crystal region and an amorphous region, and it would be difficult to evaluate crystallinity precisely by using the usual DSC. This point will be discussed in a later section.

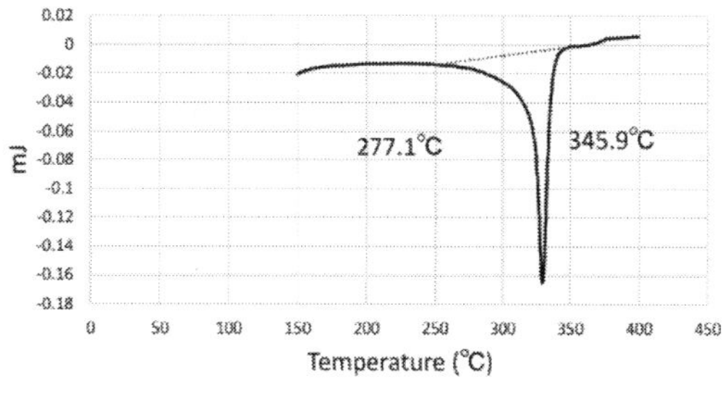

Figure 1. DSC chart for homo PTFE. The heat of fusion was evaluated to be between 277.1 °C and 345.9 °C.

3.2. Positron Annihilation Spectroscopy (PAS)

3.2.1. Positron Annihilation Lifetime Spectroscopy (PALS). There are many works related to the lifetime of *o*-Ps in PTFE; four-component analysis is known to be suitable, i.e., the lifetimes of *o*-Ps annihilations in the crystal region, τ_3, and in the amorphous region, τ_4, in addition to the lifetimes of para-Ps (the singlet of Ps), τ_1, and free annihilation, τ_2. Because our attention was focused on the temperature dependence of the structural change, the lifetimes shown in the following sections are restricted to the lifetimes of *o*-Ps. The lifetimes were also analyzed using the CONTIN program in addition to the PAFIT program to ensure the validity of the four-component analysis. Many of the lifetime spectra were confirmed to be comprised of four components; however, some spectra were decomposed into only 3 components. The number of acquired data points for each spectrum was in the range of 2 million to 3 million throughout the experiments, which might not be adequate for CONTIN. However, the longest lifetimes analyzed using CONTIN were found to well coincide with the fourth component of the PATFIT analysis. The temperature dependencies of the lifetimes of *o*-Ps are shown in Figure 2. A slight increase of τ_3 and an evident increase of τ_4 were observed between room temperature (ca. 22 °C) and 300 °C, which is consistent with the results of Drubek [3], whereas no significant differences were found for I_3 and I_4. These results indicate that the volume expansion of PTFE below the melting point is mainly due to the expansion of the amorphous region. In contrast, significant increases of τ_3 and τ_4 were observed at 335 °C and 330 °C, respectively. The increase of τ_4 occurred at a lower temperature than that of τ_3, implying that the volume expansion of the amorphous region was followed by the expansion of the crystal region. Taking into account the observation that the increase of τ_3 and τ_4 above the melting point are small and moderate, respectively, these abrupt changes in τ_3 and τ_4 would suggest the formation of positronium bubbles because the binding energy of the polymer chain is reduced greatly above the melting point and Ps is able to form a bubble by overcoming the tension of surrounding polymer chains in liquid. Above 335 °C, τ_4 increased gradually, whereas τ_3 remained almost constant, suggesting that the volume expansion is primarily due to the expansion of the amorphous region above the melting point. Based on these results, it can be deduced that at melting of homo PTFE, the relaxation of amorphous region initiates volume expansion of the amorphous region and peeling of the crystal surface that retains crystallinity [2]. Two perspectives exist regarding the annihilation of *o*-Ps in the crystal region: annihilation inside of the crystal and annihilation on the surface of the crystal. If the former is true, then the volume expansion of the crystal region indeed occurred, and the Ps bubble also formed inside of the crystal. If the latter is true, then the Ps bubble was formed on the surface, and the inter-chain distance in crystal region was not significantly changed. This issue will be discussed again in a later section. Regarding the relative intensities, both I_3 and I_4 changed abruptly at 335 °C; however, the following behavior contradicted each other, i.e., I_4 kept increasing after jumping, while I_3 slightly decreased after dropping. This behavior occurred because the volume fraction corresponding to the amorphous region increased and a number of positrons formed positronium here, with the help of an increase in the number of lower energy electrons, as shown in the next section, resulting in the decrease of positronium formation in the crystal region.

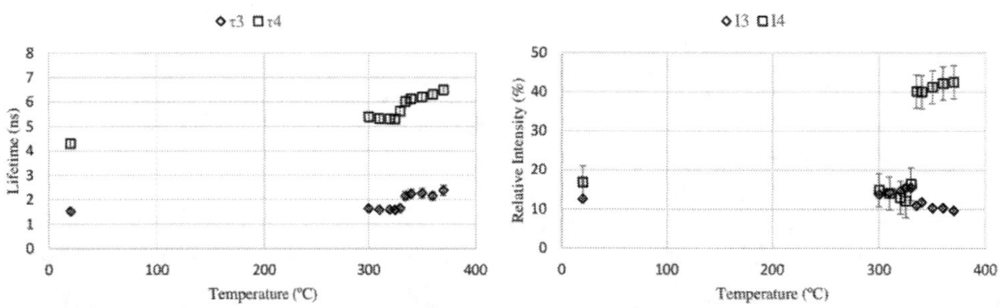

Figure 2. Temperature dependencies of lifetime and relative intensity of homo PTFE.

3.2.2. Doppler Broadening Measurements. Doppler broadened spectra were also measured at several temperatures, and the representative ones are shown in Figure 3, where the dotted, solid, and dashed lines were obtained at 24 °C, 300 °C, and 340 °C, respectively. The peaking process was mild in the range from 24 °C to 300 °C. Drastic peaking was observed at 340 °C, but no additional peaking was observed above 340 °C. This remarkable change was due to melting. To investigate the significant peaking process in detail, the spectra observed at 24 °C and 340 °C were decomposed into three Gaussians distributions of different energy spreads, representing the annihilations with core electrons, valence electrons and lower energy electrons relating to thermal radicals. The three Gaussians were characterized by the energy spreads of 3.38 keV, 1.95 keV and 0.918 keV, which were determined via fitting. Taking into account that the energy spread for the peak of Cs-137 (662 keV) was 0.845 keV, the obtained lowest energy spread is found to be close to the intrinsic energy spread due to the instrument. The fitted results are shown in Figure 4. The upper figures are the raw data (open circles) and fitted curves (solid line). The proportion of the distribution characterized by the highest energy spread to the whole distribution was the lowest and quite similar between the two spectra. The major annihilation at 24 °C was characterized by the medium energy spread (1.95 keV), and the lowest energy spread was dominant at 340 °C. The increase of the lowest energy spread component means the increase of free annihilation with low energy electrons in addition to the annihilation of *o*-Ps in the amorphous region. The increase of the number of low energy electrons, whose energy levels are almost equivalent to thermal electrons, would be useful to reprocess the materials. The intensity ratio of the lowest energy spread to the medium energy spread was 0.17 for the data of 24 °C, whereas the same ratio for 340 °C was 1.59. The annihilation rate with the lowest energy electron increased approximately 9 times at 340 °C versus 24 °C. The ratio of I_4 to I_3 in PALS was changed from 1.26 to 3.72 (3 times), but total annihilation ratio of *o*-Ps, increased 1.7 times from 24 °C to 340 °C. Thus the low energy electron was increased in amorphous region and the formation rate of positronium in there increased approximately 2.4 times. These indicate that annihilation process of positron and formation of *o*-Ps are greatly influenced by melting.

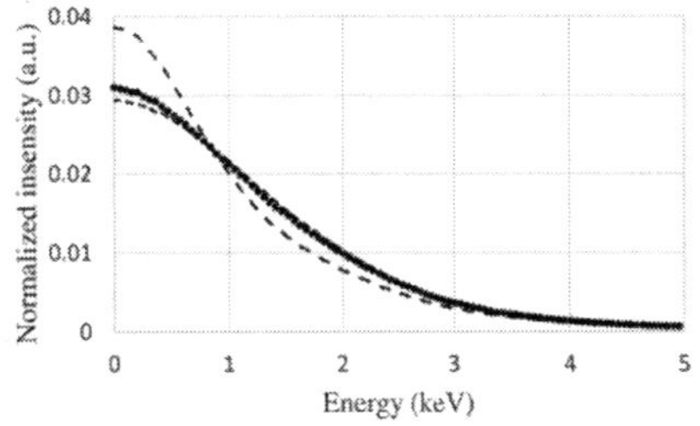

Figure 3. Annihilation photopeaks obtained at 24 °C (dotted), 300 °C (solid), and 340 °C (dashed).

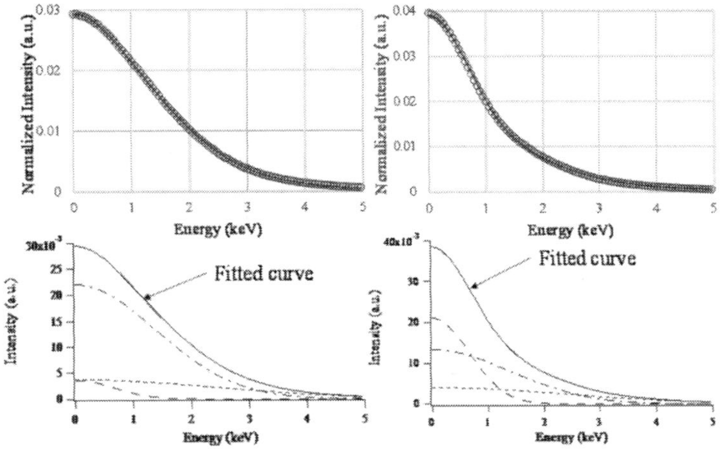

Figure 4. Curve fitting for two representative spectra: 24 °C (left column) and 340 °C (right column). The three Gaussian distributions have energy spreads of 3.38 keV (dotted), 1.95 keV (dotted-dashed) and 0.918 keV (dashed).

3.3. X-ray Diffraction (XRD).

XRD pattern obtained at 27.5 °C is shown in Figure 5. The peaks could be assigned to each reflection plane. With increasing temperature, many peaks disappeared, except the (100), (110), (200) plane peaks; these peaks were shifted toward lower angles. The temperature dependence of the diffraction peak angle for the (100) plane is shown in Figure 6. The peak shifted linearly from 18.1° to 17.5° with the increase of temperature from 27.5 °C to 326.5 °C and then decreased without shifting above the melting point. The lattice constants for the angles of 18.1° and 17.5° are 4.90 Å and 5.07 Å, respectively. To investigate the structural change of the crystal region with the increase of temperature, the crystallinity was defined as the ratio of the peak area to whole area for the (100) plane peak. The peak area was calculated by subtracting the background, which was evaluated by fitting a combination of a Lorentzian, an exponential and a constant, over the whole background area. The obtained crystallinity was approximately 90.5 % for the data of 27.5 °C. The evaluated crystallinity was quite high compared to the one evaluated using DSC. This difference is due to the reduced background, as shown in Figure 5 and is most likely a property of the semi-crystalline polymer. A detailed description regarding the differences of crystallinity among the evaluation methods is found elsewhere [4]. Semi-crystalline polymer is described by a three-phase model, and a temperature modulated DSC is better for evaluating crystallinity. However, as our attention was focused on the behavior of crystallinity with the increase of temperature, the crystallinity evaluated based on the (100) plane peak was used as an index of the crystallinity. The crystallinity remained nearly constant at approximately 88 % for temperatures increasing from room temperature to the melting point and then decreased below 80 % above the melting point. Note that the crystallinity remained at a high percentage for temperatures above the melting point.

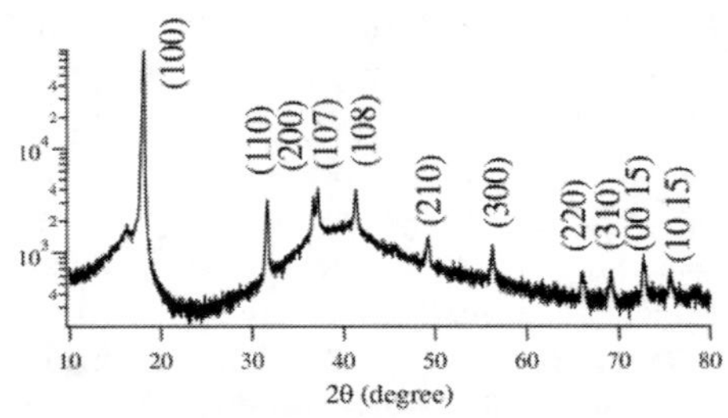

Figure 5. XRD pattern at 27.5 °C

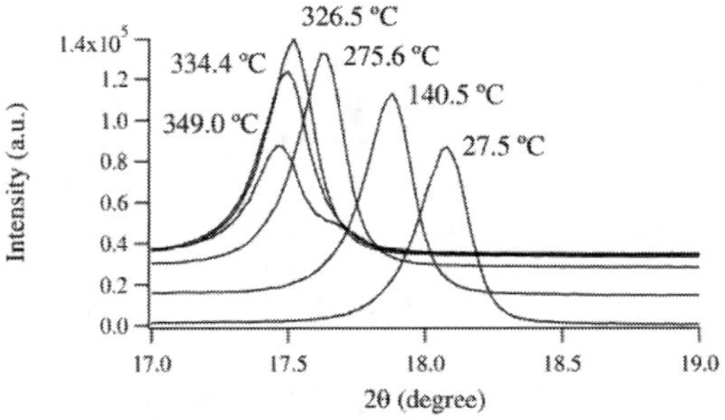

Figure 6. Temperature dependent X-ray diffraction from the (100) plane. Each curve for temperatures above 27.5 °C is shifted upward in proportion to the temperature increase.

3.4. Comparison of the PALS and XRD

In PALS, the significant increase of τ_3 from 1.64 ns to 2.15 ns was observed at 335 °C, following a mild increase from room temperature. In XRD, a gradual decrease of the (100) peak angle (2θ) was observed up to 330 °C; however,

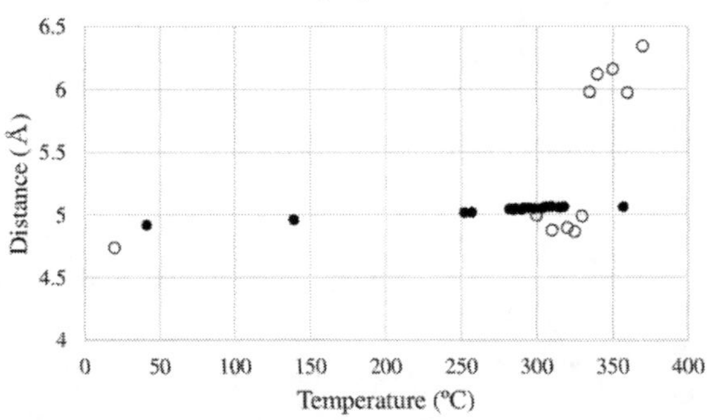

Figure 7. Temperature dependence of the inter-chain distance evaluated via XRD and the diameter of FVH based on τ_3 of PALS.

no shift of the peak was observed for temperatures above 330 °C. The lifetime of *o*-Ps is correlated with the size of the space, i.e., the so-called free volume hole (FVH), through the Tao-Eldrup equation with the interaction range of 0.166 nm, where a positron of trapped *o*-Ps in FVH annihilates with an electron. The temperature dependences of the evaluated diameter of FVH and the lattice constant calculated from the peak angle of the (100) plane are shown in Figure 7. The trends of the diameter of FVH and the lattice constant are found to be similar to each other up to 330 °C, indicating that a part of *o*-Ps indeed remained in/on crystal region. At melting, if the crystal structure retained hexagonal form and the lattice constant expanded from approximately 5 Å to 6 Å based on the result of PALS, the corresponding diffraction peak would appear at 14.8 degrees. However, no such peak shift was observed. Thus, the drastic increase of τ_3 was confirmed to be marginal with respect to an expansion of the crystal region and that the lifetime was due to the annihilation of *o*-Ps on the surface of the crystal region after forming the Ps bubble. It is unclear whether *o*-Ps was formed on the surface of crystal region or formed inside of it and leached from there; however, the dominant annihilation of *o*-Ps occurs on the surface of the crystal. It is remarkable that the evaluated inter-chain distances in the crystal region using PAS and XRD were very close to each other, despite the fact that *o*-Ps was not fully confined in the FVH but was trapped on the surface and the fact that the interaction length in Tao-Eldrup equation was the usual 0.166 nm.

4. Summary

The melting process of PTFE was investigated by PAS and XRD. Several diffraction peaks were observed in XRD at room temperature. The peaks corresponding to higher Miller indices disappeared with the increase of temperature; however, the peaks having lower Miller indices, such as (100), remained, and the peak shifts toward lower angle were observed with increase of temperature up to melting point, indicating the expansion of the crystal region. Successive peak shifting and the appearance of new peaks were not observed above the melting point. Based on the behavior of τ_3 and τ_4 in PAS for temperatures up to the melting point, volume expansion progressed both in the crystal region and in the amorphous region (the latter being dominant), and then the relaxation in the amorphous region at melting initiated volume expansion of the amorphous region while retaining the crystallinity in the crystal region. A remarkable increase and the following successive increase of τ_4 were observed at and above the melting point, respectively, indicating the formation of a Ps bubble in the amorphous region and the volume expansion of the amorphous region above the melting point. A significant increase of τ_3 was also observed at the melting point, but no successive increase was observed above the melting point. Taking into account the results of XRD above the melting point, this significant increase was certainly due to the formation of Ps bubble, indicating the annihilation of *o*-Ps on the surface of crystal. Based on these results, the melting process is initiated by the relaxation in the amorphous region, and then volume expansion starts together with peeling off the crystal surface. The decomposition of the Doppler broadened annihilation photopeak into three components of different energy spreads revealed a remarkable increase of the number of lower energy electrons at melting; the increase in the number of such electrons was found to have great influences on the formation process of *o*-Ps and the annihilation processes of positrons and *o*-Ps. Such information is useful for reprocessing PTFE.

References

[1] Honda Y, Aoyagi Y, Tojo S, Watanabe G, Akiyama Y, Nishijima S 2013 *J. Phys. Conf. Ser.* **443** 012052
[2] Atrazhev V V, Burlatsky S F, Dmitriev D V, Sultanov V I 2013 *J. Stat. Mech.* P02004
[3] Dlubek G, Gupta A S, Pionteck J, Häβler R, Krause-Rehberg R, Kaspar H, Lochhaas K H 2005 *Polymer* **46** 6075
[4] Starkweather,Jr. H W, Zoller P, Jones G A, Vega A J 1982 *J. Polym. Sci.* **20** 751

Direct correlation between free volume and dielectric constant in a fluorine-containing polyimide blend

R Ramani[1]* R Ramachandran[2] G Amarendra[2] and S Alam[1]

[1] Polymer Science Division, DMSRDE, G.T. Road, Kanpur – 208 013, India
[2] Materials Physics Division, IGCAR, Kalpakkam-603 102, India

*E-mail: ramanir70@rediffmail.com

Abstract. The dielectric constant of fluorinated polyimides and their blends is known to decrease with increase in free volume due to decrease in the number of polarizable groups per unit volume. Interestingly, we report here a polyimide which when blended with a fluoro-polymer showed a positive deviation of dielectric constant with free volume. In our experiment, we have used a blend of poly(ether imide) and poly(vinylidene fluorine-*co*-hexafluoropropylene) and the interaction between them was studied using FTIR, XRD, TGA and SEM. The blend was investigated by PALS, DB and DEA. Surprisingly, with the increase in the free volume content in this blend, the dielectric constant also increases. This change is attributed to additional space available for the polarizable groups to orient themselves to the applied electric field.

1. Introduction

The dielectric constant of a polymer is mainly influenced by two parameters. First is the molecular polarizability that can be changed by modifying the type and number of polarizable groups. Second is the free volume associated with the polymer. Since the dielectric constant (ε) of air is close to one, introducing free volume (porosity) to a polymer reduces its dielectric constant [1]. Apart from molecular polarizability and free volume of the polymer, the moisture content in a polymer can also influence the dielectric constant. Addition of hydrophobic atoms like fluorine into a polymer can diminish the moisture content and decrease the dielectric constant [1].

Polymers having fluorine atoms in their structure are classified as fluoropolymers. Amongst fluoropolymers, poly(vinylidene fluoride) (PVDF) and its copolymers are technologically important since they possess high dielectric constant and hence used in charge storage applications. Particularly, when PVDF is copolymerized with hexafluoro propylene (HFP), the pendant trifluoro methyl group ($-CF_3$) in it do not allow efficient chain packing and thus contributes to free volume [2].

Polyimides are rigid and thermally stable polymers obtained by the polycondensation reaction of diamines with dianhydrides. Apart from thermal stability, polyimides are also endowed with high glass transition temperature (T_g), good mechanical properties, low dielectric constant, high breakdown voltage. They find extensive use in aerospace and electronic applications. Poly(ether imide) (PEI) is a high performance aromatic amorphous thermoplastic polyimide aiming at replacing metals as well as targeting other specialty applications such as electronics, automotives and membranes [3].

Various strategies have been adopted in the past to decrease the dielectric constant of polymides like incorporating suitable diamine and dianhydride moieties (that minimizes polarizability and introduces high degree of free volume) as well as adding fluorine atoms in to the polyimide molecular structure which results to increase in free volume. An important contribution in understanding the molecular process that lead to dielectric properties in fluorine-containing polyimides was made at NASA-Langley [4]. The authors have studied the effect of change in free volume on the dielectric constant in them and found a correlation of high free volume content and low dielectric constant [4]. Later, other groups have also found that ε of a polyimide decreases with increasing wt% of fluorine content in different polyimides [5,6]. To the best of our knowledge, the effect of fluorine atoms on the dielectric constant and free volume properties of a polyimide has not been studied by blending it with a fluorine-containing polymer. Here we have studied the effect of fluorine on poly(ether imide) (PEI) by blending it with the fluoropolymer poly(vinylidine fluoride-*co*-hexafluoropropylene).

2. Experimental

2.1. *Blend preparation*

The copolymer of vinylidene fluoride and hexa fluoropropylene P(VDF-*co*-HFP) having M_w = 400,000 g/mol was obtained from Sigma-Aldrich in pellet form. Henceforth, this copolymer will be termed as PVH. The PEI of grade Ultem-1000 with M_w = 55,000 g/mol was obtained in powder form from SABIC innovative plastics, Europe. Blends with various weight percentage of PVH/PEI were prepared by conventional melt-mixing and the samples are designated as $PHPI_{00}$, $PHPI_{20}$, $PHPI_{40}$, $PHPI_{50}$, $PHPI_{60}$, $PHPI_{80}$, $PHPI_{85}$, $PHPI_{90}$, $PHPI_{95}$ and $PHPI_{100}$, where the subscripts denote the PEI content (wt %) in the blend.

2.2. *Characterization*

The FTIR spectrophotometer of Perkin Elmer (model Spectrum 100) was used to identify the possible interacting groups between the polymers. The XRD measurements were performed with PAN Analytical X-ray systems (model X'pert PRO) (Cu K_α radiation, λ = 1.54 Å) in the 2θ range 5° to 80°. Thermo gravimetric analysis was performed using a TGA instrument (Mettler–Toledo TGA/SDTA 851^e) in the temperature range 30 to 1000 °C. The surface morphology of the blends was examined using a Carl Zeiss EVO 50 SEM, with 10 kV operating voltage. Prior to SEM measurements, the blends were treated with acetone at room temperature for 24 h, in order to remove PVH phase from them. The DSC Q200 (TA instruments) was employed for carrying out the DSC measurements. The PALS measurements were performed at room temperature using a conventional fast-fast coincidence system having a time resolution of about 200 ps. The lifetime spectra were resolved in to three lifetime components using PATFIT program and were also analyzed using the MELT program to obtain a continuous distribution of the annihilation lifetime. The DB measurement system consists of ORTEC high-purity germanium (HpGe) detector having 1.3 keV energy resolution at 662 keV gamma line of ^{137}Cs source. The rheological measurements were performed using a Rosand capillary rheometer of Malvern make (model RH7) at 380 °C in air medium. The Dielectric measurements were conducted at a frequency of 1kHz using DS 6000 Dielectric Analyzer of M/s. Lacerta Technology, UK using circular electrodes (dia = 33 mm) and the samples used were ca. 0.4 mm thick.

3. Results and Discussion

3.1. *Interaction between PVH and PEI*

The FTIR results infer a possible dipole-dipole interaction exists between electron deficient imide group of PEI and electro-negative $>CF_2$ groups of PVH. The X-ray scattering profile of PVH reveals two sharp close diffraction peaks at 2θ = 18.2° and 20.7°, a well separated peak at 26.6° and a broad peak close to 40° which are the characteristic crystalline peaks of PVDF. These sharp peaks lose their identity at higher PEI content (≥ 80%) indicating that the dipolar interaction is significant when the PEI content is high. The introduction of fluorine is known to improve the thermo-oxidative stability of polyimides. TGA measurements on PEI exhibited two step decomposition in air medium and the

Journal of Physics: Conference Series **618** (2015) 012025 doi:10.1088/1742-6596/618/1/012025

second step corresponds to its thermo-oxidation. With a mere 20% addition of PVH, the thermo-oxidative stability of PEI enhances by ca. 53 °C which is a clear indication that the fluorine groups of PVH has interacted with PEI. The SEM photomicrograph of the blend at higher PEI content (PEI ≥ 80%) exhibits maximum interfaces which again support good interaction between the two polymers (Fig. 1). From the DSC measurements, the glass transition temperature of PVH is found to be ca. -30 °C and that of PEI is ca. 218 °C.

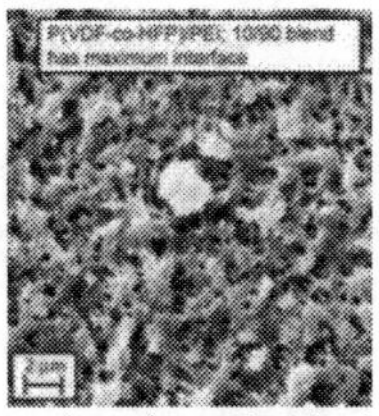

Fig.1. SEM picture showing maximum interfaces for the blend when the PEI content is ≥80%.

3.2. Free volume and Dielectric constant

The positron annihilation lifetime results for this blend series reveal three lifetime components and the variation of the long-lived lifetime component that corresponds to *o*-Ps pick-off lifetime (τ_3) was used to obtain the free volume hole size using the Tao-Eldrup relation. The average free volume size was calculated using the formula $V_{f3} = (4/3)\pi R^3$. The relative free volume fraction (F_{vr}) is then found as $F_{vr} = V_{f3} * I_3$, where I_3 being the *o*-Ps intensity.

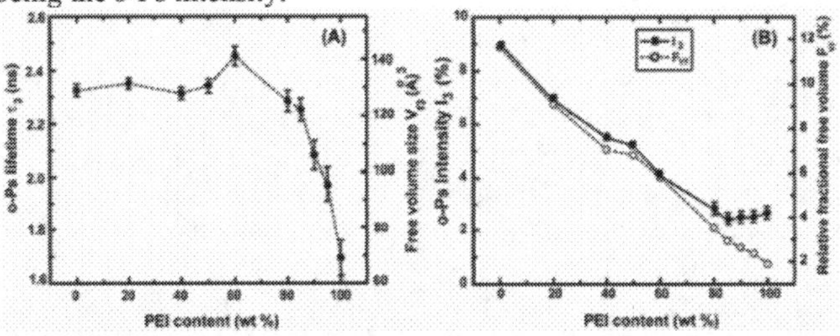

Fig.2. Variation of (A) *o*-Ps lifetime (τ_3) and free volume size (V_{f3}) and (B) *o*-Ps intensity (I_3) and the relative fractional free volume (F_{vr}) as a function of PEI content. The lines through the data point are for visual aid.

The *o*-Ps lifetime shows an initial constancy and attains a maximum when the PEI content is 60% and starts to decrease significantly at later stages (PEI ≥ 80%). The increase in lifetime with PEI content is due to increase in free volume which is an indication of non-compatibility between the two polymers (Fig. 2A). But, when the PEI content is ≥ 80%, the free volume decreases significantly that goes in agreement with the dipolar interaction between the two polymers at higher PEI content as given by FTIR, XRD and TGA measurements. Also, the SEM showed increased interfaces in this range of PEI content. The free volume distribution pattern of the blends lie within the distribution limits of their pure polymers at higher PEI contents (PEI ≥ 80%), revealing good interaction between the two polymers. The I_3 continuously decreases with increasing PEI content (I_3 increases with increase in PVH content) till the PEI content is ≥80% and thereafter remains almost constant. The increase of I_3 with increase in PVH (fluorine) content indicates anti-inhibition effect. The constancy of I_3 at higher PEI content indicates that the *o*-Ps formation is not that significant at the interfaces. The DB results reveal that the 'S' parameter decreases with increase in *o*-Ps intensity, inferring that the high kinetic energy of the p-electrons of fluorine participates in the annihilation process and thus reveals the information regarding 'chemical environment' inside these free volume 'holes'. To verify the free volume results, shear viscosity measurements were done which showed high viscosity for PEI compared to PVH which is in agreement with the free volume results.

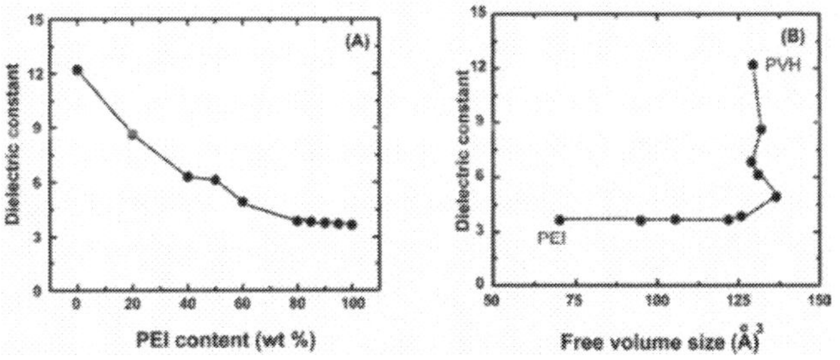

Fig.3. Variation of (A) dielectric constant as a function of PEI content in the blend and (B) dielectric constant as a function of free volume size. The lines through the data point are for visual aid.

The dielectric constant (ε) decreases with increase in the PEI content but at higher PEI content, the change in ε becomes insignificant (Fig. 3A). In case of heterogeneous systems like polymer blends, dielectric constant arises mainly due electronic polarization, interfacial polarization and / or dipole orientation polarization [1,2,5]. The interfacial polarization arises due to accumulation of mobile charges at the interfaces constituted by unlike phases. The almost constant value of ε even when the interfaces are maximum (as revealed by SEM when PEI content is ≥80%), indicates that interfacial polarization is not contributing significantly to the observed change in dielectric constant. The dipole orientation polarization arises due to redistribution of charges when a group of atoms with a net permanent dipole moment reorient itself in space fast enough to keep in pace with the oscillations of an applied alternating electric field. The dielectric constant remains almost constant with increase in free volume size till the size reaches ca. 125 $Å^3$ and thereafter it steeply increases (see Fig. 3B). The process of dipole orientation polarization accompanies the movement of polymer chain segments and the movement requires considerable free volume in the matrix. If the two polymers are compatible, it is difficult for the dipoles to orient and relax in the applied field [5] and for this reason we observe almost constant value of ε when the PEI content is ≥80% (see Fig 3B). The process of dipole orientation polarization becomes significant when the PEI content in the blend is < 80% and at this composition, we know that the interaction of PEI and PVH is insignificant. This incompatibility results to more free volume in the blend which in turn increases the chain mobility and the associated dipole orientation polarization.

4. Conclusions

The origin of dielectric constant in this blend is identified to be mainly from dipole orientation polarization owing to the availability of more free volume. The positive variation of dielectric constant with free volume content is distinct from the generally observed behaviour in polyimides and their blends studied so far. The present results would be of help in more effective design of polymeric electrical charge storage devices.

Acknowledgements: The authors thank Dr. A.K. Saxena, Director, DMSRDE, Kanpur for his encouragement and support and for permitting to present this work.

References

[1] Hougham G Tesoro G and Viehbeck A 1996 *Macromolecules* **29** 3453.
[2] Simpson J O and St. Clair A K 1997 *Thin Solid Films* **308-309** 480.
[3] Xia J Liu S Pallathadka P K Chng M L and Chung T S 2010 *Ind. Eng. Chem. Res.* **49** 12014.
[4] Singh J J St. Clair T L Holt W H and Mock W Jr.1985 *NASA Tech. Memorandum* No. 86431.
[5] Zha J W Jia H J Wang H Y and Dang Z M 2012 *J. Phys. Chem. C* **116** 23676.
[6] Meador M A B McMillon E Sandberg A Barrios E Wilmoth N G Mueller C H and Miranda F A 2014 *ACS Appl. Mater. Interfaces* **6** 6062.

DBAR investigation on films of polypyrrole incorporated polyvinylalcohol doped with ferric chloride

Blaise Lobo[1*], B.M. Baraker[1], P. B. Hammannavar[1], R. F. Bhajantri[2], M. R. Ranganath[1], M. Hurkadli[1] and V Ravindrachary[3]

[1]Department of Physics, Karnatak Science College, Karnatak University, Dharwad 580001, Karnataka, India
[2]Department of Physics, Karnatak University, Dharwad 580003, Karnataka, India
[3]Department of Physics, Mangalore University, Mangalore 574199, Karnataka, India

* E-mail: blaise.lobo@gmail.com

Abstract. Flexible films of pyrrole (Py) sorbed, ferric chloride ($FeCl_3$) doped polyvinylalcohol (PVA) were prepared by solution casting. The films were characterized by XRD, UV-Visible spectrometry, Thermal Analysis (DSC, DTA/TGA), FTIR and electrical measurements. In this paper, the results of Doppler Broadening of Annihilation Radiation (DBAR) spectra in the doping range, from 4 wt% up to 18 wt%, are discussed. The XRD and DSC scans complement the DBAR results. The computed S- parameter and W –parameter reflect changes in the degree of crystallinity and the average crystallite size, respectively, of polypyrrole (PPy) incorporated PVA samples doped with ferric chloride.

1. Introduction

Polyvinylalcohol (PVA) is a polymeric material, with many versatile applications. It has been doped with different reduction/oxidation (redox) agents like iodine, ferric chloride and other salts, and these doped polymeric materials show significant modifications in their micro-structural features, as well as their thermal, optical and electrical properties [1]. The blend of PVA with other polymers has been extensively studied [2]. Flexible films of pyrrole (Py) sorbed, ferric chloride ($FeCl_3$) doped PVA were prepared by solution casting, after in-situ polymerization in aqueous solution [3]. Ferric chloride acts both as a dopant (redox agent) as well as an initiator for the polymerization of pyrrole, resulting in the formation of a conducting polymeric blend [4]. The films were characterized using optical spectrometry, powder X-Ray Diffraction (XRD), Differential Scanning Calorimetry (DSC), and the Doppler Broadening of Annihilation Radiation (DBAR) technique of Positron Annihilation Spectroscopy (PAS). The DBAR spectrum was analysed using line-shape parameters, namely, S-parameter and W- parameter. The two important portions of the Doppler broadened 511 kilo- electron Volt (keV) line are the area under the curve immediately around the peak, and the area under the tails of the curve, on both sides [5]. The S parameter can be defined as the ratio of area around the 511 keV peak normalised over the total area, and the W parameter is defined as the normalised area under the tails [6]. The energy of the electron – positron pair undergoing two gamma annihilation is not equally shared by them, in the laboratory frame of reference. The energy of one of these annihilation photons is increased (Doppler shifted upwards) and the other annihilation photon is decreased (Doppler shifted downwards) by an amount given by $\Delta E = (c \times P_L)/2$, where '$P_L$' is the longitudinal component of momentum of the electron-positron pair in the direction of gamma ray emission and 'c' is the velocity of light in vacuum. In terms of P_L, every S and W parameter was obtained by the fraction of counts which appeared at $0 \sim 4\times10^{-3}m_oc$ and $15 \sim 30\times10^{-3}m_oc$ from the centre to the total counts for the annihilation peak. Here, 'm_o' refers to the rest mass of the electron (or positron).

2. Experimental

The incorporation of PPy in PVA matrix is achieved by in-situ polymerization of pyrrole (Py). Aqueous solutions of PVA doped with different concentrations (from 4 wt% up to 18 wt %) of ferric chloride ($FeCl_3$) were exposed to Py vapour for sixteen hours. The resulting mixture was dried in an air cooled, temperature controlled furnace, maintained at 40°C, to obtain the $FeCl_3$ doped, PPy incorporated PVA films. These films were characterized using XRD, DSC, Attenuated Total Reflectance- Fourier Transform Infra-red Spectrometry (ATR-FTIR) and Ultraviolet (UV)-Visible Spectrometry. Fig. 1 shows the variation of average crystallite size of the samples (extracted from the XRD data) for different doping levels of $FeCl_3$ in PPy- PVA blend. The XRD curves were analysed using the PowderX software [7]. The ATR-FTIR scans reveals that an absorption peak appears at 1640 cm^{-1} (attributed to amine vibrations within the pyrrole ring) [8], which increases in intensity on increasing the doping level (Fig. 2). In addition, the NH absorption peak is observed at 3400 cm^{-1}, thereby confirming the formation of PPy in the PVA matrix. For DBAR studies, a high resolution High Purity Germanium (HPGe) detector was used to record the energy spectrum of annihilation gamma rays from the sample. The energy resolution was measured to be 1.4 keV at 662 keV photo-peak of Cesium-137 source. A 300 kilo-becquerel (kBq) sodium -22 (Na^{22}) source, deposited on a thin kapton foil of thickness 7 micro-metres (μm), was used as a source of positrons. This source was sandwiched on either side by the $FeCl_3$ doped PPy-PVA films (stacked to 1 mm thickness, on either side of the source). The source-sample sandwich assembly was kept at a distance of 10 cm from the beryllium window of the HPGe detector. Each spectrum was analysed using a programme developed by us to extract the line-shape parameters, namely, the S- parameter (Fig. 3) and the W- parameter (Fig. 4).

3. Results and discussion

The modifications in the micro-structure of $FeCl_3$ doped PVA caused by incorporation of PPy are discussed in this paper. It is noted from Fig. 1 that the average size of crystallites (obtained from XRD scans) increases at low and moderate doping levels (from 0.5 wt% up to 5.7 wt% of $FeCl_3$ in PPy incorporated PVA), beyond which it decreases. Note that the W- parameter of DBAR (see Fig. 4) also shows a similar trend (increase from 4 wt% up to 10.7 wt%, followed by a decrease). The S-parameter (Fig. 3) shows an increase at higher dopant levels (from 5.7 wt% up to 18 wt %) indicating enhanced annihilation of positrons with low momentum electrons (at higher dopant concentrations), which is expected when the material becomes more amorphous, as indicated by the XRD results (Fig. 5). The DSC curves (Fig. 6) for these samples supports this line of reasoning. At low levels of doping, there is molecular level distribution of the dopant ($FeCl_3$) in the blend, and the chlorine ions will interact with the PVA chains, thereby aligning these chains, and increasing the average crystallite size of the sample. As the dopant concentration increases, the polymerization process of pyrrole (in-situ polymerization in aqueous solution before casting) increases, and the polypyrrole regions start accumulating in the sample, thereby drastically reducing the average crystallite size and the degree of crystallinity of the sample. This explanation can account for the fact that, as the doping concentration is increased from 4 wt% up to 10.7 wt%, there is increase in W- parameter, followed by a decrease (reversal in trend) at higher doping concentrations. Note that (see Fig. 1 and Fig. 4) there is a correlation between the W- parameter (DBAR) and the average crystallite size (XRD). Also, note the inverse relationship between the S- parameter (DBAR) and the degree of crystallinity (XRD) of these samples (Fig. 3 and Fig. 5). The XRD results are supported by the DSC curves (Fig. 6).

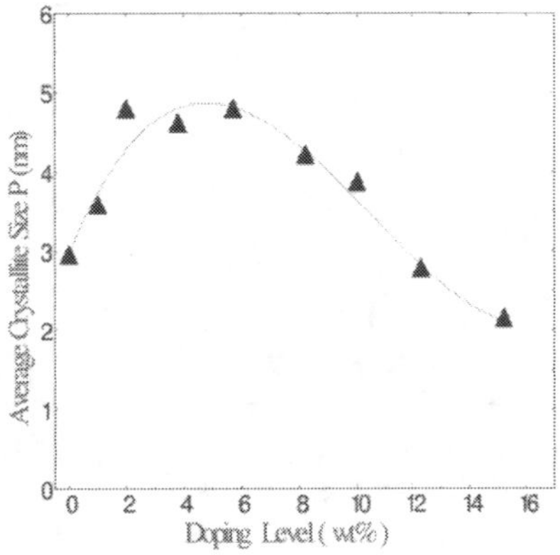

Figure 1. Average crystallite size of the $FeCl_3$ doped, PPy incorporated PVA ,

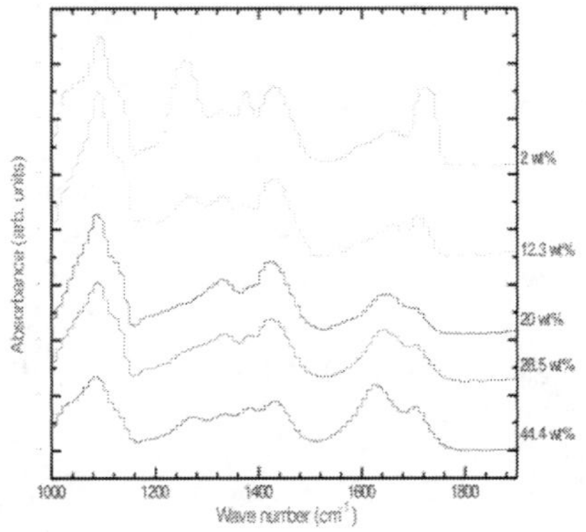

Figure 2. FTIR Scans of the $FeCl_3$ doped, PPy incorporated PVA films in the finger print region.

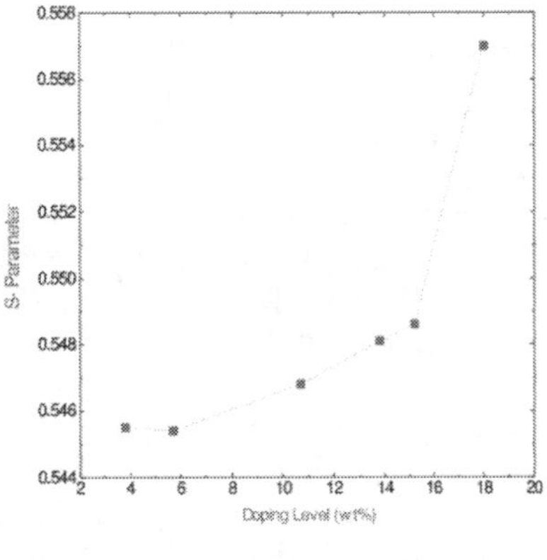

Figure 3. Variation of S-parameter of DBS relative to $FeCl_3$ doping level in PPy incorporated PVA

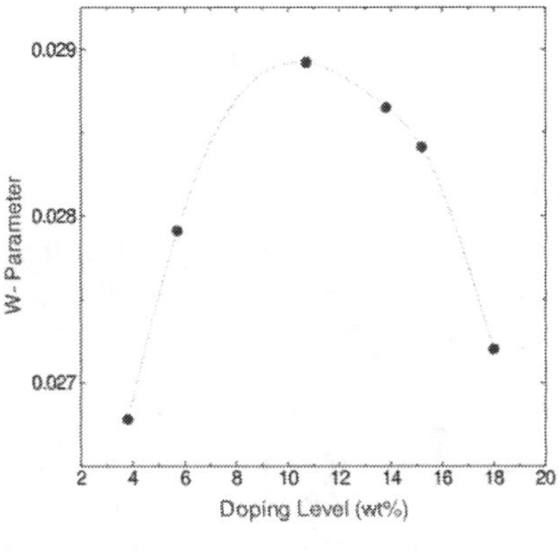

Figure 4. Variation of W-parameter of DBS relative to $FeCl_3$ doping level in PPy incorporated PVA

4. Conclusion

The variation of line-shape parameters of DBAR spectra of PPy incorporated PVA films with variation in $FeCl_3$ doping level is discussed. The S- parameter shows an increase at higher dopant levels (from 5.7 wt% up to 18 wt %) indicating enhanced annihilation of positrons with low momentum electrons (at higher dopant concentrations), which is expected when the material becomes more amorphous, as indicated by the inverse trend shown by degree of crystallinity of the PPy-PVA blend, extracted from the XRD data. This conclusion is supported by the DSC curves. The W-parameter of DBAR shows a trend similar to the average crystallite size, extracted from the XRD data.

Acknowledgements: The facilities at the University Science Instruments Centre, Karnatak University, Dharwad was utilized for DBAR, UV-Vis and ATR-FTIR study. In addition, the prepared films were studied using XRD (Panalytical XPERT-PRO, Department of Physics, Alagappa University, Karaikudi) and DSC (Netzsch DSC 204, SAIF, IITM, Chennai). We acknowledge that Dr. Cheng Dong of NLSC, Chinese Academy of Sciences, P.R. China provided the PowderX software to analyse the XRD data. A part of this work was supported by UGC (Government of India) Major Research Project F.38-103/2009(SR).

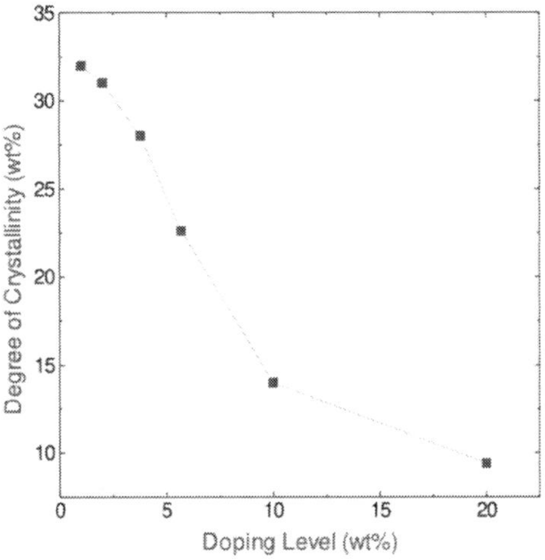

Figure 5. Variation of the degree of crystallinity of $FeCl_3$ doped PPy incorporated PVA films, for different concentrations (in wt %) of the dopant

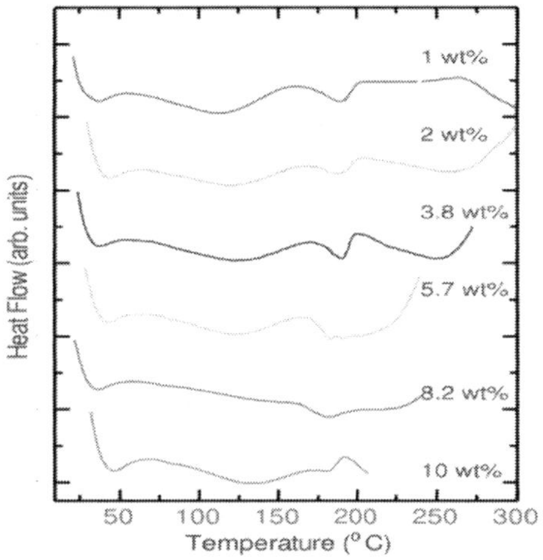

Figure 6. DSC scans of $FeCl_3$ doped PPy incorporated PVA films, for different concentrations (in wt %) of the dopant

References

[1] Lobo. B, Ranganath. M. R, Ravi Chandran T. S. G, Rao. G. V, Ravindrachary.V , and Gopal. S 1999 *Phys. Rev. B* **59** 13693.
Tawansi. A, El-Khodary. A, and Abdelnaby. M. M 2005 *Current Applied Physics* **5** 572.
Bhajantri R F, Ravindrachary V, Harisha A, Ismayil and Ranganathaiah C 2009 *Physica Status Solidi C* **6** 2429.

[2] Patil.R.V, Ranganath. M. R and Lobo. B 2014 *AIP Conf. Proc.* **1591** 183.
Patil.R.V, Ranganath. M. R and Lobo. B 2013 *AIP Conf. Proc.* **1512** 578.
Ravindrachary. V, Bhajantri. R. F, Harisha. A, Ismayil and Ranganathaiah. C 2009 *Physica Status Solidi* C **6** 2438.

[3] Ranganath. M. R and Lobo. B 2009 *Solid State Physics (India)* **54** 535.

[4] Benseddik. E, Bonnet. A and Lefrant.S 1998 *J. Appl. Polym. Sc.* **68** 709.
Dhawan. S. K and Trivedi. D. C 1993 *Bulletin of Materials Science* **16** 371.

[5] Choudalakis. G, Gotsis. A. D, Schut. H and Picken S. J 2011 *European Polymer Journal* **47** 264.

[6] Mohammed. H. F. M and Abd El-Aziz . N. S 2001 *Polymer* **42** 8013.

[7] Dong. C 1999 *J. Appl. Cryst.* **32** 838.

[8] Cruz. G. J, Morales . J and Olayo. R 1999 *Thin Solid Films* **342** 119.

Journal of Physics: Conference Series **618** (2015) 012027 doi:10.1088/1742-6596/618/1/012027

The study of synthetic food dyes by positron annihilation lifetime spectroscopy.

A A Pivtsaev[1] and V I Razov[1]

[1] Far Eastern Federal University, st. Sukhanova 8, Vladivostok,690950, Russia

E-mail: aleksey_pivtsaev@mail.ru

Abstract: By method of positron annihilation lifetime spectroscopy (PALS), substances are food dyes were studied: E-102 (Tartrazine), E-124 (Ponso 4R), E 132 (Indigo carmine), E-133 (Brilliant Blue), E-151 (Black Shiny). They are examined for the presence of carcinogenic properties. The difference between dyes having explicit carcinogenic properties and mutagenic properties (non-explicit carcinogens) is established.

1. Introduction.

Technical progress promotes the appearance of large amounts of new chemical types which present cheaper analogues for natural products. At the present time, the issues of human health protection against negative effects from scientific and technological advance have become substantial. The pollution of food with extraneous chemicals causes the violation of the body's adaptive functions and creates threats to human health, up to the development of a malignant neoplasm.

It is known that any carcinogenic effect has a cumulative character, which means a necessity to decrease bodily contact with carcinogens. As far as it is impossible to influence natural factors (solar radiation, natural radiation background), it is necessary to limit the ingress of artificial ones (air pollution with heavy metals, chemicals [1]).In the present day, the food industry has practically completely replaced natural food additives by their synthetic analogues, which have negative characteristics, due to their cheapness and utility.

Food dyes have come into wide use in the food industry, in the capacity of color reducing agents or color enhancers. Long-term experiments on animals have already revealed that synthetic dyes are carcinogens [2, 3, 4] and therefore their use is allowed in small concentrations (50-500 mg/l) [5]. However this calculation does not take into account the cumulative character of carcinogens.

Previously in our research, we have studied the method of positron annihilation lifetime spectroscopy (PALS) as a fast and effective method for the revelation of the carcinogenic effects of substances by the value of a positron long-lived time component τ_3 [6]. It has been discovered that value τ_3=(1,005±0,005) ns is the border between carcinogens and non-carcinogens. Studying synthetic dyes by the PALS method has been suggested in order to:

1. Establish whether this method is more optimal (fast and effective) for studying chemical additives (using the example of dyes) than experiments on animals;
2. Study annihilation spectra of solutions with different concentrations of dyes;
3. Confirm or disprove the safety of use of these concentrations in the food industry.

Getting positron into a substance together with medium electrons creates a hydrogen-like atom - positron Ps. Ps lifetime depends on medium characteristics - the more electrons there are, the less its life lasts [7]. Carcinogens are characterized by a large concentration of electrons, as opposed to elementary substances having no carcinogenic properties. Researching chemical carcinogens has revealed that the substance has no carcinogenic properties if the value of τ_3 is more than (1,005±0,005) ns; if it is less, it is carcinogen. And the lower the time of the long-lived time component Ps (τ_3), the higher the carcinogenic level of the substance [6].

2. Material and methods.

In order to solve the given tasks, we have measured annihilation spectra of synthetic food dyes (in the form they are used in the food industry: the main substance is dye and the rest is admixture), demonstrating both explicit carcinogenic properties (E-133, E-124) and non-explicit carcinogenic properties (mutagenic: E-132, E-151, E-102 and their 10%, 5%, 2.5% and 1% solutions) [2, 4, 8]. Those spectra are gathered within 6 hours and then processed by a program. Using the programs Taufit and PALSfit v2.23 [9], the annihilation characteristics of the studied substances have been calculated: τ_i (Ps lifetime) and I_i (intensity). The results received after processing are given in fig.1 and in table 1.

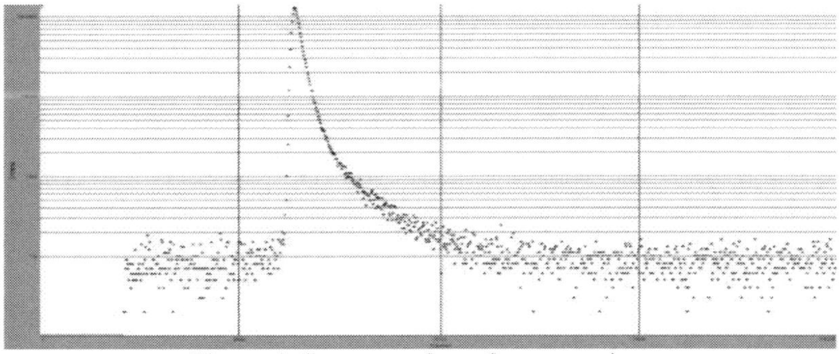

Figure 1. Spectrum view after processing.

Using the data from table 1, time component (τ_3)-dye concentraction charts have been plotted, fig.2.

Table I. Characteristics of received annihilation spectra.

Dye	Concentration, %	τ3, ns	I3, %
	100	453	5.16
	10	0.542	4.6
E102	5	0.561	5.94
	2.5	0.925	8.51
	1	0.981	8.56
	100	0.481	6.41
	10	0.567	2.11
E124	5	0.613	8.26
	2.5	0.888	7.82
	1	1.017	9.39
	100	0.556	5.35
	10	0.570	7.85
E133	5	0.675	7.63
	2.5	0.910	7.84
	1	1.109	8.29
	100	0.508	4.34
	10	0.730	4.55
E151	5	0.839	5.88
	2.5	0.964	8
	1	1.009	9.25
	100	0.506	5.9
	10	0.880	12.04
E132	5	0.952	11.31
	2.5	1.006	9.27
	1	1.171	9.02

3. Results.

Based on the data received and shown in table 1 and fig.2 it is arguable that all dyes in their initial conditions are carcinogens (τ_3<1,005 ns). It is permitted to use them only in very small concentrations because by decreasing the concentration, the level of carcinogenicity decreases as well.

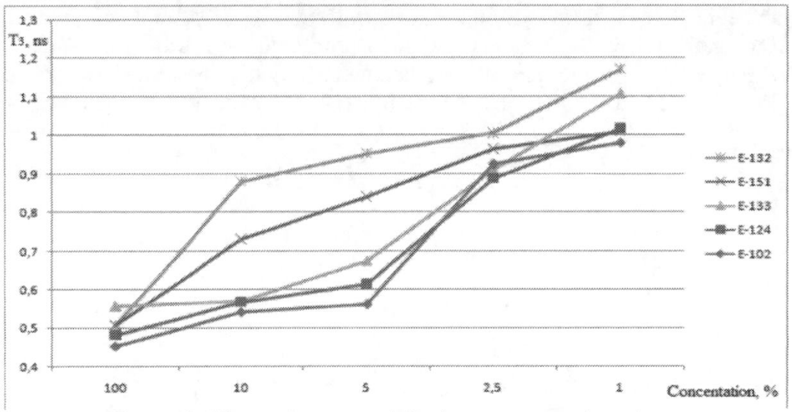

Figure 2. Time component (τ3)-dye concentration chart.

4. Discussion.

According to fig.2, at a concentration 5%, there is an obvious difference between mutagenic dyes E151, E132 and other dyes having explicit carcinogenic properties, and E102 (arguments about its carcinogenic properties still continue). As the concentration is reduced from 100% to 5%, the value of long-lived time component Ps (τ_3) rises sharply for mutagenic dyes and very slowly for carcinogenic dyes. Upon further dilution of the solutions, the τ3 value rises, and in a 1% solution it overruns 1.0 ns for practically all of them, which transfers them into the non-carcinogenic group. This is confirmed by experiments on animals. But it is not the case!

In a 1% solution, carcinogenic properties do not disappear, they just reduce their activity. But if it is remembered that damage caused to the body by any carcinogens has a cumulative character, because the free radicals which appear due to the reciprocity of a carcinogen with the body do not vanish but continue to have an effect, and this effect accumulates during the whole lifetime, then further calculation is simple. Dyes are added to food in a concentration of 0,005÷0,05% (50÷500 mg/l), therefore eating 100 kg of food (drinks, meat products, etc) includes a 0,5÷5% concentration of dyes. As shown in fig.2 this volume enters the carcinogen area, and the dyes E-133, E-124, E-102 appear as strong carcinogens. Moreover, the concentration increase of carcinogenic dyes leads to a sharp decrease of positron (τ_3) lifetime, i.e. to an increase of carcinogenic activity. As for mutagenic dyes, while the concentration increases, τ_3 decreases slowly, which means that carcinogenic properties will be evidently demonstrated in most concentrations. This shows the necessity of reviewing the possibility of using additives in given concentrations for some synthetic food dyes, or the prohibition of adding them to food, because their influence on a human body causes a strong negative effect over the course of time, during which a definite quantity of food would have been consumed. And it must surely be remembered that food products contain a lot of «E» numbers besides dyes - stabilizers, conservatives, flavors, etc. - whose effect on the body is similar to the effect of dyes, i.e. lots of them have carcinogenic properties [10, 11]. As well as this, most chemical additives are used in baby food, and, taking into account children's fast cell metabolism, those effects may have a fast and inconvertible character.

In this way, spectra of synthetic food dyes have been measured and analyzed by means of the PALS method. This method is a faster method for the determination of the presence of carcinogenic properties in chemical dyes. The method of positron annihilation spectroscopy shows a full picture of carcinogenic solutions and points to the negative effect of the use of the above concentrations of synthetic dyes in the food industry.

References

[1] Report on Carcinogens, Twelfth Edition 2011, U.S.Department of Health and Human Services National Toxicology Program.

[2] Andrianova M.M. Kancerogennye svojstva krasnyh pishhevyh krasitelej amaranta, puncovogo SX i puncovogo 4R. Voprosy pitanija. —1970. —N 5. —S. 61—65

Journal of Physics: Conference Series **618** (2015) 012027 doi:10.1088/1742-6596/618/1/012027

[3] Volkova N.A., Kup'ev V.V., Popov V.I. Gigienicheskaja ocenka nekotoryh pishhevyh krasitelej. Materialy 17-j nauchnoj konferencii instituta pitanija, 1—3 ijunja 1971. —M., 1971. —S. 35—36.
[4] Karpljuk I.A., Volkova N.A., Okuneva L.A., Gogol' A.T., Rybakova K.D. Izuchenie mutagennogo dejstvija pishhevyh krasitelej tartrazina i indigokarmina. Voprosy pitanija. 1984. —N 2. —S. 58—61.
[5] SanPiN 2.3.2.1293-03 Gigienicheskie trebovanija po primeneniju pishhevyh dobavok (utv. Glavnym gosudarstvennym sanitarnym vrachom RF 18 aprelja 2003 g.)
[6] Pivtsaev A. A., Razov V. I. Issledovanie himicheskih kancerogenov metodom pozitronnoj annigiljacionnoj spektroskopii s vremennym razresheniem. Zhurnal prikladnoj spektroskopii. T. 80, № 5, 2013. 806 – 809.
[7] Buchihin A.P., Gol'danskij V.I., Tatur A.O., Shantarovich V.P. Atom pozitronija v organicheskih zhidkostjah. Zhurnal jeksperimental'noj i teoreticheskoj fiziki, - 1971. – 60, №3. – S. 1136 – 1140.
[8] Sysoev A.B. Toksikologo-gigienicheskoe izuchenie sinteticheskogo pishhevogo krasitelja solnechnogo zheltogo zakata FCF i ego analoga. Avtoref. diss. na soiskanie uchenoj stepeni k.m.n. —M., 1974. —25 s.
[9] Available at: http://www.palsfit.dk
[10] Nejman I. Kancerogeny i pishhevye produkty. Medicina – 1972.
[11] Nikitina E. V., Reshetnik O.A. Biobezopasnost' pishhevyh produktov: Uchebnoe posobie. Kazanskij gosudarstvennyj tehnologicheskij universitet – Kazan', 2006, 92 s.

Journal of Physics: Conference Series **618** (2015) 012028 doi:10.1088/1742-6596/618/1/012028

Investigation of PCM microcapsules at various temperatures and pressures by PALS method

B Zgardzińska, M Tydda, A Błażewicz

Institute of Physics, Maria Curie-Sklodowska University, 20-031 Lublin, Poland

E-mail: bozena.zgardzinska@poczta.umcs.lublin.pl

Abstract. The positron annihilation lifetime spectroscopy (PALS) was used to investigate the properties of Phase Change Material (PCM): the Micronal® DS 5001 X microcapsules. The microcapsules are composed of a polymer capsule shell and a paraffin core. Two groups of free volumes were observed: those producing the long-lived component can be ascribed to the polymer, the other to the mixed alkanes. Both *o*-Ps components show the known effect of the intensity rise with positron irradiation time. At temperatures below -3°C we have observed the short lifetime as for alkane rigid phase, then the rotator-like phase preceded liquid phase. This form of temperature dependence of PAL spectra is discussed. There was no destructive effect of temperature. The PALS results were compared with those obtained by scanning electron microscopy.

1. Introduction

The lifetime of *ortho*-Ps substate depends on the size of free volume (void) where the *o-Ps* is trapped, thus *Ps* is considered as a convenient probe of medium structure.

Phase Change Materials (PCM) are able to absorb, accumulate and release a large amount of energy in the range of the phase transition temperature. The PCM in the form of microcapsules of core-shell morphology were the subject of this study. The shell material of microcapsules (polymer) determines their physical properties (e.g. strength, toxicity, reactivity), while the core material (paraffin) the thermal ones (e.g. operating temperature, capacity of accumulation and release of energy). When the environment temperature increases, the filler material absorbs the energy until the core becomes molten. When the temperature falls, the filler material solidifies, releasing back the energy to the environment.

The microcapsules Micronal® DS 5001 X (MPCM) were obtained from the producer and distributor the BASF Chemical Company. Their manufacturing process is described in [1]. The manufacturer declares the melting temperature 26°C. They are a kind of prefabricated material for building industry; for example they are added to the building materials: paints, plasters, fillings or cement mortar and can replace the air conditioning systems.

The filling material is usually alkane or their mixture. The pure hydrocarbons as well as the mixtures of alkanes were successfully investigated as a function of temperature or pressure by the PALS technique [2-5]. At low temperatures (usually below 200 K) in pure n-alkanes and some of their mixtures, the *o*-Ps intensity increases with time by several percent with time constant θ of several hours, while the *o*-Ps lifetime ($1.1 \div 1.2$ ns) does not change. The growth of intensity is associated with trapping the post-ionization electrons. The stepwise change of intensity and lifetime (as a function of T or p) appears at the phase transition points.

The existing data about alkanes allow to single-out the share of them in the complex structure of microcapsule. It was particularly interesting to find out how the spatial limitation (encapsulation) affects the properties of the filling material (the formation of crystalline structures, phase transitions points, other effects typical for pure alkanes, e.g. low-temperature e- trapping).

2. Experimental

Positron annihilation lifetime spectra (PALS) were measured using two standard fast-slow spectrometers with different chambers: one to measure spectra as a function of temperature (*T*) and-the other as a function of pressure (*p*). The chamber for *f (T)* measurements gives the opportunity to carry out the measurements in the range of 120 ÷ 500 K. The other chamber was connected to the Unipress U11 gas compressor which allows to achieve the pressures from 0 to 500 MPa. The resolution time (FWHM) was about 260 ps and 240 ps, respectively. The ^{22}Na positron sources of activity 0.80 MBq in the Kapton envelopes were used ($8 \cdot 10^5$ coincidences were collected during 1 h). The spectra were processed by the LT 9 programme [6]. The PAL spectra were assumed to consist of four exponential components convoluted with the instrumental resolution curve and random coincidence background. Two shortest-lived components were ascribed to the para-positronium (*p*-Ps) and to the annihilation of free positrons (*e*+). The remaining longest-lived ones belonged to the decay of ortho-positronium (*o*-Ps) in free volumes of different sizes. The correction for positron absorption in the Kapton foil was also applied. The structure of microcapsules was observed using a high resolution scanning electron-ion microscope Quanta 3D FEG form FEI company. For our investigations the MPCM with grain size in the range 180-250 μm were selected.

3. Results and discussion

In the microcapsules two *o*-Ps components with different lifetimes exist. In the Figure 1, the growth of *o*-Ps intensities at 123 K in time is shown; the lifetimes, 1.1 and 1.7 ns were stable in time. This means that in the microcapsules electron trapping effect is still present, like in the previously described alkanes. The time constant of this process for the longer-lived *o*-Ps component is 9.6 h, while for the shorter-lived one is 4 h. The maximal increase of intensities is in both cases similar (4.6 and 4.8 %) and the scale of that rise was greatly reduced compared to neat alkanes. We suppose that the trapping effect can occur if the crystal structure of the filler material is the same as of macroscopic samples of the alkanes.

As one can see in the Figure 2, the lifetime of shorter-lived *o*-Ps component τ_3 is practically independent on temperature. The intensity I_3 decreases initially by 4%, then the rate of change slows, and above 269 K I_3 stabilizes at a relatively low value ~ 6%. The changes in the longer-lived component (τ_4, I_4) as a function of temperature are more pronounced. The lifetime τ_4 increases almost linearly up to 2.4 ns, and at 269 K the first stepwise change is visible. Above 298 K, the τ_4 stabilizes at the value of 2.95 ns. The intensity I_4 reaches a maximum at 143 K, but above this temperature the excess electrons are already throne out of the traps, that leads to the reduction of intensity down to 13%. Two steps of I_4 are seen, at 269 K and at 298 K. Such a dependence of τ_4, I_4 on temperature suggests that the 4th component should be attributed to the filling material of the microcapsules. The intensity I_4, significantly higher than I_3, suggests that in the mass composition of the microcapsules the filling is predominant. Stepwise changes of the τ_4 and I_4 allow us to specify the phase transitions points in MPCM, estimated as 269 K and 298 K. Above 298 K, the filling material is liquid. Below 269 K it should be in the rigid crystal phase. Between these two points we see the rotator phase. The *o*-Ps lifetime in liquid alkanes and their mixtures a few K above the melting point is about 3.15 ns, and is reduced to about 2.9 ns, it can be the result of the *ortho-para* conversion in non-degassed samples. The *o*-Ps lifetime τ_4 in MPCM is 2.95 ns – just like in the alkanes in the air, but after degasing the τ_4 increases to 3.1 ns (Fig.2), which confirms that the 4th component corresponds to the paraffin filling.

Melting point of 298 K is quite close to that of *n*-octadecane (301 K), however, in so short-chain even-numbered alkane the rotator phase does not appear, moreover, in pure alkanes the lifetime in rigid phase is 1.2 ns only. Increased lifetime in MPCM suggests that they are filled with a mixture of

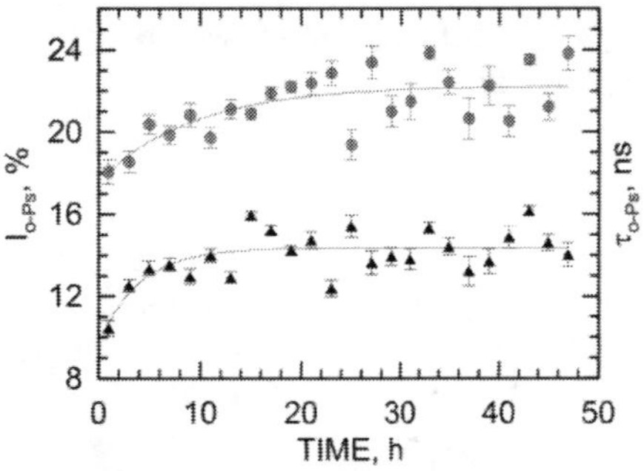

Figure 1.(up) Increase of *o*-Ps intensities (I_3 – triangles, I_4 – dots) with time at 123 K.

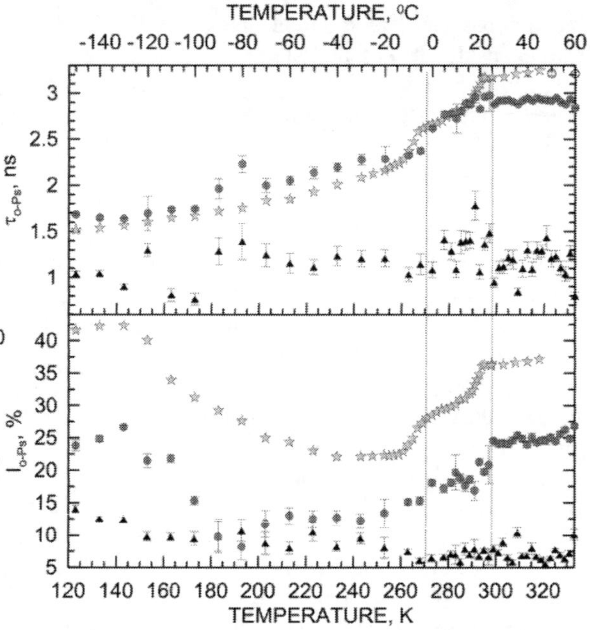

Figure 2.(right) Temperature dependence of *o*-Ps lifetimes and intensities in MPCM (τ_3, I_3 – triangles, τ_4, I_4 – dots) and alkane mixture (stars). Open circle – τ_4 in degased MPCM.

alkanes, the longer of them is *n*-octadecane (because the mp. in mixture is an average of mp. of their components). Among several mixtures of alkanes investigated by us analogous behaviour was found in the mixture of 50% $C_{16}H_{34}$ +50% $C_{18}H_{38}$. As one can see in the Figure 2 (stars) the *o*-Ps lifetime in the mixture is similar to that measured in MPCM (except the liquid phase, because the mixture was degassed and the microcapsules were not). In the mixture the rotator phase appears and the phase transition points are roughly reproduced (deviations may be due to small difference in the percentage composition of the mixture).

In the MPCM, the intensity is smaller than in alkane mixture, because we have here two different kinds of free volumes. Comparing the intensities in the liquid, both MPCM and mixed alkanes, we observe the reduction to about 75% of that for pure mixture. Thus, one can expect that the polymer content on the capsule is about 25%. In the literature [7] one can find the information that the shell material is about 24% of microcapsule mass, well consistent with our estimate. Average size of our microcapsules is about 5 μm, thus the respective shell wall thickness should be 0.226 μm. It can be easily verified in the SEM images; the shell thickness of selected microcapsule with the diameter 4.75 μm is estimated as 0.225 μm.

After cooling the MPCM the measurements as a function of temperature were repeated on the same sample and found fully reproducible - there was no destructive effect of temperature, which is confirmed by SEM pictures.

The phase change (for example from liquid to rigid crystal) can be realized by decreasing the temperature or increasing the pressure. Microcapsules were investigated as a function of increasing pressure at constant temperature 293 K, when the MPCM were in the rotator phase. The increase of pressure by 1 MPa corresponds in alkanes to a decrease in temperature of about 0.22 K. Thus it was expected to observe the transition to the rigid phase at $p \cong 80$ MPa. The results illustrated in Figure 3 show that even small pressure significantly influences the PAL spectra parameters. At zero pressure *o*-Ps lifetimes are identical as in temperature measurements, but already above 6 MPa they are shorter than in the measurements as a function of *T*. Below 20 MPa the intensity I_3 increases - such an effect is rather unexpected. The lifetime shortening can be easily explained as the transition to rigid phase, the increase of intensity suggests microcapsules crushing, which leads to the relative increase of the share of *o*-Ps annihilating in polymer. We have verified this hypothesis by performing a SEM photograph of microcapsules treated with high pressure (Fig. 4). The grain capsules are deformed, some of the grains

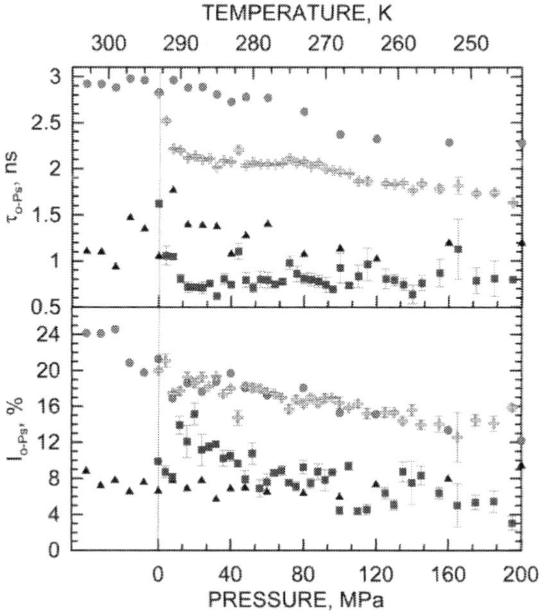

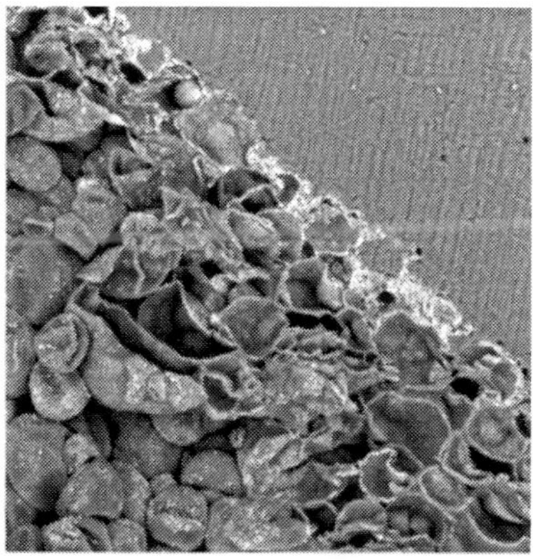

Figure 3. (left) Pressure dependence of *o*-Ps lifetimes and intensities in MPCM (τ_3, I_3 – squares, τ_4, I_4 – crosses). The o-Ps lifetimes and intensities as a function of temperature are added (symbols as on Fig.2).

Figure 4. (up) SEM image of MPCM after high pressure treatment. Figure width 45 μm.

has been broken, split into smaller fragments. As can be seen in the Figure 4 the microcapsules were squeezed but on the grain boundary region only. Inside the grain deformed microcapsules were observed.

4. Conclusions

Positronium is a useful probe in the research of microcapsules. Using the PALS technique we can state that in microcapsules the electron trapping effect occurs, suggesting that the crystalline structure formed inside the microcapsules is the same as the lamellar structure of pure paraffins. The phase transitions points in MPCM were determined, and the presence of additional phase (rotator) has been found. Comparing the changes of PAL spectra in MPCM and mixtures allows to determine the composition of the microcapsules. The filling material was identified as a mixture of *n*-hexadecane and *n*-octadecane in equal proportions. The percentage weight of the core was estimated as 75%. The devastating effect of pressure was found. The SEM images confirmed the conclusions from the PALS measurements.

5. Acknowledgments

This study is supported by the grant 2013/09/D/ST2/03712 of National Science Centre in Poland. The authors thank the BASF Chemical Company for providing samples for testing.

References

[1] Schmidt M 2008 *Phase Change Materials-latent heat storage for interior climate control. BASF Micronal*, Energiforum Danmark.
[2] Zgardzińska B, Goworek T 2012 *Chem. Phys. Letters* 547, 35.
[3] Zgardzińska B, Gorgol M 2014 *Acta Phys. Polon A* 125, 816.
[4] Goworek T, Zgardzińska B, Pietrow M, Wawryszczuk J 2013 *Mat. Sci. Forum* 733, 67.
[5] Zgardzińska B, Goworek T, Wawryszczuk J, Standzikowski K 2011 *Chem. Phys.* 379, 1.
[6] Kansy J 1996 *Nucl. Instr. & Meth. A* 374, 235.
[7] Giro-Paloma J, Oncins G, Barreneche C, Martínez M, Fernández A I, Cabeza L F 2013 *Applied Energy* 109, 441–448.

Journal of Physics: Conference Series **618** (2015) 012029 doi:10.1088/1742-6596/618/1/012029

Positron annihilation in benzene and cyclohexane: a comparison between gas and liquid phase.

Kamil Fedus

Institute of Physics, Faculty of Physics, Astronomy and Informatics, Nicolaus Copernicus University, Grudziadzka 5, 87-100 Torun, Poland

E-mail: `kamil@fizyka.umk.pl`

Abstract. A comparative study about positron annihilation in gas and liquid phases of two non-polar ring molecules: benzene (C_6H_6) and cyclohexane (C_6H_{12}) is presented including the most recent experimental and theoretical achievements. In addition the preliminary results of positron annihilation lifetime measurements in a liquid phase at room temperature for these two molecules are reported.

1. Introduction

The problem of linking the gas-phase and the condensed-phase interactions is one of the most fundamental concepts in physics. The goal is to describe how much the mechanisms of individual interactions present in the gas-phase are distorted and/or retained in the condensed phases. In particular this issue has been raised some time ago [1] by community studying behavior of slow electrons at interfaces of the gaseous and the condensed phases of matter. It is well known that the high-energy electrons, whose wavelength is short in comparison with the "diameter" of the elementary constituents of condensed matter, preserve the unique nature of two-body electron-target interaction present in the gas-phase. This concept is no longer valid at low energies, where the electron wavelength is of the order of the interatomic or intermolecular distances. In this case, the electron interacts collectively with many neighboring atoms and/or molecules and thus the mechanisms of interactions are significantly modified when compare to a pure two-body case.

Nowadays the low-energy electron studies in a gas and condensed phases are quite well developed. On the other hand the attempts to find similar links have been rarely undertaken for an antimatter counterpart – the positron (see pages 121-140 in Ref.[1]). The recent progress in studies on slow positron scattering and annihilation in a gas phase matter combined with a very fast development of positron chemistry focused on the behavior of antielectron in condensed matter, makes possible to construct some first conclusive connections between positron interactions in gas and condensed phases. For example, the latest paper by Zubiga et al.[2] shows that a theory for positron interaction with single atoms can be transferable to the condensed phase. Such study is particularly important in the context of numerical modeling of positron behavior in liquids, solids and living tissues [3] where the gas-phase results can serve as an input data.

The main purpose of the present paper is to shortly describe the current state of the art of the research on low-energy positron interaction with two non-polar ring molecules: benzene

(C_6H_6) and cyclohexane (C_6H_{12}) in gas and liquid phases. In addition the preliminary results of positron annihilation lifetime measurements in a liquid phase for these two molecules are reported.

2. Gas phase

Recently, a substantial breakthrough in understanding interactions of low-energy positrons with materials in gas phase has been achieved (see for example [4]), where both scattering and annihilation occur. The most recent measurements of total cross sections for positron scattering from atoms and molecules in gas phase show a sharp rise of total cross sections when positron energy goes to zero. This effect has been explained by a dominating long-range polarization interaction at low energies [5]. Furthermore, the increase of the scattering cross sections is accompanied by the enhancement of annihilation rates in the low energy limit. It has been proved by Gribakin [6] that the annihilation rates are directly related to the elastic scattering cross sections in thermal energy range for simple targets such as noble gases and small molecules. However the situation appears to be much more complex for large molecules such benzene (C_6H_6) and cyclohexane (C_6H_{12}).

Fig.1 shows all available experimental and theoretical data of total cross sections (TCS) for positron scattering from C_6H_6 and C_6H_{12} near and below the positronium formation threshold. The experimental results were obtained using two positron beam systems: from Trento University in Italy (Karwasz et al.[7], Zecca et al.[8]) and from Yamaguchi University in Japan (Sueoka [9], Makochekanwa et al.[10]). Note that the Yamaguchi group reported two completely different datasets for benzene. The older cross-sections (1988) by Sueoka [9], obtained with very low magnetic field applied in scattering region to guide the positron beam, are in quite good agreement with both results from Trento. Whereas a more recent dataset (2003) by Makochekanwa et al.[10], obtained with much higher magnetic field, largely underestimates the cross-sections in the low energy limit. The same effect is also valid for c-hexane data by Sueoka et al.[11]. Both datasets from Trento, Karwasz et al.[7] (2005) and Zecca et al.[8] (2007), are similar - the small difference comes from different energy scaling in both experiments.

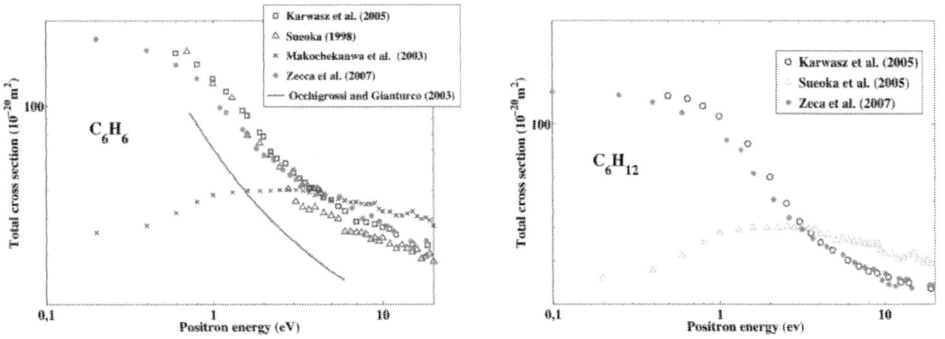

Figure 1. Total cross-sections for positron scattering on benzene (C_6H_6) and cyclohexane (C_6H_{12}). Experimental data are from Karwasz et al. [7], Sueoka et al. [9, 11], Makochekanwa et al. [10] and Zecca et al. [8]. Theory from Occhigrossi and Gianturco [13] is also presented.

Experimental TCS for benzene are higher by about 20-30% when compare to data for c-hexane at around 1eV. However the measurements of positron binding energies carried out by Surko and co-workers using positron trap techniques [6] show that the zero-energy cross-sections ($\sigma(E = 0)$) are exactly in an opposite relationship. To visualize this reversal of TCS (see Fig.2) in low energy range for both molecules, the experimental cross-sections by

Journal of Physics: Conference Series **618** (2015) 012029 doi:10.1088/1742-6596/618/1/012029

Karwasz et al.[7] have been extrapolated down to zero energy using a simple model based on the Modified Effective Range Theory (MERT, see for more details [5]). Unlike in our previous paper on this subject [12], here both negative and positive scattering lengths $A = \pm 9.52a_0$ and $A = \pm 13.04a_0$ are assumed for benzene and c-hexane, respectively. The latter were determined from relation $E_b = 1/2A^2$ [6] (atomic units) using experimental positron binding energies ($E_b = 150$ meV and 80 meV for C_6H_6 and C_6H_{12}, respectively) and they were used to calculate the zero-energy cross sections from $\sigma(0) = 4\pi A^2$. The results of extrapolation are shown in Fig.2. It is clear that regardless the sign of the scattering length, the cross sections for c-hexane are larger than for benzene in the thermal energy range, $E < 100$meV, where the annihilation process is highly probable. Note however that MERT extrapolation has to be treated as a rough approximation since the model considers only elastic scattering neglecting the contribution of the inelastic processes related to vibrational excitations.
Experimentally determined annihilation rates (thermally averaged) at room temperature expressed by a dimensionless quantity $\langle Z_{eff} \rangle$ [6] indicate higher annihilation rates for positron interaction with c-hexane (≈20000) than with benzene (≈15000). This is qualitatively consistent with MERT-derived cross-section in thermal energy range - in analogy to noble gases and small molecules: higher cross-sections are followed by larger annihilation rates [6]. However the measured annihilation rates in hydrocarbones are few order of magnitude larger than the corresponding coefficeints in simpler targets. Energy-resolved positron trap [6] experiments show an appearance of numerous resonances in the annihilation at thermal and sub-thermal energies, which are attributed to the Feschbach resonances associated with low vibrational modes of C-H bonds [6]. Hence in the case of large molecules the annihilation rates at room temperature should be a function of both, elastic and vibrational cross-sections. However no *ab initio* calculations have been done so far to prove this prediction.

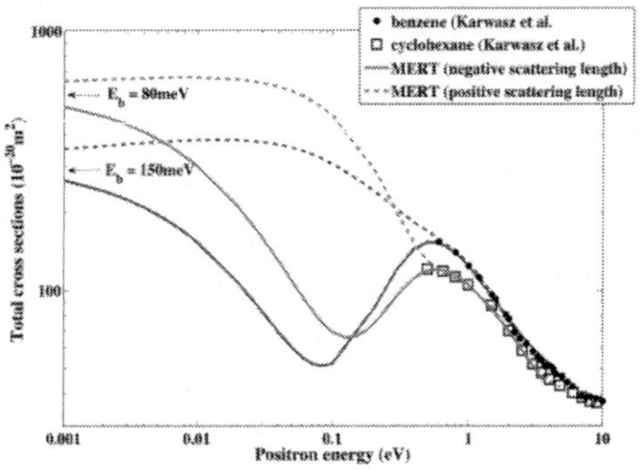

Figure 2. MERT extrapolation of experimental total cross-sections [7] for positron scattering from C_6H_6 and $C_{6}H_{12}$. See text for more details.

3. Liquid phase

Positrons implanted into the condensed matter are quickly slowed down to the thermal energies by inelastic processes. Then they diffuse through the material interacting with the environment before being eventually annihilated. The main difference when comapred to a gas phase is related to the Ps formation. In liquids the mechanisms responsible for this process are much more complicated than in gases where a simple Ore picture is valid [14]. Different theories including the modified Ore, spur and blob models, have been proposed to explain the results of annihilation experiments in condensed matter. The comparison of these models is out of the scope of this paper and it can be found elsewhere (see for example Refs.[14, 15] and references therein).

We measured the positron annihilation lifetime spectra for both C_6H_6 and $C_{6}H_{12}$. The experiments were carried out at room temperature using the fast-fast coincidence ORTEC PLS system described in Ref.[16]. At least 10^6 counts of annihilation events were collected in each experimental acquisition. The obtained spectra were analyzed with two methods: (i) the MELT procedure [17] and (ii) the conventional multiexponential model. Both methods give similar results revealing the presence of three lifetime components corresponding to three different annihilation events. This is consistent with results reported for other liquid hydrocarbons [14]. The derived averaged lifetimes are summarized in Tab.1 where they are compared with some data from literature [14, 18]. Note that the quantitative agreement between the present data and the literature is relatively poor. This is because our liquid samples were not degassed before the measurements. It has been shown recently by Zgardzinska and Goworek [18] that the presence of air significantly changes the room temperature lifetime spectra of liquid c-hexane, giving the longest component τ_3 =2.40 ns in non-degassed samples and 3.20 ns in degassed ones. The first results is in good agreement with our present data. The measurements with the air-free samples are in progress. Despite the quantitative inconsistency, all spectra from Tab.1 have a common feature - all lifetimes in C_6H_{12} are always higher than in C_6H_6. Hence the positron annihilation rates in liquid c-hexane are smaller than in liquid benzene. This results is in direct contradiction to the output from the gas-phase experiments [6]. Note however that a simple three exponential model typically used to analyze PALS data does not take into account the complexity of the Ps formation process occuring in the liquids. Therefore probably a more sophisticated approach is needed to get more accurate annihilation rates in liquids and to do more reliable comparison with a gas phase.

Table 1. Positron annihilation lifetimes in C_6H_6 and C_6H_{12}.

C_6H_6	τ_1(ns)	τ_2(ns)	τ_3(ns)
present	0.261	0.542	2.30
Ref.[14]	0.142	0.440	3.18
Ref.[18]	-	-	3.00
C_6H_{12}	τ_1(ns)	τ_2(ns)	τ_3(ns)
present	0.284	0.601	2.57
Ref.[14]	0.214	0.469	3.25
Ref.[18]	-	-	3.20

4. Conclusions

The most recent experimental and theoretical results describing the positron interaction with molecules of benzene and cyclohexane at room temperature in gas and liquid phases have been shortly summarized. In particular it was noticed that the gas-phase experiments give higher annihilation rates for C_6H_{12} when compare to C_6H_6, while the liquid-phase experiments give an opposite result when analyzing PALS data with the conventional three exponential model. This is confirmed by positron annihilation lifetime measurements in liquid phase presently reported.

References

[1] Christophorou L G, Illenberger E and Schmidt W F 1994 *Linking the Gaseous and Condensed Phases of Matter, The Behavior of Slow Electrons* vol 326 (NATO ASI Series, Springer)
[2] Zubiaga A, Tuomisto F and Puska M J 2014 *Phys. Rev. A* **89** 052707
[3] Munoz A, Oller J C , Blanco F , Gorfinkiel J D , Limao-Vieira P, Maira-Vidal A, Borge J G, Tengblad O, Huerga C, Tllez M and Garcia G 2008 *Journal of Physics: Conference Series* **133** 012002
[4] Karwasz G P 2005 *Eur. Phys. J. D* **35** 267
[5] Fedus K, Karwasz G P and Idziaszek Z 2013 *Phys. Rev. A* **88** 012704
[6] Gribakin G F, Young J A and Surko C M 2010 *Rev. Mod. Phys.* **82** 2557
[7] Karwasz G P , Pliszka D, Brusa R S, Perazzolli C 2005 *Acta Phys. Pol.* **107** 666
[8] Zecca A, Moser N , Perazzolli C, Salemi A and Brunger M J 2007 *Phys. Rev. A* **76** 022708
[9] Sueoka O 1988 *J. Phys. B* **21** L631
[10] Makochekanwa C, Sueoka O and Kimura M 2003 *Phys. Rev. A* **68** 032707
[11] Sueoka O, Makochekanwa C, Tanino H and Kimura M 2005 *Phys. Rev. A* **72** 042705
[12] Fedus K and Karwasz G 2014 *Acta Phys. Pol. A* **125** 829
[13] Occhigrossi A and Gianturci F A 2003*J. Phys. B* **36** 1383
[14] Mogensen O E 1995 *Positron Annihilation in Chemistry* (Springer-Verlag, Berlin)
[15] Stepanov S V and Byakov M V 2002 *J. Chem. Phys.* **116** 6178
[16] Karbowski A, Fidelus J D, Karwasz G P 2011 *Materials Science Forum* **666** 155
[17] Shukla A, Peter M and Hofmann I 1993 *Nucl. Instrum. Methods Phys. Res. A* **335** 310
[18] Zgardzinska B and Goworek T 2013 *Chem. Phys.* **421** 10 + private communication

Journal of Physics: Conference Series **618** (2015) 012030 doi:10.1088/1742-6596/618/1/012030

PLT and DBAR Investigations on MPDMAPP Doped PVA/PVP Blend

R. F. Bhajantri [1]*, V. Ravindrachary [2], Blaise Lobo [3], P. K. Pujari [4], Sunil G. Rathod [2], Jagadish Naik [2], Vidyashree Hebbar [1], Chandrappa H [1]

[1] Department of Physics, Karnatak University, Pavate Nagar, Dharwad - 580 003, Karnataka, India
[2] Department of Physics, Mangalore University, Mangalagangotri - 574 199, Karnataka, India
[3] Department of Physics, Karnatak Science College, Karnatak University, Dharwad 580 001, Karnataka, India
[4] Radiochemistry Division, Bhabha Atomic Research Centre, Trombay, Mumbai – 400 085, Maharashtra, India

* Email: rfbhajantri@gmail.com, rfbhajantri@kud.ac.in

Abstract. Poly(vinylalcohol) (PVA)/Poly(vinylpyrrolidone) (PVP) blend films, doped with chalcone derivative (1-(4-methylphenyl)-3-(4-N,N,dimethylaminophenyl)-2-propen-1-one) (MPDMAPP) from 0.025 wt% up to 1 wt% were prepared using solution casting technique. The o-Ps lifetime τ_3 is found to change little, from 1.61 ns at 0.025 wt% dopant concentration to 1.63 ns at 0.5 wt% dopant level, but drops to 1.4 ns at 1 wt% dopant concentration, indicating the onset of phase separation. The S-parameter of DBAR was found to be linearly related to the ortho-Positronium (o-Ps) intensity I_3. The S-parameter drops significantly from 0.1 wt% up to 1 wt% doping concentration. This is supported by the XRD scans.

1. Introduction

The blending of different polymers, and doping polymers with organic molecules, are strategic routes to improve the performance of polymeric materials [1-2]. The properties of polymeric blends are mainly dependent on the characteristics of the parent homo-polymers, the blend composition and the polymer-polymer miscibility [3]. Poly(vinylalcohol) (PVA) and poly(vinylpyrrolidone) (PVP) are perfectly compatible and miscible polymers; via the hydrogen bond interactions between the carbonyl groups of PVP and hydroxyl groups of PVA in the amorphous regions of the blend [4-5]. The chalcone derivatives are organic materials which have attracted attention due to their excellent blue-green light transmittance [6-7]. They are typically polarizable molecules, consisting of π-conjugated system with donor and acceptor groups.

The main objective of this study is to measure the *o*-Ps lifetime τ_3 and calculate the bulk lifetime τ_b, mean lifetime τ_m, positron trapping rate κ and S-parameter in (1-(4-methylphenyl)-3-(4-N,N,dimethylaminophenyl)-2-propen-1-one) (MPDMAPP) doped PVA/PVP (50:50) blend using Positron Lifetime Technique (PLT) and Doppler Broadening of Annihilation Radiation (DBAR) technique. In molecular media including polymers, the ortho-Positronium (*o*-Ps) localizes (gets trapped) in the free volume holes and there is a finite probability that the positron of *o*-Ps bound state may annihilate with a molecular electron of opposite spin, rather than its bound partner [8]. It is known that the positrons are selectively trapped by vacancy type defects, micro-voids, negatively charged ions and radicals. The chromophore dopants in polymers are trapping sites for the positrons, which are created as a consequence of spatially localized dopant ions. This creates lattice distortion, and negatively charged domains that localises the positrons [9-13]. The evolution of defects and positron trapping rates in chalcone chromophore doped miscible blend systems remains unexplored, to the best of our knowledge. DBAR measurements on the other hand can provide information about the polar groups and chemical surrounding of nanoholes in polymers, and helps to investigate the local chemical environment around the positron trapping sites.

2. Experimental

The films of PVA/PVP (50%:50%) blend, doped with MPDMAPP in different concentrations were prepared by solvent casting method using DMF as solvent, as explained in our earlier papers [7, 10]. The details of positron lifetime set up are explained elsewhere [10]. The energy spectrum is recorded using a high purity Germanium (HPGe) detector having resolution of 1.4 keV at 662 keV photopeak of ^{137}Cs. The width and shape of the 511 keV annihilation line yields the information about the momentum distribution of electrons with which the positrons annihilate. The *S* and *W* parameters were obtained by fraction of the counts which appeared at 0 to $4\times10^{-3}m_oc$ and 15 to $30\times10^{-3}m_oc$ respectively from the centre to the total counts for the annihilation peak. XRD studies were carried out using Rigaku Miniflex-II benchtop X-ray diffractometer with Copper K_α radiation of wavelength λ=0.15406 nm, with graphite monochromator.

3. Results and Discussion

The measured PLT spectra were resolved into three components. The first life time component (τ_1, I_1) is attributed to the self-annihilation of *p*-Ps and the free annihilation of positrons in the bulk. The intermediate lifetime component τ_2, having intensity I_2, is considered to be associated with annihilation of positrons trapped by defects present in the crystalline regions and positrons trapped in the crystalline–amorphous interface regions of the polymeric material.

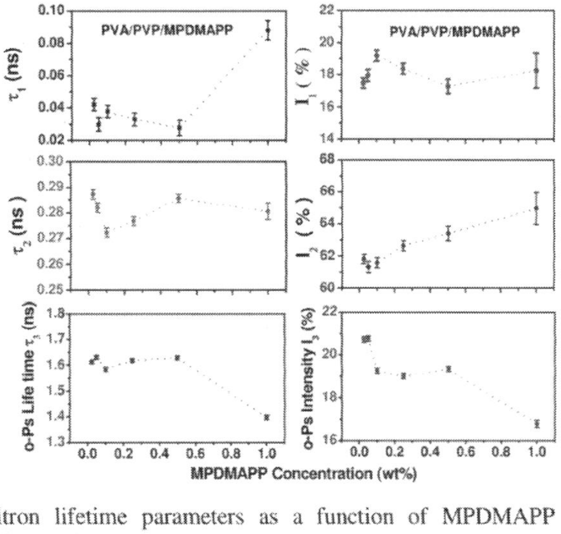

Figure 1. Positron lifetime parameters as a function of MPDMAPP doping concentration in PVA/PVP (50:50) blend

The *o*-Ps lifetime (τ_3) is a measure of the size of free-volume holes (in the amorphous regions of the polymer). The intensity (I_3) is a measure of Ps formation probability, and not the concentration of free volume holes. From Fig 1, we notice that τ_3 remains constant at 1.6 ns from 0.025 wt% doping level (DL) up to 0.5 wt% DL. It however decreases significantly to 1.4 ns at 1 wt% DL. The chalcone MPDMAPP contains methyl group on one side and dimethyl amino group on the other side, the C_6H_4–$N(CH_3)_2$ group acting as donor site due to its strong mesomeric electron releasing behaviour. On the other hand, the methyl group attached to the phenyl ring containing CO group, i.e. CO–C_6H_4–CH_3 acts as an acceptor group. The lone pair of nitrogen in the donor is transferred to the acceptor group to form a highly polar charge transfer state and when such the aromatic compounds are added to polymers, the electrons can be scavenged by these aromatic compounds, mainly because of the CTCs (defects) which are expected to trap the positrons. By accepting two-state positron trapping model [1, 9, 13], the numerical parameters of positron trapping (mean lifetime τ_m, bulk lifetime τ_b and positron trapping rate κ) can be calculated using equations (1), (2), and (3).

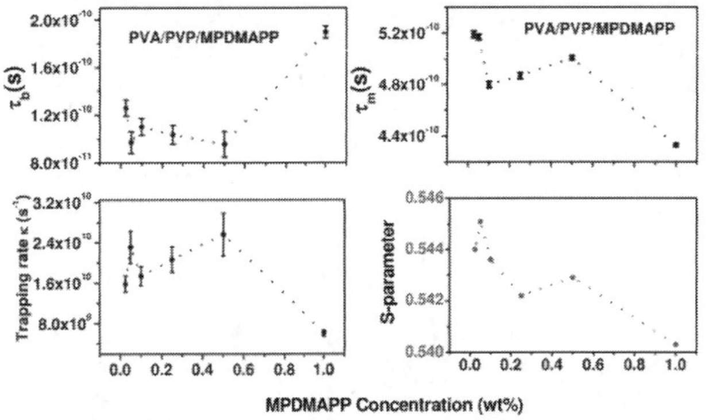

$$\tau_b = \frac{(I_1 + I_2)}{\left(\frac{I_1}{\tau_1} + \frac{I_2}{\tau_2}\right)} \quad (1)$$

$$\tau_m = \frac{(\tau_1 I_1 + \tau_2 I_2 + \tau_3 I_3)}{(I_1 + I_2 + I_3)} \quad (2)$$

$$\kappa = \left[\frac{1}{\tau_1} - \frac{1}{\tau_b}\right] \quad (3)$$

Figure 2. Variation of bulk lifetime τ_b, mean lifetime τ_m, positron trapping rate κ and S-parameter as a function of MPDMAPP doping concentration in PVA/PVP (50:50) blend.

The Fig. 2 shows the variations of bulk lifetime τ_b, mean lifetime τ_m, positron trapping rate κ and S-parameter as a function of MPDMAPP doping concentration in PVA/PVP (50:50) blend. The first component τ_1 can be an average due to both the bulk positron lifetime (τ_b) and the Bloch state residence time (τ_B), as a result of which, $\tau_1 < \tau_b$ [13]. It may also be expected that the OH group of PVA should trap the positrons. This is clear evidence of positron trapping at the polar groups and suggests that the positron can act as a sensitive chemical probe of polar structures in chromophore doped polymers. One may expect the negatively charged parts of polar groups in molecular substances to act as trapping centers as well [1].

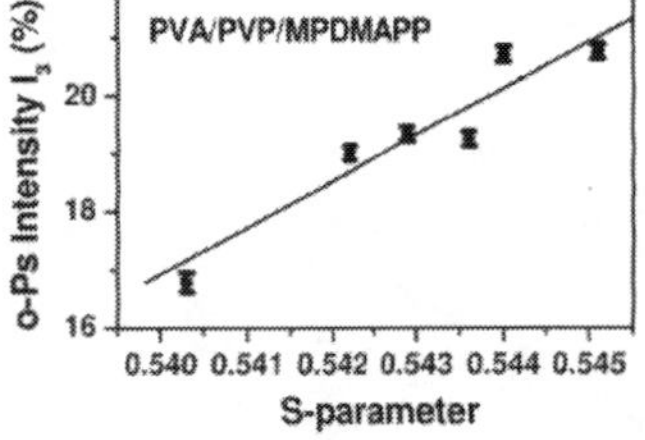

Figure 3. Variation of o-Ps intensity I_3 with S-parameter in MPDMAPP doped PVA/PVP (50:50) blend

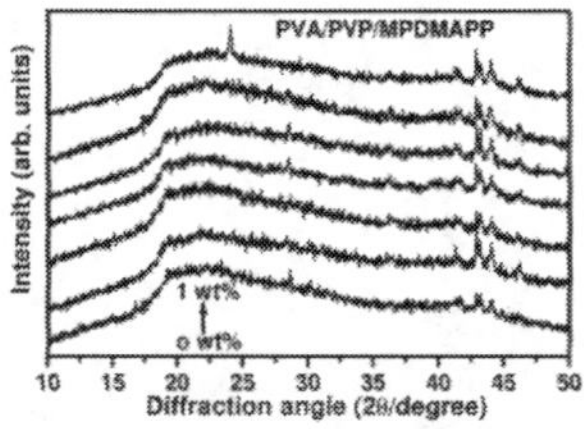

Figure 4. XRD patterns of MPDMAPP doped PVA/PVP (50:50) blends

The *S*-parameter is 0.543 for the sample with 0.5 wt% MPDMAPP in PVA/PVP blend. The variation of *o*-Ps intensity relative to the *S*-parameter shows a linear relationship as shown in Fig. 3. The initial trend can be explained as the partial occupation of the larger free-volume holes of by the doping species, thereby inhibiting o-Ps formation for highest dopant concentration. The *o*-Ps lifetime τ_3 and the positron lifetime τ_2 correspond to Ps and positron trapping in different regions of the polymeric material. The lifetime τ_3 corresponds to pick-off annihilation of Ps trapped in the constraint free amorphous regions of the polymeric material. However, the lifetime τ_2 comprises of contributions from positrons trapped in micro-voids of the crystalline region as well as a constrained region of the amorphous zone (constrained by the crystalline areas) [4, 14]. The XRD patterns of MPDMAPP doped PVA/PVP blend films are shown in Fig. 4. It is observed from the figure that the XRD patterns of the films reveals a dominant amorphous nature at low dopant concentration, and manifests the semicrystalline nature at moderate and high concentration of the dopant. There is a sudden drop in o-Ps lifetime τ_3 and intensity I_3 at 1 wt% doping level, due to aggregation of dopant in the amorphous regions.This results in increase in crystallinity of the sample.

4. Conclusions

The PLT results of the prepared MPDMAPP doped PVA/PVP blend films shows that the intermediate lifetime component τ_2, with intensity I_2 are considered to be associated with the annihilation of positrons trapped by defects present in the crystalline regions or trapped at the crystalline–amorphous interface regions. At low dopant concentrations, there is molecular level homogeneous distribution of dopant. The *o*-Ps intensity I_3 varies in a manner similar to the *S*-parameter of DBS spectrum. Positron trapping rates of the MPDMAPP doped blend were calculated. Distortions induced by aggregation of dopant (at higher doping) results in increase in trapping rate up to 0.5 wt% dopant concentration.

Acknowledgements

The authors are thankful to DAE-BRNS for sanctioning the project (2010/37C/7/BRNS/832) and DST, Govt. of India (SR/FTP/PS-011/2010); DST-FIST, Dept. of Physics, Mangalore University for XRD facility; Prof. Vincent Crasta, St. Joseph Engg. College, Mangalore for providing sample.

References

[1] Nagai Y, Nonaka T, Hasegawa M, Kobayashi Y, Wang C L, Zheng W and Zhang C 1999 *Phys. Rev.* B **60** 11863
[2] Huang C M, Yuan J -P, Cao H, Zhang R, Jean Y C, Suzuki R, Ohdaira T and Nielsen B 2000 *Rad Phys. Chem.* **58** 571
[3] Ramya P, Guagliardo P, Pasang T, Ranganathaiah C, Samarin S and Williams J F 2013 *Phys. Rev.* E **87** 052602
[4] Sharma S K, Prakash J, Bahadur J, Sudarshan K, Maheshwari P, Mazumder S and Pujari P K 2014 *Phys. Chem. Chem. Phys.* **16** 1399
[5] Lu J, Nguyen Q, Zhou J and Zheng-Hua P 2003 *J. Appl. Polym. Sci.* **89** 2808
[6] Ravindrachary V, Crasta V, Bhajantri R F and Poojari B 2005 *J. Crys Growth* **275** e313
[7] Bhajantri R F, Ravindrachary V, Poojary B, Ismayil, Harisha A and Crasta V 2009 *Polym. Eng. Sci.* **49** 903
[8] Tao S J 1972 *J. Chem. Phys* **56** 5499
[9] Boyko O, Shpotyuk Y and Filipecki J 2013 *Phys. Status Solidi* C **10** 121
[10] Bhajantri R F, Ravindrachary V, Harisha A, Ismayil and Ranganathaiah C 2009 *Phys. Status Solidi* C **6** 2429
[11] Mohamed H F M, Kobayashi Y, Kuroda S and Ohira A 2012 *Chem. Phys. Lett.* **544** 49
[12] Kansy J, Consolati G and Dauwe C 2000 *Rad. Phys. Chem.* **58** 427
[13] Rana U, Nambissan P M G, Malik S, Chakrabarti K 2014 *Phys. Chem. Chem. Phys.* **16** 3292.
[14] Cerrada M L, Perez E, Peren J M, Benavente R, Misheva M and Grigorov T 2005 *Macromolecules* **38** 8430

Journal of Physics: Conference Series **618** (2015) 012031 doi:10.1088/1742-6596/618/1/012031

Per-fluorinated sulfonic acid/PTFE copolymer studied by positron annihilation lifetime and gas permeation techniques

Hamdy F M Mohamed[1*], E E Abdel-Hady[1] and A Ohira[2]

[1] Physics Department, Faculty of Science, Minia University, B.O.61519, Minia, Egypt
[2] Fuel Cell Cutting-Edge Center (FC-Cubic), Technology Research Association, 2-3-26 Aomi, Koto-ku, Tokyo 135-0064, Japan

*E-mail: hamdyfm@gmail.com

Abstract. The mechanism of gas permeation in per-fluorinated sulfonic acid/PTFE copolymer Fumapem® membranes for polymer electrolyte fuel cells has been investigated from the viewpoint of free volume. Three different samples, Fumapem® F-950, F-1050 and F-14100 membranes with ion exchange capacity (IEC) = 1.05, 0.95 and 0.71 meq/g, respectively were used after drying. Free volume was quantified using the positron annihilation lifetime (PAL) technique and gas permeabilities were measured for O_2 and H_2 as function of temperature. Good linear correlation between the logarithm of the permeabilities at different temperatures and reciprocal free volume indicate that gas permeation in dry Fumapem® is governed by the free volume. Nevertheless permeabilities are much smaller than the corresponding flexible chain polymer with a similar free volume size due to stiff chains of the perfluoroethylene backbone.

1. Introduction

The polymer electrolyte fuel cell (PEFC) has been considered as one of the most promising power sources for stationary and portable applications [1] due to their high energy conversion efficiency and near-zero pollutant emission. The key element of PEFC is a proton conductive polymer electrolyte membrane (PEM). Currently, per fluorinated membranes and proton conductive polymer, Fumapem® is a commercially available per-fluorinated sulfonic acid/PTFE copolymer membrane used as the electrolyte in polymer electrolyte fuel cells. The proton conductivity must be affected by the electronic state near the ion exchange site, namely hydrophilic site, as well as the structure of nanoscale spaces (free volume). The free volume holes in polymers play a crucial role in determining its physical properties e.g., conductivity, permeability, thermal stability, etc [2]. The Positron Annihilation Lifetime (PAL) technique has been established as a powerful probe for microstructures of polymers; in particular, angstrom-sized free volume holes [3]. Positrons introduced from a radioactive source into a polymer annihilate either from the free state or positronium (Ps), the positron-electron bound state. Positronium can be formed in either a triplet state, ortho-positronium (*o*-Ps), or a singlet state, *para*-positronium (*p*-Ps), depending on the spin combination of the electron and the positron. A correlation between the radius R of a spherical free volume hole and the observed *o*-Ps lifetimes, τ_3, has been proposed by Tao [3] and Eldrup et al. [4] as follows;

$$\tau_3 = 0.5\left\{1 - \frac{R}{R_o} + \frac{1}{2\pi}\sin\left(\frac{2\pi R}{R_o}\right)\right\}^{-1} \quad \text{(ns)}, \tag{1}$$

where $R_o = R + \Delta R$ and $\Delta R = 0.166$ nm is the thickness of the homogenous electron layer in which the positron in o-Ps annihilates [5]. The free volume size $V_{FV,Ps}$ in nm^3 is calculable as;

$$V_{FV,Ps} = 4\pi R^3/3. \tag{2}$$

The free volume hole size obtained by Equations (1) and (2) has been successfully applied to explain diffusion and permeation of various gases in polymers. Muramatsu *et al.* [6] showed that $V_{FV,Ps}$ can satisfactory describe the variation of O_2 permeability in both dry and hydrated ethylene vinyl alcohol copolymer. To date, no data regarding free volume and gas permeation is available, so the aim of this work is to study the free volume in dry per-fluorinated sulfonic acid/PTFE copolymer Fumapem® membranes by the PAL technique and relate it to O_2 and H_2 permeabilities. Further we discuss the mechanism of gas permeation based on the correlations between the permeabilities and $V_{FV,Ps}$.

2. Experimental

Pperfluorinated sulfonic acid/PTFE copolymers Fumapem® F-14100, Fumapem® F-1050 and Fumapem® F-950 with ion exchange capacity (IEC) = 0.71, 0.95 and 1.05 meq/g, respectively, were purchased from FuMA-Tech, St. Ingbert, Germany. The chemical structure of the membrane is shown in Figure 1.

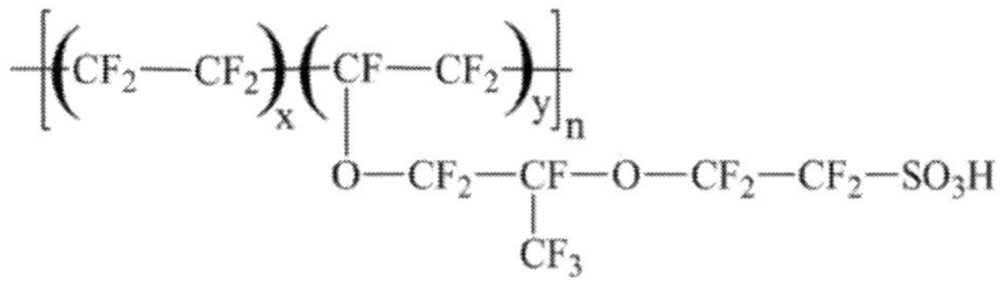

Figure 1. Chemical structure of Fumapem® F-950, Fumapem® F-1050 and Fumapem® F-14100.

The positron annihilation lifetime (PAL) measurements were performed using a conventional fast-fast coincidence system with a time resolution of 280 ps. Positron lifetime spectra were recorded at different temperatures up to 140 °C with 10 degree steps. Before the measurements, the samples were dried by evacuation for one day at room temperature (~25 °C). The source correction was about 9% and contributed to the short lifetime components. Since we were interested in the long-lived component, this absorption was not separated in the analysis of the lifetime spectra. The PAL spectra, having about one million total counts, were analyzed into three lifetime components using the PALSfit program [7] without constraints or source correction.

Gas permeabilities were measured with a GTR-Tech 30XFST apparatus equipped with a Yanaco G2700T gas chromatograph by monitoring the amounts of O_2 and H_2 that permeated across the membrane from one side to the other. Temperature dependence of the permeation coefficients P of O_2 and H_2 was also measured from 40 to 120 °C. Ar and He were used as carrier gases of H_2 and O_2, respectively.

3. Results and Discussion

Figure 2 shows the temperature dependence of the *o*-Ps lifetime τ_3 and its relative intensity I_3 for dry Fumapem® F-14100, F-1050 and F-950. The right hand ordinate of the left figure represents the free volume hole size quantified by Eqs. (1) and (2). It is seen that the free volume of all the samples is considerably enlarged by thermal expansion. Heating scans show an abrupt change of the thermal expansion coefficient of the free volume 75 - 110 °C which may be attributed to α-transition caused by the glass transition of Fumapem®. This transition is associated with the onset of long-range mobility of both the main- and side-chains, which is facilitated by a profound weakening of the electrostatic interactions within the ionic aggregates. It is interesting to note that Fumapem® F-950 with higher IEC and little larger free volume than the other samples. The larger slope above the α-transition temperature for all the samples suggests that the backbone segmental motions significantly contribute to the free volume at high temperatures. The *o*-Ps intensity I_3 for all the samples varies only slightly with temperature, indicating that Ps formation does not significantly depend on temperature. It is clear that sample with low IEC (Fumapem® F-14100) has high intensity compared with that in high IEC (Fumapem® F-950 and/or Fumapem® F-1050). A chemical structure or compound that captures an electron-positron pair, which would otherwise recombine with each other to Ps, can inhibit the formation of Ps.

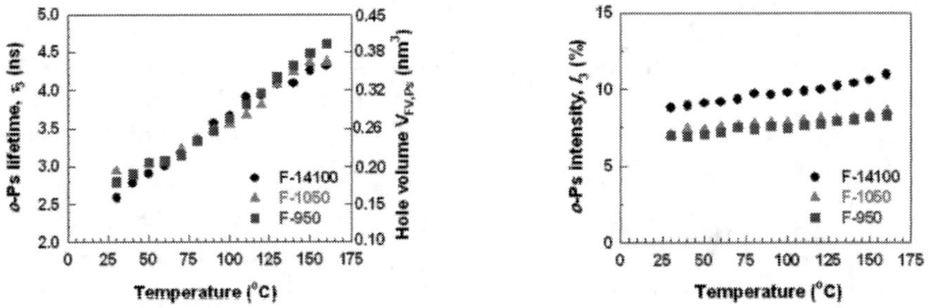

Figure 2. Lifetime τ_3 (left) and its intensity I_3 (right) of *o*-Ps as a function of temperature for dried Fumapem® F-950, Fumapem® F-1050 and Fumapem® F-14100. The scale of the right ordinate in left figure is the size of free volume hole $V_{FV,Ps}$ deduced from Eqs. (1) and (2).

Figures 3 and 4 show the temperature dependence of H_2 and O_2 permeabilities of dried Fumapem® F-950, Fumapem® F-1050 and Fumapem® F-14100. It is noticed that both H_2 and O_2 permeablities increase by orders of magnitudes with increasing temperature and also they increase with increasing IEC at constant temperature. For example, at 80 °C the O_2 permeability is 1.44×10^{-13} and 3.57×10^{-13} cm^3 cm cm^{-2} s^{-1} Pa^{-1} in the Fumapem® F-14100 and Fumapem® F-950, respectively. At all the temperatures studied, H_2 is much more permeable than O_2. All the samples show similar tendency in H_2 and O_2 permeations with temperature. The IEC dependence of mass transport parameters can be simply interpreted on the basis of the water content of the fully hydrated membranes. The microstructure of perfluorosulfonate membranes such as Fumapem® is comprised of an amorphous hydrophilic phase containing the hydrated sulfonic acid sites, and partly crystalline hydrophobic Teflon backbone regions. Both of these hydrophobic and hydrophilic phases have been described in prior research similar to Nafion® as having a strong impact on the gases transport process [8]: the aqueous domain (hydrophilic phase) is predominately involved in the O_2 diffusion pathway, hence more water uptake enables a higher diffusion coefficient of oxygen in the membrane.

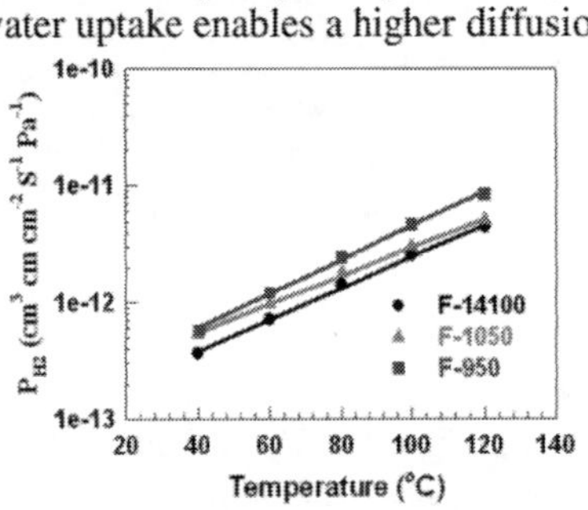

Figure 3. H_2 permeability as a function of temperature for dried Fumapem® F-950, Fumapem® F-1050 and Fumapem® F-14100. The straight lines are drawn just for guiding the eyes.

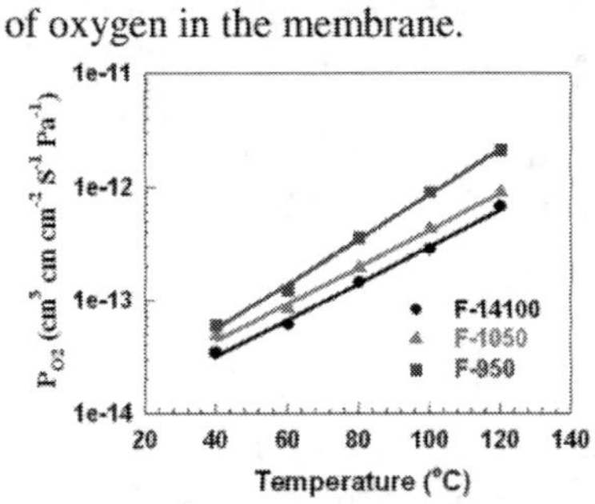

Figure 4. O_2 permeability as a function of temperature for dried Fumapem® F-950, Fumapem® F-1050 and Fumapem® F-14100. The straight lines are drawn just for guiding the eyes.

To correlate the hole volume $V_{FV,Ps}$ to gas permeability P, the following equation was used [9];

$$P = CT \exp(-b/V_{FV,Ps}), \tag{3}$$

where b is a constant depending on the size of the permeant molecule, C is another constant depending on solubility and T is the absolute temperature. Figures 5 and 6 show the plots of log P/T versus $1/V_{FV,Ps}$ for the permeation of H_2 and O_2, respectively, in dried three Fumapem® samples at different temperatures. A plot of O_2 permeability for relatively flexible chain polymers at different temperatures in vacuum and at different relative humidities at room temperature [6] is also included in Figure 6. The considerable downward deviation of the correlations for Fumapem® may be attributed to the stiff backbones of the ionomer in comparison with the other polymers. Reasonably good correlations in line with Eq. (3) are observed indicating the important role played by the free volume in O_2 permeation as shown in Figures 4 and 5. The considerable downward deviation of the correlations form Fumapem® F-950 to Fumapem® F-14100 is attributed to the stiff backbones of the ionomer by decreasing IEC (*i.e.* the concentration of the sulfonic groups). The stiffness of the ionomer is enhanced by decreasing IEC.

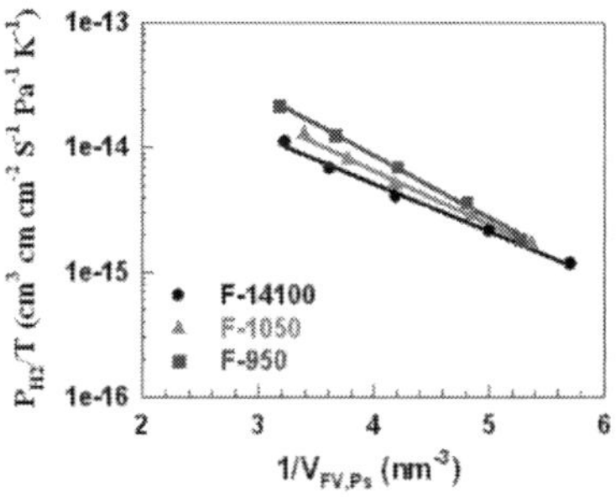

Figure 5. Correlation between the permeability of H_2 and reciprocal free volume hole size for dried Fumapem® F-950, Fumapem® F-1050 and Fumapem® F-14100 at different temperatures.

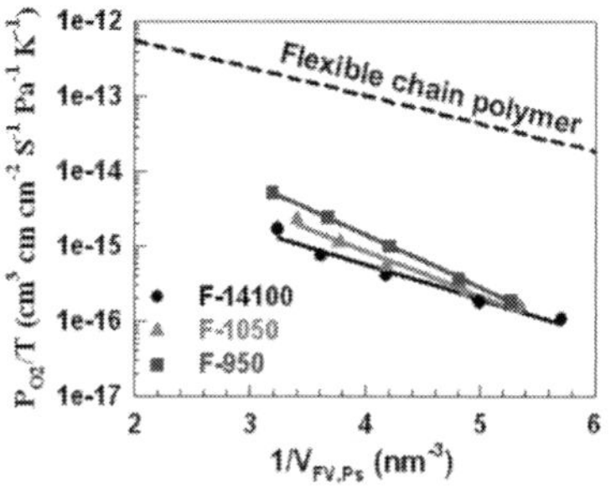

Figure 6. Correlation between the permeability of O_2 and reciprocal free volume hole size for dry ion-exchanged forms of Fumapem® F-950, Fumapem® F-1050 and Fumapem® F-14100 at different temperatures. The small dashed line shows the correlation for relatively flexible chain polymers [6].

4. Conclusion

For the development of a polymer electrolyte membrane with a supreme gas barrier property, molecular design optimizing not only free volume but also polymer stiffness is required.

Acknowledgments: One of the authors (A. Ohira) is grateful to the New Energy and Industrial Technology Development Organization (NEDO), Japan, for their financially support.

References

[1] Costamagna P and Srinivasan S (2001) *J. Power Sources* **102** 253.
[2] Mohamed H F M, Kobayashi Y, Kuroda C S and Ohira A (2010) J Phys Conference Series **225** 012038.
[3] Toa S T (1972) *J. Chem. Phys.* **56** 5499.
[4] Eldrup M, Lightbody D and Sherwood J N (1981) *Chem. Phys.* **63** 51.
[5] Nakanishi H, Wang S J and Jean Y C 1988 *Microscopic surface tension studied by positron annihilation*, in: Sharma S C (Ed.) *Positron annihilation studies of fluids* Singapore, World Scientific pp. 292.
[6] Muramatsu M, Okura M, Kuboyama K, Ougizawa T, Yamamoto T, Nishihara Y, Saito Y, Ito K, Hirata K and Kobayashi Y (2003) *Radiat. Phys. Chem.* **68** 561.
[7] Olsen J V, Kirkegaard P, Pedersen N J and Eldrup M 2007 *Phys. Stat. Sol. C* **4** 4004.
[8] Zhang L, Hampel C and Mukerjee S (2005) *J. Electrochem. Soc.* **152** A1208.
[9] Mohamed H F M, Kobayashi Y, Takimoto N, Takeoka Y and Ohira A (2008) *Polymer* **49** 3091.

Electron beam induced microstructural changes and electrical conductivity in Bakelite polymer RPC detector material: A positron lifetime study

K V Aneesh Kumar[1], S Ningaraju[1], L M Munirathnamma[1], H B Ravikumar[1] and C Ranganathaiah[1,2]
[1] Department of Studies in Physics, University of Mysore, Mysore-570006, India.
[2] Govt. Research Centre, Sahyadri College of Engineering & Management, Adyar, Mangalore – 575 007, India

E-mail: cr@physics.uni-mysore.ac.in:crang1@rediffmail.com

Abstract. In order to explore the structural modification induced electrical conductivity, samples of Bakelite RPC polymer detector materials were exposed to 8 MeV of electron beam with the irradiation dose from 20 kGy to100 kGy in steps of 20 kGy. The microstructural changes upon electron beam irradiation have been studied using Positron Annihilation Lifetime Spectroscopy (PALS) and Fourier Transform Infrared (FTIR) Spectroscopy. Positron lifetime parameters viz., o-Ps lifetime and its intensity show chain scission at lower doses (20 kGy, 40 kGy) followed by cross-linking beyond 40 kGy due to the radical reactions. The reduction in electrical conductivity of Bakelite material beyond 60 kGy is correlated to the conducting pathways and cross-links in the polymer matrix. The appropriate doses of electron beam irradiation of Bakelite material may reduce the leakage current and hence improves the performance of the detector.

1. Introduction

Resistive plate chambers (RPCs) [1] have found extensive applications in high energy and astrophysics experiments as the active detectors for muon detection. The RPCs made up of high resistive materials like Bakelite and glass are being used as the detectors for iron calorimeter (ICAL) to study atmospheric neutrinos in the proposed India based neutrino observatory (INO) project in India [2]. As the RPC detectors are continuously exposed to charged particles like cosmic ray muons, there will be some modification in the microstructure of RPC material in the long run. The vital problem faced in such experiments is that Bakelite RPCs exhibit undesirable high leakage current compared to the glass RPCs [3]. However, Bakelite has several advantages over glass detector material like low cost, easy to handle in the fabrication of RPCs and as such it is very important to understand the origin of this leakage current. This problem has not been rectified in the past and it is interesting to find whether the high leakage current owes its origin to the microstructural changes of the Bakelite material upon exposure to charged radiation. In recent years, irradiation is treated as an effective tool for structural modification of polymers and plays a significant role in the material modifications. In the present study, authors carried out experimental investigations on the effects of electron beam irradiation on the microstructure of the polymer based Bakelite RPC detector material by one of the well established techniques viz., Positron Annihilation Lifetime Spectroscopy (PALS) [4] and an attempt has been made to correlate the free volume change due to the electron beam irradiation on the electrical conductivity.

2. Experimental

2.1. Sample Preparation and Electron Beam Irradiation

Bakelite samples (P-120, Matt finished NEMA LI-1989, Grade XXX) of density 1.22 g/cm^3, manufactured by Bakelite Hylam, India were procured from VECC Kolkata (INO-Lab), India. The chemical structure of Bakelite is shown in figure 1. The pairs of rectangular samples having dimensions 1 cm x 1 cm x 0.175 cm were exposed to the electron beam of energy 8 MeV in air at Microtron Centre, Mangalore University, India, for different doses up to 100 kGy in the interval of 20 kGy. These samples were used for PALS, electrical conductivity and FTIR measurements.

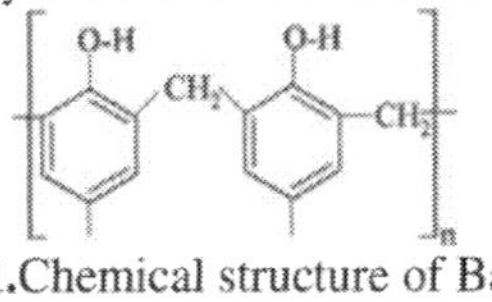

Figure 1. Chemical structure of Bakelite.

2.2. Positron Annihilation Lifetime Measurements

Positron annihilation lifetime spectra were recorded for the as received and electron beam irradiated Bakelite samples using positron lifetime spectrometer. The detailed description of PALS technique was given in our earlier paper [5]. The o-Ps lifetime (τ_3) is related to the free volume hole size by a simple relation given by Nakanishi *et al.* [6], which was developed on the basis of theoretical models originally proposed by Tao [7] for molecular liquids and later by Eldrup *et al.* [8].

2.3. FTIR Characterization

The Fourier Transform Infrared (FTIR) Spectra for the as received and the electron beam irradiated Bakelite samples at different doses were recorded using KBr pellet method in the range of 4000-400 cm^{-1} with JASCO-460 Plus, Japan FTIR Spectrometer with a resolution of 4 cm^{-1}.

2.4. Electrical Conductivity Measurements

The electrical conductivity of the as received and electron beam irradiated Bakelite samples at different doses were measured using a standard set up with Keithley 2636A model with Source meter and computer program Lab Tracer 2.0.

3. Results and Discussions

3.1. Positron Annihilation Lifetime Spectroscopy

PALS results reveal that the free volume size increases from 64.90 $Å^3$ to 69.46 $Å^3$ on electron beam irradiation up to 40 kGy and then decreases continuously to 58.93 $Å^3$ at 100 kGy (figure 2). This can be explained as follows; Positronium (Ps) formation takes place preferentially at the free volume cavities, which are the regions of low electron density exist mainly in the amorphous domains of Polymers. Upon irradiation, the scission of polymer chains will increase the size of the free volume cavities and hence the o-Ps lifetime (τ_3) [7]. Above 40 kGy, there is a remarkable decrease of o-Ps lifetime (τ_3) about 71 ps and free volume size (V_f) by about 6 $Å^3$ starting from 40 to 100 kGy irradiation dose in comparison to as received sample. This is attributed to the cross-linking of Bakelite polymer chains due to the radicals formed by the scission of polymeric chains in the initial stages of irradiation; this will hinder the polymer chain mobility [9]. On the other hand, the o-Ps intensity (I_3) decreases up to 60 kGy and a small increase at 80 kGy (only about 0.5%) followed by a decline up to 100 kGy dose as seen from figure 3.The decrease in I_3 may be due to inhibition of positronium (Ps) formation. The cleavage of bonds in Bakelite yield phenyl radicals, these radicals undergo reactions with the electrons of the spur created during slowing down of positrons, thereby inhibits the formation of positronium (Ps) [10]. Also, there is a possibility of the interaction of free radicals with o-Ps, this may cause ionization or oxidation. All such interactions may contribute to the inhibition of o-Ps formation. The variations of o-Ps intensity from 60 to 100 kGy is due to the structural modification of Bakelite polymer induced by the chain scission followed by the cross linking of the polymer chains. These arguments are adequately supported by FTIR results.

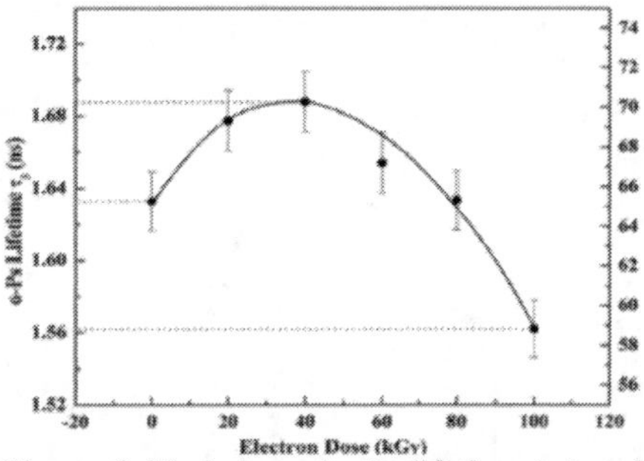

Figure 2. Variation of o-Ps lifetime (τ_3) and free volume size (V_f) as a function of electron dose.

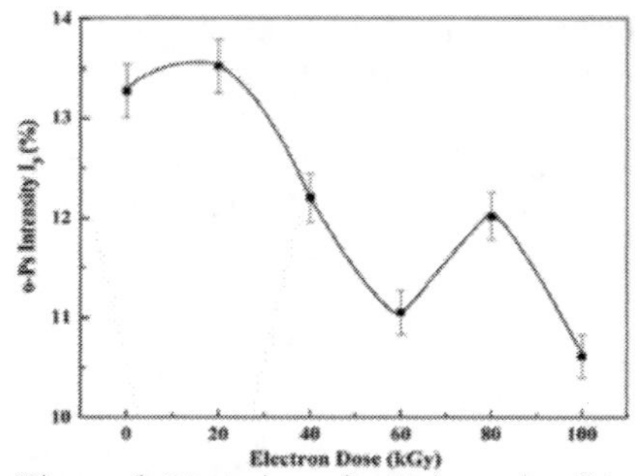

Figure 3. Variation of o-Ps intensity (I_3) as a function of electron dose.

3.2. FTIR Studies

The chemical modifications caused by electron beam irradiation on Bakelite are understood using FTIR scans shown in figure 4. The absorbance bands at wave numbers 1457, 1647, and 2923 cm^{-1} correspond to the functional groups viz., C-H aliphatic bridge, C=C aromatic ring and stretching vibrations of CH_2 alkane respectively [5, 11]. Since no changes in these wave numbers observed upon electron beam irradiation suggest that they do not influence structural modification of Bakelite. On the other hand, the absorption band at 3394 cm^{-1} corresponds to Phenol/OH in the as received sample gets shifted to 3420 cm^{-1} at 20 kGy and remains at this value for higher electron doses. As the radiation dose increases, the % of transmittance also increases which can be attributed to the chain scission resulting to the formation of more number of phenolic radicals. This may further leads to the formation of OH^- and H^+ ions [12, 13]. The decrease in % of transmittance of O-H group at higher electron doses suggests the cross linking of polymeric chains of Bakelite. This is possibly due to the formation of hydrogen bonds by the radicals released during the scission of phenolic groups [11, 14]. The absorbance band at 1053 cm^{-1} corresponding to single bond C-O stretching vibrations of $-CH_2OH-$ group shifted to 1036 cm^{-1} at 20 kGy and then increases to 1055 cm^{-1} and remains at 1055 cm^{-1} for higher dosages. These results indicate that the chemical changes due to the rearrangement of free radicals after the chain cleavage seems to contribute to the microstructural modifications in the Bakelite sample. These results agree well with the PALS results.

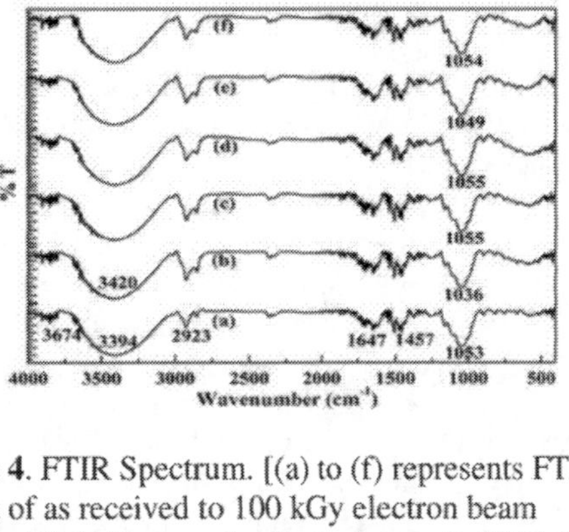

Figure 4. FTIR Spectrum. [(a) to (f) represents FTIR spectra of as received to 100 kGy electron beam irradiated Bakelite samples in the interval of 20 kGy.]

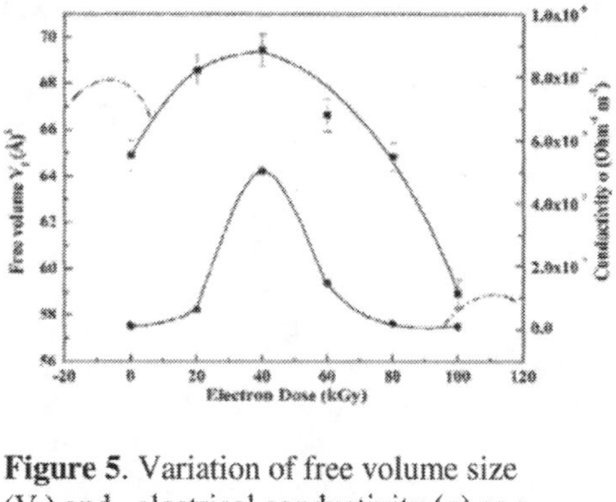

Figure 5. Variation of free volume size (V_f) and electrical conductivity (σ) as a function of electron dose.

3.3. Electrical Conductivity Results

Figure 5 shows the variation of free volume size and electrical conductivity as a function of electron dose and they exhibit behaviour upon electron beam irradiation that as the free volume increases, conductivity increases and vice-versa. The electrical conductivity (σ) of the as received sample was $1.23x10^{-8}$ $(\Omega m)^{-1}$. It was observed that the electrical conductivity increases on irradiation of electron beam dosage up to 40 kGy and then decreases to 1.08 x 10^{-8} $(\Omega m)^{-1}$ at 100 kGy. The increase in conductivity for the low electron doses may be due to the scission of hydrogen bonded phenolic groups lead to the formation of OH^{-} and H^{+} radicals [5, 15]. At higher doses (above 60 kGy) cross linking is the predominant process and the formation of cross-links of the network hinders the movement of chains inside the polymer. This indicate the formation of cross-links of Bakelite polymeric chains and reduces the formation of free radicals and reduce their mobility and hence reduction in electrical conductivity [16, 17]. This is also evident from the reduced free volume size (V_f) obtained from the PALS study.

4. Conclusions

PALS results revealed that chain scission and cross linking are the predominant processes under electron beam irradiation on polymer based Bakelite RPC detector material. FTIR spectroscopy indicates the scission of hydrogen bonded phenolic groups leading to the formation of OH^{-} and H^{+} radicals in the low dosage region. The electrical conductivity for 40 kGy electron irradiated sample is more compared to the as received sample indicative of increased mobility of a large number of OH^{-} and H^{+} ions produced from the cleavage of hydrogen bonded phenolic groups. The increased cross-links at higher electron dose may possibly reduce the leakage current although it depends on the surface properties of the materials as well. We propose that improvement in the performance of Bakelite RPC detector is possible with electron beam irradiation resulting in reduced leakage current.

Acknowledgement

The authors are grateful to DST, Govt of India for sanctioning the INO-DST Project to UOM, Mysore.

References

[1] Santonico R and Cardarelli R 1981 *Nucl. Instr. Meth.* **187** 377-380
[2] Biswas S, Bhattacharya S , Bose S, Chattopadhyay S, Saha S, Sharan M K and Viyogi Y P 2009 *Nucl. Instr. Meth. Phys. Res. Sect. A* **602** 749-753
[3] Carlson P, Crotty *et al.* 2001 Study and optimization of RPCs for high rate applications *IEEE Nuclear Sci. Symp. Medical imaging Conf.* (USA, 4-10 November 2001) pp 1-17
[4] Shaojin J, Zhicheng Z, Yangmeib F, Huimingb W, Xianfengb Z, Rongdiana H 2002 *Eur. Polym. J.* **38** 2433
[5] Aneeshkumar K V, Ravikumar H B and Ranganathaiah C 2013 *J. Appl. Polym. Sci.***130** 793-800
[6] Nakanishi H, Wang S J and Jean Y C 1988 Microscopic surface tension studied by positron annihilation, in: Sharma S C (Ed.) Positron annihilation in fluids Singapore, *World Scientific* pp 292- 93
[7] Tao S J 1972 *J.Chem. Phys.* **56** 5499-5510
[8] Eldrup M, Lightbody D and Sherwood 1981 *Chem. Phys.* **63** 51-58
[9] Ismayil, Ravindrachary V, Bhajantri R F, Praveena S D, Poojary B, Dutta D and Pujari P K 2010 *Polym. Degrad.Stab.* **95** 1083-91
[10] Shariff G, Sathyanarayana P M, Thimmegowda M C and Ranganathaiah C 2002 *Polymer.***43** 2819
[11] Lee Y, Kim D, Kim H J, Hwang T S and Rafailovich M 2003 *J. Appl. Polym. Sci.* **89** 2589-96
[12] Loo J S C, Ooi C P and Boey F Y 2005 *Biomaterials.***26** 1359-67
[13] Chen Y, Chen Z, Xiao S and Liu H 2008 *Thermochim. Acta.* **476** 39-43
[14] Roczniak K, Biernacka T and Skarzynski M 1983 *J. Appl. Polym. Sci.* **28** 531-542
[15] Va'vra J 2003 *Nuclear Symp. Conf. Record 2003 IEEE.* **5** 3704-08
[16] Siti A, Mohd N, Azizan A, Mohd Y A R and Ibrahim A T T 2010 J. *Nat. Sci.* **2** 190
[17] Liu H, Hu X B, Wang J Y and Boughton R I 2002 *Macromolecules.***35** 9414-19

Physical Selectivity of Molecularly Imprinted polymers evaluated through free volume size distributions derived from Positron Lifetime Spectroscopy

T. Pasang[1,2] and C Ranganathaiah[1,2]

[1]Govt. Research Centre, Sahyadri College of Engineering and Management, Adyar, Mangalore-575007, India

[2]Department of Studies in Physics, University of Mysore, Manasagangotri, Mysore-570006, India

Email: crang1@rediffmail.com

Abstract. The technique of imprinting molecules of various sizes in a stable structure of polymer matrix has derived multitudes of applications. Once the template molecule is extracted from the polymer matrix, it leaves behind a cavity which is physically (size and shape) and chemically (functional binding site) compatible to the particular template molecule. Positron Annihilation Lifetime Spectroscopy (PALS) is a well known technique to measure cavity sizes precisely in the nanoscale and is not being used in the field of MIPs effectively. This method is capable of measuring nanopores and hence suitable to understand the physical selectivity of the MIPs better. With this idea in mind, we have prepared molecular imprinted polymers (MIPs) with methacrylic acid (MAA) as monomer and EGDMA as cross linker in different molar ratio for three different size template molecules, viz. 4-Chlorophenol (4CP)(2.29 Å), 2-Nephthol (2NP) (3.36 Å) and Phenolphthalein (PP) (4.47Å). FTIR and the dye chemical reactions are used to confirm the complete extraction of the template molecules from the polymer matrix. The free volume size and its distribution have been derived from the measured o-Ps lifetime spectra. Based on the free volume distribution analysis, the percentage of functional cavities for the three template molecules are determined. Percentage of functional binding cavities for 4-CP molecules has been found out to be 70.2% and the rest are native cavities. Similarly for 2NP it is 81.5% and nearly 100% for PP. Therefore, PALS method proves to be very precise and accurate for determining the physical selectivity of MIPs.

1. Introduction

The Molecular Imprinted Polymers (MIPs) are man tailored polymers with specific recognition cavities complementary in shape, size and functionality to the target molecule [1]. Research in MIPs has emerged as an attractive field owing to their wide variety of applications like chemical and biosensors, solid-phase extraction (SPE), drug delivery systems etc.

Selectivity of an MIP is the prime character from the view point of its numerous applications particularly as sensors. However, in such designs there is a probability of non-selective binding sites in the presence of foreign molecule with similar functional group as that of the template molecule or with the similar size of the template. Therefore, to achieve high selectivity for a given molecule by the MIP, both functional group selectivity (chemical) and pore-size selectivity (physical) must be properly understood and estimated. With this view point, MIPs with MAA-co-EGDMA as a matrix polymer and three different molecules (viz 2-Nephthol, 2-naphthol and Phenolphthalein) as the templates have been synthesized by non covalent binding method and characterized with Positron lifetime spectroscopy with a special emphasis on physical selectivity of the MIPs.

2. Experiment

2.1. Materials

The chemicals for the synthesis of MIPs namely methacrylic acid (MAA), ethylene glycol dimethacrylate (EGDMA) and 4-Chlorophenol (4-CP) were procured from Sigma Aldrich, Bangalore, India. 2-Naphthol(2NP) and Phenolphthalein (PP) were procured from Loba Chemie, India. The radical initiator 2,2'-azobisisobutyronitrile (AIBN) and solvent acetonitrile (ACN) were obtained from Merck (Germany). All the chemicals were of lab grade and used as received.

2.2. Synthesis of MIP

MIPs were synthesized by employing thermal bulk polymerization [1]. The pre-polymerization complex comprises of the mixture of monomer MAA and the template taken in a 25ml thick walled glass tube. The mixture was dissolved in the solvent (ACN). After proper mixing of pre-polymerization complex, cross liker (EGDMA) of known molar ratio to monomer and radical initiator (AIBN) were added to the solution. The solution was kept in an ice bath for cooling and nitrogen gas was purged for removal of oxygen. Then the solution was placed in a water bath at a constant temperature of 60°C. The thermal polymerization was allowed to proceed for nearly 19 hours [1,2] to get hard monolith product. Later the template molecules were chemically extracted from this using a soxhlet extractor following the standard procedure. Then the hard sample was sliced to make positron measurement samples.

The control polymer without template molecule was prepared using same method and is labeled as NIP5 for monomer cross linker ratio 1:5. When the template is incorporated, it will be addressed as NIP5+template; on extraction of template from the polymer matrix, it is called MIP5 and finally when the template is reloaded to the polymer matrix, it is labeled as RMIP5.

2.3. Positron Lifetime Spectroscopy.

In polymers, the positron captures an electron with parallel spin to form *ortho*-Positronium (*o*-Ps) and localized in free volume or voids is the process of interst. *o*-Ps annihilation [3] process is a microprobe for free volume cavities since its lifetime and intensity are characteristics of free-volume size and their relative concentration in polymers. In the present work positron lifetime measurements were made using a standard fast–fast coincidence system with time resolution of 220 ps. All measurements were made at room temperature. More details of the experimental procedure and analysis can be found in the earlier works [2, 4].

The lifetime spectra so acquired were analyzed into three average lifetime components using PATFIT-88 [5], with proper source correction applied. The reproducibility of the PALS results was found to be satisfactory within the experimental error limits. To get the distribution of free volume cavity sizes from the same measured positron lifetime spectrum, we used another computer program CONTIN-PALS2 [6]. CONTIN - PALS2 generates annihilation rate probabilities that produce distribution peaks from which free volume cavity radius and size (volume) distributions can be generated.

2.4. FTIR

The Fourier transform infrared spectra were recorded using FTIR, JASCO-460 PLUS, Japan, with a resolution of 4 cm^{-1}.

3. Results and Discussion

The formation of co-polymerization of MAA-*co*-EGDMA have been verified using FTIR. FTIR scans also confirm the incorporation, extraction and rebinding of the template molecules [2,4]. Using the Bondi's method [7], the radius of the template molecules 4-CP, 2NP and PP were found to be 2.99, 3.36 and 4.47 Å respectively. The results obtained from both PATFIT-88 and CONTIN-PALS2 are found to be consistent. It is know that the free volume cavities (*o*-Ps lifetime) come with a distribution rather a single value and maximum probability corresponds to the peak of this distribution as can be seen in the graphs below.

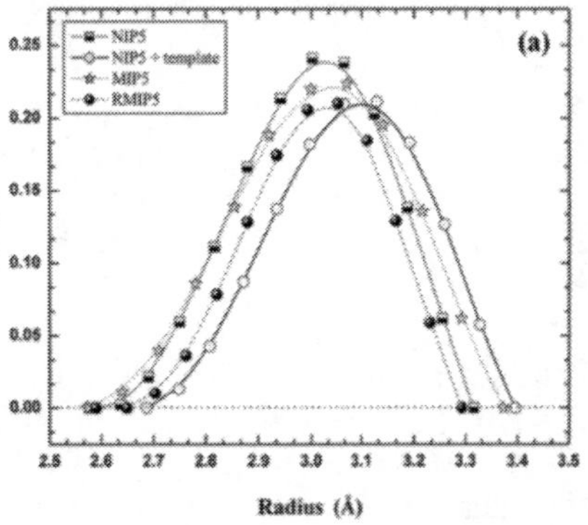

Figure 1: Pore Radius distribution curve of different stages in 4-CP imprinting polymer

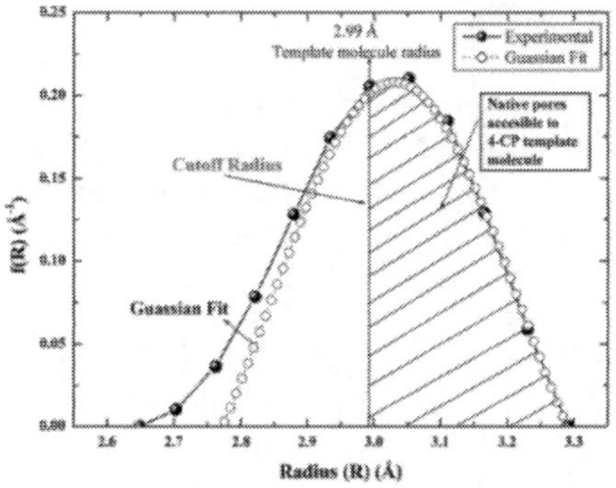

Figure 2: Free volume cavity radius distribution of MIP5 with cutoff radius line for 4-CP

As we can see from figure 1 for 4-CP imprinted polymer, pore size radius distribution in each stage of the imprinting gets slightly displaced. NIP5 will have only the pores which are due to chain folding and cross-linking of the polymer chains. When the monomer is polymerized in the presence of template molecules, the template molecules get into the polymer matrix thus affecting the microstructure of the polymer. These changes can be clearly seen in the distribution of radius distribution (*o*-Ps lifetimes). It is to be noted that some of the template molecules may get in to the preexisting pores big enough to accommodate them and the rest will be embedded to polymer matrix when the monomer is cross-linked around these templates. The pores present in the NIP5+template sample are taken as "native pores".

When the template molecules in the matrix are extracted chemically, it is expected that they leave cavities behind which are complementary to their size, shape and functionally. Thus, the MIP samples will have natives pore in addition to the cavities left behind by the template molecules. In the process of template rebinding, the templates will go in to the cavities created by the template and few to the native pores with similar size. Thus, the RMIP will have all the pores in MIP minus pores filled by template molecules [4].

Table 1: Percentage area accessible and non-accessible to the 4-CP template molecules in 1:5 ratio samples

Sample	Area under the radius distribution PDF curve (%)	Area under the curve above cut-off accessible to 4-CP template (%)	Area under the curve below the cut-off line non-accessible to 4-CP template (%)
NIP5	9.1	4.79	4.31
NIP5+template	8.12	4.43(100)	3.69
MIP5	9.45	5.28(119.2)	4.17
RMIP5	7.15	4.07(91.9)	3.08
Physical selectivity according to equation 1 is **70%**			

Now, with the understanding that the template molecules can be accommodated by a pore with radius greater or equal to its size, the free volume cavities/pores with respect to imprinting, extracted and rebinding, we use the following formulation to compute physical selectivity of the MIPs as.

$$\text{Physical selectivity} = \frac{X-Y}{X-Z} \times 100\% \qquad \ldots 1$$

Where, X=Area under the Gaussian curve above the cutoff radius in MIP5 ;Y= Area under the Gaussian curve above the cutoff radius in NIP5+template; Z= Area under the Gaussian curve above the cutoff radius in RMIP5 sample. The data in Table 1are computed by calculating the total area under free volume radius distribution curve and the area above the cut-off line (radius of template) for different samples.

By considering the available native pores in NIP5+template sample accessible to 4-CP as 100%, the available accessible cavities in MIP5 (after template extraction) has increased to 119.2% which is 19.2% higher. When cavities in MIP5 are rebound with template molecules (RMIP5 is produced) the available accessible pores decreases to 91.9%. This is less by 27.3 % as compared to MIP5. This is also less compared to NIP5 +template by 8.1%. From these we understand that the difference of 8.1% native pores which may have been occupied by the template molecules are non-intentional or we can label them as non functional binding sites for the template molecules. This is very interesting and found for the first time. Therefore, the present MIP5 is characterized to have (19.2/27.2)100=70.2 ~70% physical selectivity for 4-CP molecules and this is tabulated in Table 1.

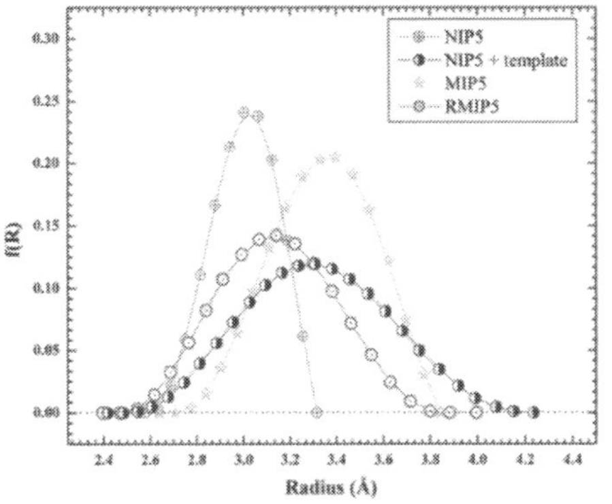

Figure 3: Pore Radius distribution curve of different stages in 2NP imprinting polymer

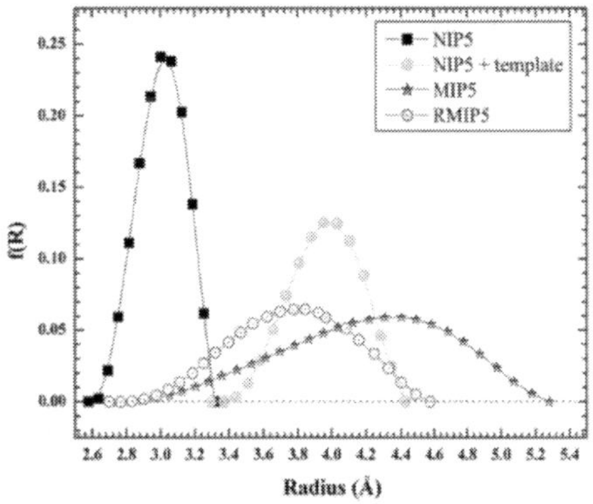

Figure 4: Pore Radius distribution curve of different stages in PP imprinting polymer

Similarly, we have evaluated the physical selectivity for 2NP and PP imprinted polymers. The radius distribution of different stages of 2NP and PP imprinted polymers are shown in figure 3 and figure 4 respectively. We found the physical selectivity for 2NP and PP imprinted polymers according the simple equation1 to be 81.5% and close to 100% respectively.

4. Conclusion

Based on the free volume PDF analysis and results presented here, we propose that not all pores in the MIP are accessible to template molecules when reloaded. This work portrait that PALS is a useful technique to evaluate the physical selectivity of MIPs. We also see from the results that the physical selectivity clearly depends on the size of template molecule used in imprinting. This method could be further tried for other MIPs with different base polymer for determining the selectivity.

References

[1] Alexandrer C, Adersson H S, Adersson L I, Ansell R J, Kirsch N, Nicholls I A, O'Mahony J and Whitechomb M J 2006 *J. Mol. Recognit.* **19** 106
[2] Pasang T and Ranganathaiah C, 2014 *Polym. Engg. Sci.* **54** 667
[3] Tao S J 1972 *J. Chem. Phys.* **56** 5499
[4] Pasang T and Ranganathaiah C 2013 *Ind. Eng. Chem. Res.* **52** 7445
[5] Kirkegaard P, Pederson N J and Eldrup M 1989 PATFIT-88: Riso National Lab. Rep., PM-2724, Denmark.
[6] Gregory R B 1993 CONTIN(PALS-2): Kent State University, USA
[7] Bondi A 1964 *J. Phy. Chem.*, **68(3)** 441

DBS investigation on films of cobalt chloride doped PVA-PVP blend

Preeti. B. Hammannavar[1], Basavarajeshwari .M. Baraker[1], R. F. Bhajantri[2], V Ravindrachary[3] and Blaise Lobo[1*]

[1]Department of Physics, Karnatak Science College, Karnatak University, Dharwad 580001, Karnataka, India
[2]Department of Physics, Karnatak University, Dharwad 580003, Karnataka, India
[3]Department of Physics, Mangalore University, Mangalore 574199, Karnataka, India

* E-mail: blaise.lobo@gmail.com

Abstract. Films of Cobalt Chloride ($CoCl_2$) doped polyvinylalcohol (PVA)-polyvinylpyrrolidone (PVP) blend (doped from 0.5 wt% up to 28 wt%) were prepared by solution casting, and characterized by XRD, DSC, UV-Visible Spectrometry TGA, FTIR and electrical measurements. In this paper, the results of Doppler Broadening Spectroscopy (DBS) in $CoCl_2$ doped PVA-PVP blend is discussed. An increase in crystallinity of PVA-PVP blend, is observed, on doping it with $CoCl_2$. The DBS results are complemented by XRD and DSC scans.

1. Introduction

Polyvinylalcohol (PVA) is a polymeric material which has gained interest of researchers due to its potential applications, the scope for easy modification and formation of miscible blends with many other polymers. PVA doped with different redox agents have been studied, and these doped polymeric materials show significant modification of micro-structural, optical and electrical properties [1-5]. The doping of a miscible blend of PVA with polyvinylpyrrolidone (PVP) involves inter-molecular and intra-molecular hydrogen bonds [6]. Cobalt chloride ($CoCl_2$) has a potential to significantly modify the dielectric and optical properties of a material [7]. The modifications in the electronic band-structure caused by incorporation of cobalt chloride ($CoCl_2$) as dopant in polyvinylalcohol – polyvinylpyrrolidone (PVA-PVP) blend has been studied by analyzing the experimental optical spectra and electrical behaviour of the material [8]. The information provided by thermal analysis and X-ray diffraction (XRD) scans yield complementary data, which helps in understanding the micro-structural nature of semi-crystalline polymeric materials, especially the changes caused by doping [9].

Doppler Broadening Spectroscopy (DBS) or Doppler Broadening of Annihilation Radiation (DBAR) is one of the efficient techniques of Positron Annihilation Spectroscopy (PAS), giving information about the electron momentum distribution in materials [10]. The DBS spectrum can be analysed using line-shape parameters, namely S- parameter and W- parameter. The S-parameter is a measure of positrons annihilating with low momentum electrons from the molecules in amorphous (disordered) regions of the polymeric material. The W-parameter is a measure of positrons annihilating with high momentum electrons from the crystalline (ordered) regions of the polymeric material. Hence, an increase in the degree of crystallinity and width of crystalline domains is expected to be reflected in a decrease in S- parameter, and an increase in the W- parameter.

Journal of Physics: Conference Series **618** (2015) 012034 doi:10.1088/1742-6596/618/1/012034

2. Experimental

The films of PVA-PVP-$CoCl_2$ blend were prepared by solution casting method. Films having different concentrations (doping levels from 1 wt% up to 28 wt%) of $CoCl_2$ were prepared. The method of preparation of these films is discussed elsewhere [8]. A Hitachi U 3310 UV-Visible spectrometer was used in the wavelength range 200-1000 nm for obtaining the Optical absorption (UV-Visible) spectra. The Urbach energy which is the width of band tails in the forbidden gap, is found to decrease significantly with increase in doping level of cobalt chloride in PVA-PVP blend films. It is 0.9 eV at 1 wt% doping level, 0.1 eV at 4.3 wt% doping level and becomes negligible at 28 wt% doping level, suggesting an increase in degree of structural order in $CoCl_2$ doped PVA-PVP blend, when concentration of $CoCl_2$ is increased. An arrangement for temperature variation (PID Controlled oven/ furnace) of polymer samples has been used with a temperature controller accurate to one degree Celsius, for the study of temperature variation of electrical conductivity, using two-probe method (SES, Roorkee). The powder XRD spectra were recorded using Rigaku X-ray Spectrometer, with Cu K_α x-rays of wavelength 1.5406 angstrom.

An 8 micro-curie sodium-22 source, deposited on a thin kapton foil, was used as source of positrons. This source was sandwiched on either side by $CoCl_2$ doped PVA-PVP films (stacked to 1 mm thickness, on either side). This assembly was kept at a distance of 10 cm from the window of the detector. A high resolution HPGe detector was used to record the energy spectrum of annihilation gamma rays from the sample. The energy resolution was measured to be 1.4 keV at 662 keV photo-peak of Cesium-137 source.

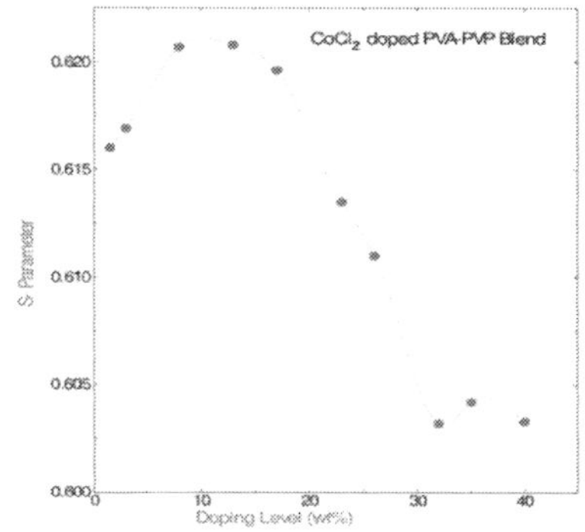

Figure 1. The variation of S- parameter w.r.t doping level of $CoCl_2$ in PVA-PVP blend.

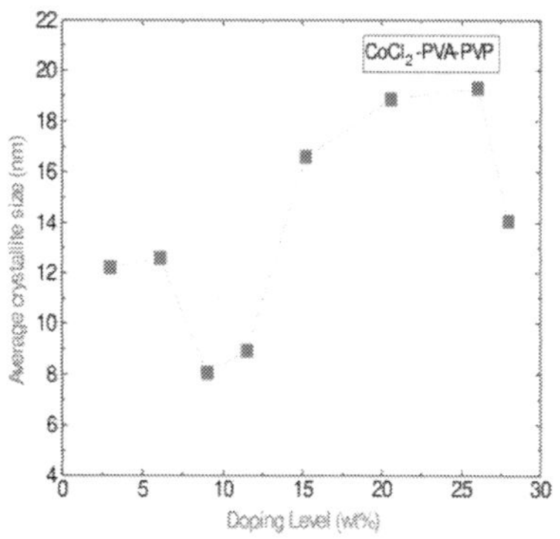

Figure 2. The variation of average crystallite size with doping level, for $CoCl_2$ doped PVA-PVP blend.

3. Results and discussion

It is noted that the computed S- parameter and W –parameter reflect changes in the degree of crystallinity and average crystallite size of $CoCl_2$ doped PVA-PVP blend samples. The Variation of S-parameter of DBAR spectrum of $CoCl_2$ doped PVA-PVP blend w.r.t doping level is shown in Fig. 1. The variation of W-parameter with doping level is complementary to that of the S- parameter. Fig. 2 shows the variation of average crystallite size (determined from XRD scans) with doping level. There

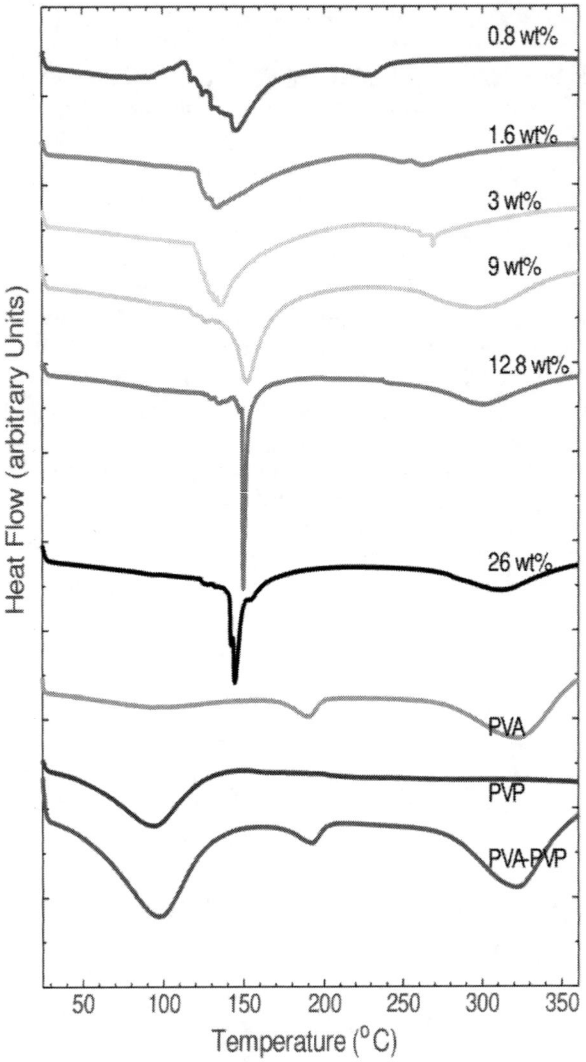

Figure 3. DSC curves of PVA, PVP, PVA-PVP blend, and different dopant levels (in wt%) of CoCl2 in PVA-PVP blend.

is complementary evidence from XRD for increase in average size of crystallites. The $CoCl_2$ doped PVA-PVP blend films were semi-crystalline in nature. The variation of average crystallite size versus doping level is shown in Fig. 2. The average crystallite size increased from 7.8 nm at 9 wt% up to 19 nm at 26 wt%, indicating an increase in structural order caused by doping PVA-PVP blend with $CoCl_2$. This is supported by the variation of S-parameter with doping level, shown in Fig. 1 and also the decrease in Urbach energy derived from the UV-Visible spectra (described in the experimental section). Note that the S-parameter, on low levels of doping, first increases and then saturates, dropping significantly on further increase in doping concentration. The S-parameter, being a measure of annihilation of positrons (or positrons bound in Ps) with low momentum electrons, shows an increase if there is an increase in amorphous nature of the sample. An increase in crystallite size will result in a drop in the S- parameter.

The DSC curves of PVA, PVP and PVA-PVP blend in Fig. 3, confirms the formation of miscible blend, with a single glass transition temperature. On doping with $CoCl_2$, at low dopant concentrations, a number of just resolved endotherms are observed at lower melting temperatures, indicating the formation of crystalline lamellae of different thicknesses.

As the doping concentration increases, the area under the melting endotherm increases, indicating an increase in the degree of crystallinity. This is in agreement with the XRD and DBS results.

This is supported by the presence of an absorption band at 1142cm^{-1} (crystallinity indicator), in the FTIR scans of all $CoCl_2$ doped PVA-PVP blend samples. In addition, from DSC scans, it is observed that the temperature of onset of melting increases, and the melting endotherm becomes sharper on increased dopant concentration, indicating formation of larger crystalline domains.

4. Conclusion

Significant structural changes are caused by incorporation of cobalt chloride ($CoCl_2$) as dopant in polyvinylalcohol –polyvinylpyrrolidone (PVA-PVP) blend. The XRD and DSC scans reveal an increase in average crystallite size and the degree of crystallinity of the blend. The line-shape parameters of DBS complement this observation.

Acknowledgements: The facilities at the University Science Instruments Centre, Karnatak University, Dharwad was utilized for DBS, UV-Vis, FTIR and DSC study. The XRD facility was utilized at the Department of Physics, Gulbarga University, Gulbarga. A part of this work was supported by UGC (Government of India)Major Research Project F.38-103/2009(SR).

References

[1] Lobo B, Ranganath M R, Ravi Chandran T S G, Rao G V, Ravindrachary V and Gopal S 1999 *Phys. Rev. B* **59** 13693.

[2] Bhajantri R F, Ravindrachary V, Harisha A, Ismayil and Ranganathaiah C 2009 *Physica Status Solidi C* **6** 2429.

[3] Ravindrachary V, Sreepad H R, Chandrashekara A, Ranganathaiah C and Gopal S 1992 *Phys. Rev. B* **46** 11471.
Tawansi A, El-Khodary A and Abdelnaby M M 2005 *Current Applied Physics* **5** 572.

[4] Bhajantri R F, Ravindrachary V, Harisha A, Crasta V, Nayak S P and Poojary B 2006 *Polymer* **47** 3591.
Hammannavar P B, Hurkadli M Y, Hundekar C B, Ravindrachary V and Lobo B 2014 *Int. J. ChemTech Res. (ISSN : 0974-4290)* **6** 1867.

[5] Bhajantri R F, Ravindrachary V, Poojary B, Ismayil, Harisha A and Crasta V 2009 *Polym. Eng. Sci.* **49** 903.

[6] Umadevi C, Mohan K R, Achari B S, Sharma A K and Rao V V R N 2010 *Ionics* **16** 751.
Sudheesh P, Sharafudeen K N, Vijayakumar S and Chandrasekharan K 2011 *Journal of Optics* **40** 193.

[7] El-Shahawy M A 1993 *Polymer Bulletin* **31** 199.
Patil R V, Ranganath M R and Lobo B 2014 *Int. J. ChemTech Res. (ISSN : 0974-4290)* **6** 1852.

[8] Patil R V, Ranganath M R and Lobo B 2014 *AIP Conf. Proc.* **1591**, 183.

[9] Somashekarappa H, Prakash Y, Mahadevaiah, Hemalatha K and Somashekar R 2013 *AIP Conf. Proc.* 1512 532.
Murthy N S and Minor H 1990 *Polymer* **31** 996.

[10] Sato K, Ito K, Hirata K, Yu R S and Kobayashi Y 2005 *Phys. Rev. B* **71** 012201.
Hunt A W, Cassidy D B, Sterne P A, Cowan T E, Howell R H, Lynn K G and Golovchenko J A 2001 *Phys. Rev. Lett.* **86** 5612.
Szpala S, Kumar A, Nielsen B, Peng J P, Hayakawa S, Lynn K G and Gossmann H J 1996 *Phys. Rev. B* **54** 4722.

Journal of Physics: Conference Series **618** (2015) 012035 doi:10.1088/1742-6596/618/1/012035

Free volume study on the miscibility of PEEK/PEI blend using positron annihilation and dynamic mechanical thermal analysis

R Ramani* and S Alam

[1] Polymer Science Division, DMSRDE, G.T. Road, Kanpur – 208 013, India

*E-mail: ramanir70@rediffmail.com

Abstract. High performance polymer blend of poly(ether ether ketone) (PEEK) and poly(ether imide) (PEI) was examined for their free volume behaviour using positron annihilation lifetime spectroscopy and dynamic mechanical thermal analysis methods. The fractional free volume obtained from PALS shows a negative deviation from linear additivity rule implying good miscibility between PEEK and PEI. The dynamic modulus and loss tangent were obtained for the blends at three different frequencies 1, 10 and 100 Hz at temperatures close to and above their glass transition temperature. Applying Time-Temperature-Superposition (TTS) principle to the DMTA results, master curves were obtained at a reference temperature T_o and the WLF coefficients c^0_1 and c^0_2 were evaluated. Both the methods give similar results for the dependence of fractional free volume on PEI content in this blend. The results reveal that free volume plays an important role in determining the visco-elastic properties in miscible polymer blends.

1. Introduction

The polymers poly(ether ether ketone) (PEEK) is a high performance engineering semi-crystalline thermoplastic having excellent chemical resistance and superior mechanical properties. But, the glass transition temperature (T_g) of this polymer is low (ca. 152 °C) and hence its modulus decreases beyond this temperature. The poly(ether imide) (PEI) is an amorphous polymer with comparatively high T_g of ca. 215 °C. Blending with PEI improves the T_g of PEEK and thus PEEK/PEI blend turns out to be a suitable matrix resin in the preparation of high performance composites for aerospace applications. Now, it is known that PEEK and PEI are molecularly miscible in the amorphous phase [1].

The experimental determination of free volume in polymers is of great importance both from the theoretical point of view and for practical applications. The free volume of a polymer influences its molecular motion and physical properties. This in turn will have an influence on their mechanical and visco-elastic properties. The positron annihilation lifetime spectroscopy (PALS) is used for characterizing the free volumes in polymers for decades [2] and later, it is also been used to study the polymer blends [3,4]. Recently, we have studied the miscibility of PEEK and PEI in their blend by studying their free volume behaviour using PALS and identified that when the PEI content in the blend is ≥ 50%, there is a high degree of miscibility [4].

The linear visco-elastic properties of the polymers are both temperature and time (frequency) dependant. At ambient temperature, the relaxation process in a polymer occurs over considerably longer times than at higher temperatures. The Williams-Landel-Ferry (WLF) relation is one of the most popular methods based on time-temperature-superposition (TTS) principle to evaluate the fractional free volume in polymers and is applicable in the temperature range $T_g < T < T_g+100$ °C. The WLF relation is given as

$$\mathrm{Log}_{10}\, a_T = -\, c^0_1\, (T-T_o) \,/\, (c^0_2 + T-T_o)$$

where a_T is the shift factor along the frequency (time) axis and T_o is the reference temperature. The dynamic mechanical thermal analysis (DMTA) is a powerful technique to study the visco-elastic behaviour of the sample. Upon performing DMTA measurements in multi-frequency mode as a function of temperature and applying TTS, a master curve can be obtained at T_o. Using the shift factor derived from the master curve, it is possible to evaluate the WLF constants, which in turn is related to the fractional free volume [5].

In this work, the free volume behaviour of the PEEK/PEI blend for different compositions has been investigated by both PALS and DMTA. When the fractional free volume obtained from PALS results is compared with that of the WLF coefficient obtained from DMTA results, a good correlation is found.

2. Experimental

2.1. Blend preparation

Granular PEEK-grade Victrex 450G having molecular weight of 40,000 g/mol, and a polydispersity of 2.8 was purchased from Victrex, U.K. The PEI-grade Ultem 1000 with molecular weight of 30,000 g/mol and a polydispersity of 2.5 was obtained from General Electric Plastics, Europe. Blends with weight ratios of PEEK/PEI 100/0, 90/10, 80/20, 70/30, 50/50, 30/70 and 0/100 were prepared by melt mixing [4]. These blends are designated respectively as PP_0, PP_{10}, PP_{20}, PP_{30}, PP_{50}, PP_{70} and PP_{100} (where the subscripts represent the wt% of PEI).

2.2. Characterization

The PALS measurements were performed at room temperature using a conventional fast-fast coincidence system having a time resolution of about 200 ps. The lifetime spectra were resolved in to three lifetime components using PATFIT program. The DMTA measurements were carried out using a GABO EPLEXOR 150N using rectangular samples (10 x 45 x 3 mm^3) in a three point bending geometry. The storage modulus (E′) and the mechanical loss (tan δ) evolution with temperature were registered. To obtain master curves, fresh set of DMTA measurements were conducted between 1 and 100 Hz in isothermal conditions in the glass transition region of the samples. From the master curves, the horizontal shift factor (a_T) was found by shifting the values obtained along the frequency (time) scale using the EPLEXOR-8 software.

3. Results and Discussion

3.1. PALS results

The positron annihilation lifetime results for PEEK/PEI blend series reveal three lifetime components and the variation of the long-lived lifetime component that corresponds to *o*-Ps pick-off lifetime (τ_3) was used to obtain the free volume hole size using the Tao-Eldrup relation. The average free volume size was calculated using the formula $V_{f3} = (4/3)\pi R^3$. The relative fractional free volume (F_{vr}) is then found as $F_{vr} = V_{f3} * I_3$, where I_3 being the *o*-Ps intensity.

The average free volume size remains almost constant (ca. 70 Å^3) when the PEI content is ≤ 10% (Fig. 1A). It increases to a value of ca. 78 Å^3 when the PEI content is 30%, and with further addition of PEI, it gradually reverts back close to its original value (ca. 71 Å^3). Generally, it is expected that in a miscible blend, the free volume to decrease with the increase in composition of the other component. At the initial stages of PEI addition may be due to stearic hindrance, the free volume shows a little increase. However, an increase in free volume in a miscible blend with the change in composition of

Journal of Physics: Conference Series **618** (2015) 012035 doi:10.1088/1742-6596/618/1/012035

its other partner is not something new. The free volume of miscible blends may be additive or decrease and is influenced more by segmental conformation and packing than by specific interactions [6]. The decrease in free volume when the PEI content is ≥ 50% could be due to close packing of PEEK and PEI chains contributing to greater extent of miscibility [4].

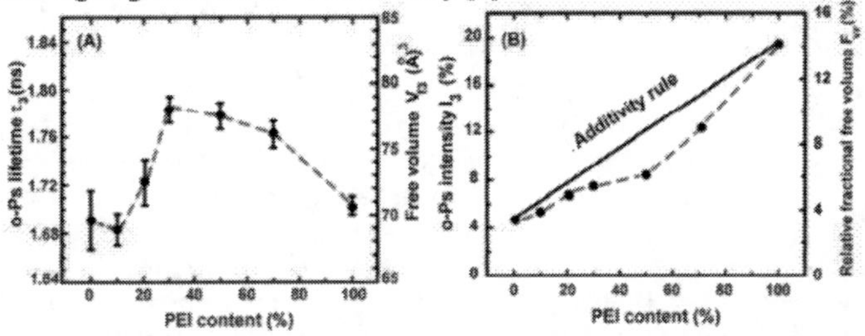

Fig.1. Variation of (A) *o*-Ps lifetime (τ_3) and average free volume size (V_{f3}) and (B) *o*-Ps intensity (I_3) and the relative fractional free volume (F_{vr}) as a function of PEI content. The dotted lines are guide to eyes.

Although the change in *o*-Ps lifetime (τ_3) is not that significant, the *o*-Ps intensity (I_3) shows a prominent change with PEI content (Fig. 1B). For pure PEEK, I_3 is just ca. 4.7% but with the increase in PEI content, the I_3 value also increases gradually and reaches a value of ca. 20% for pure PEI. The I_3 together with F_{vr} shows a negative deviation from linearity indicating that the blend is miscible at the molecular level and the deviation is more when the PEI content in the blend is ≥ 50%. The large negative deviation of F_{vr} when the PEI content is ≥ 50% also reveals that the PEEK/PEI blend achieves a high degree of miscibility at higher PEI content.

3.2. DMTA results

The DMTA measurements are used to evaluate the stiffness of the polymer and molecular segmental mobility under dynamic conditions. The storage modulus (E') of PEEK is higher than that of PEI. This could be due to the semi-crystalline nature of PEEK while PEI is completely amorphous. The study on mechanical deformation as a function of temperature also helps to find the glass transition temperature (T_g). The T_g is generally identified from the temperature at which the E' value drops significantly or at which the tan-δ is maximum. In the present case, all the blends showed a single transition in terms of drop in E' value as well as a single tan-δ peak that shifts to high temperatures with increase in PEI content. This infers that the blend possess a single T_g and is completely miscible in all the proportions studied, in agreement with earlier findings [1].

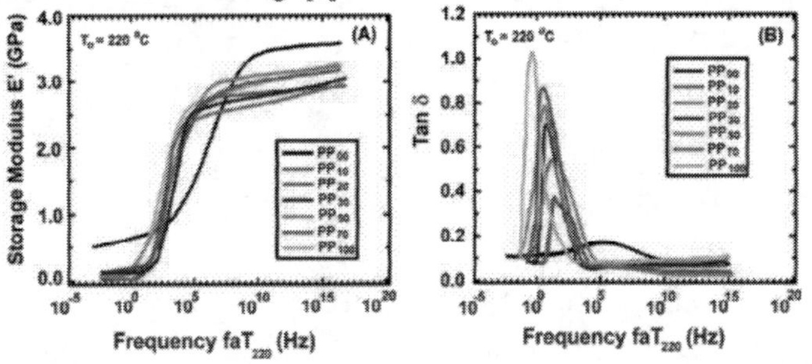

Fig.2. The master curves for (A) storage modulus (E') and (B) Tan-δ for various PEI content in the blend.

The glass transition events in these blends were further explored by the construction of master curves applying TTS using separate isothermal measurements performed at 1, 10 and 100 Hz in the glass transition region. The generated master curves of E' and tan-δ at T_o = 220 °C (T_o is selected close to or slightly above T_g for the entire blend compositions studied) are shown in Fig. 2(A) and (B). The resulting master curve makes the estimate of E' and tan-δ possible for a wide frequency range. The data superpose very well and the uniformity of the curves indicates that the TTS principle is valid for all the blend samples studied. Both the master curves of E' and tan-δ show a regular shift of T_g to lower frequencies with increase in PEI content. Such a shift of the master curve to low frequencies reveals the presence of long relaxation times due to increased T_g [7].

Using the values of the shift factor (a_T) deduced from E' and tan-δ master curves, a plot of ($T-T_o$)/log a_T against ($T-T_o$) is made for each of the blends that showed a linear behavior which confirms that the viscoelastic properties in this blend follow the TTS principle. Using these linear plots, the WLF coefficient c^0_1 has been obtained from the reciprocal of the slope and the coefficient c^0_2 from the intercept.

$c^0_1 = B/2.303\, f$; $c^0_2 = f/\alpha_f$; and $\alpha_f = B/2.303\, c^0_1\, c^0_2$

where α_f is the thermal expansion of the free volume (V_f) that is related to the total volume (V) as $f = V_f / V$, where 'f' is the fractional free volume.

3.3. *Correlation between PALS and DMTA results*

The plot of F_{vr} of PALS results against the fractional free volume of DMTA results, a good correlation is obtained (see Fig. 3). So, PALS measurements can be used to obtain information on dynamic processes in miscible polymer blends.

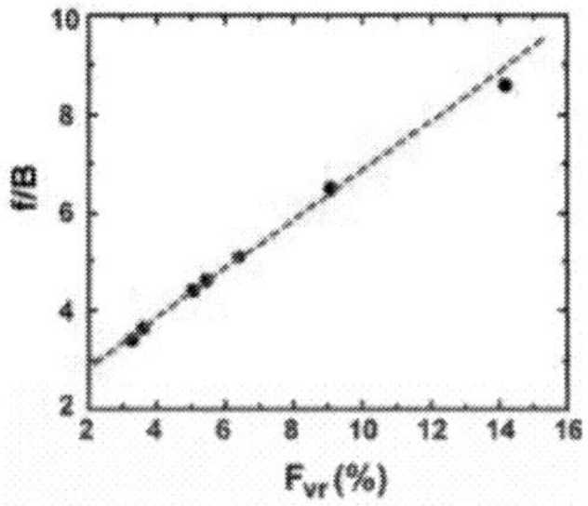

Fig.3. Correlation between relative fractional free volume (F_{vr}) from PALS and *f/B* from DMTA results.

4. Conclusions

The results reveal that PEEK/PEI blend is miscible in all the proportions studied. A good correlation is obtained between the F_{vr} found from PALS and the fractional free volume derived from DMTA results. This reveals that the two techniques provide essentially the same information on free volume in this high performance polymer blend.

Acknowledgements: The authors thank Dr. A.K. Saxena, Director, DMSRDE, Kanpur for his encouragement and support and for permitting to present this work. The authors also thank Prof. C. Ranganathaiah, Department of Physics, University of Mysore, Mysore, India for extending the positron lifetime facility for this work.

References

[1] Crevecoeur G and Groeninckx G 1991 *Macromolocules* **24** 1190.
[2] Jean Y C 1990 *Microchem. J.* **42** 72.
[3] Wästlund C and Maurer F H J 1997 *Macromolecules* **30** 5870.
[4] Ramani R and Alam S 2012 *J. Appl. Polym. Sci.* **125** 3200.
[5] Marzocca A J Cerveny S Salgueiro W Somoza A and Gonzalez L 2002 *Phys. Rev. E.* **65** 021801
[6] Hill A J Zipper M D Tant M R Stack G M Tordan T C and Shultz A R 1996 *J. Phys. Cond. Matter.* **8** 381.
[7] Ramani R and Alam S 2014 *High Performance Polymers* **26** 347.

Gas transport and free volume study in polyethylene based epoxy membranes

Pushkar N Patil[1,2*], R Checchetto[1], R Ferragut[3], S Aghion[3], A Miotello[1] and R S Brusa[1]

[1]Department of Physics, University of Trento, via Sommarive 14, I-38123, Povo, Italy
[2]Institute for Polymers, Composites and Biomaterials, National Research Council, Piazzale Tecchio 80, I-80125, Napoli, Italy
[3]Department of Physics and L-NESS, Politecnico di Milano, via Anzani 42, 22100 Como, Italy

E-mail: pushkhem6@gmail.com

Abstract. The mechanism of gas diffusivity in amine modified epoxy membranes is studied in the frame of free volume properties. Epoxy membranes with two different crosslinking densities and composite with Graphene nanoplatelets were prepared by solvent casting method. The free volume parameters measured by Positron Annihilation Spectroscopy (PAS) show inverse correlation with crosslinking density of the samples. The gas permeability and diffusivity for CO_2 have been studied by gas permeation measurements. The study reveals that the gas permeability and diffusivity change significantly as a function of crosslinking density, free volumes and structural relaxations of the molecular chains.

1. Introduction

Polymeric membranes are essential in applications where gas separation is required such as in the carbon dioxide capture and removal from natural gas mixtures [1]. Polyethylene based epoxy membranes are of interest due to their excellent mechanical strength and high thermal stability [2], ease processing and good compatibility with inorganic fillers [3,4] as well as for easy alteration of the microstructure by changing the crosslinking density. The effect of microstructural changes of the polymeric membranes is of fundamental importance to understand the gas transport mechanisms and the gas separation properties through the membranes. Aliphatic epoxy resins with low viscosity are of interest as gas separation membranes due to their ability to produce flexible polymeric networks with high thermal fluctuations [5]. The microstructure of epoxy networks can be tuned by changing the crosslinking density and the flexibility of the backbone chains [2]. In this paper, epoxy membranes with two different crosslinking densities as well as a composite membrane with Graphene nanoplatelets (GNP), have been prepared with the aim to study the influence of nanolevel molecular packing on the CO_2 gas permeability.

2. Experimental

The aliphatic epoxy resin, neo-pentyl glycol diglycidyl ether (NPGDE) with an epoxy equivalent weight of 145±5 g eq^{-1} was purchased from Sigma-Aldrich *S.r.l.* (Italy). The polyether diamines, namely Jeffamine D-230 (M. W. = 230 g; amine hydrogen equivalent weight = 59 g $eq.^{-1}$), and

Jeffamine ED-900 (M.W. = 900 g; amine hydrogen equivalent weight = 220 g eq.$^{-1}$) were procured from Huntsman Chemical Co and used as curing agents.

The membranes were prepared by mixing 1:1 stoichiometric ratio of NPGDE epoxy and polyether diamines and hand-mixed vigorously for 10 minutes to ensure uniform mixing. For GNP composite sample (NPGD-900-5), 5 weight % of GNP platelets was added using mechanical stirrer at 100 rpm for 30 min, followed by sonication for 20 min to get fine dispersion in epoxy matrix. The mixture was then degassed for 15 min under vacuum to remove air bubbles. During the curing process, the samples were first cured at 323 K for 4 h, and then post-cured at 353 K for next 12 h. Samples with thickness in the order of 300μm and 2mm were used for gas permeation measurements and other structural characterization analysis, respectively. Fourier transform IR (FTIR) spectra were measured using TENSOR 27 apparatus to ensure about complete curing process. Differential scanning calorimetry (Perkin-Elmer DSC-8500) measurements were carried out to determine the glass transition temperature (T_g). X-ray diffraction (XRD) and Scanning electron microscopy (SEM) have been used to observe the morphology of GNP composite membrane. In the discussion, epoxy-Jeffamine D-230 and Jeffamine ED-900 networks will be referred as NPGD-230 and NPGED-900, respectively.

The PALS measurements were performed at 313 K using conventional fast–fast coincidence system with time resolution of 260ps. The positron source ^{22}Na (15 uCi), deposited on a 7 um thick Kapton film, was sandwiched between two stacks of epoxy sample. The membrane samples were 1.2cm X 1cm X 2mm in dimension. The positron lifetime spectra (3 X 10^6 counts each) were fitted in three discrete and continuous lifetime components by using routine PATFIT-88 [6] and LT9.0 [7] programs, respectively.

Samples shaped in form of discs with diameter of 1.1 cm were used for studying the gas transport process using gas phase permeation technique. The details about gas measurements and experimental apparatus are reported elsewhere [3,8]. Measurements were done at 313 K using CO_2 (purity > 99.99% mol) as test gas kept at pressure 600 mbar.

3. Results and discussion

The completion of curing process in all amine modified epoxy membranes were confirmed by FTIR measurements as shown in Figure 1. The disappearence of peaks at 757 (-CH_2), 838 (C-O-C), 908 cm^{-1} (C-O) from epoxy ring as well as appearence of peaks at 1650 (deformed N-H) and 3440 cm^{-1} (O-H) from diamine confirms the epoxy ring opening and complete crosslinking process. Nano level molecular packing in the membranes has been altered by using curing agents, namely D-230 and ED-900. The backbone chain length of these curing agents is in the order D-230 < ED-900 and hence the crosslinking density is assumed to vary inversely. The crosslinking densities have been calculated by

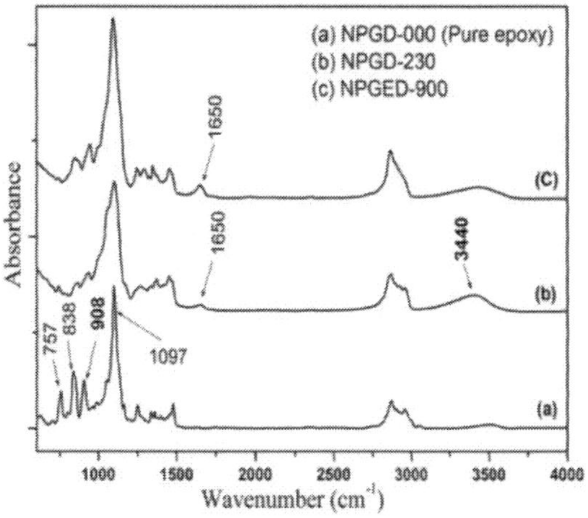

Figure 1. FTIR spectrum of amine-modified epoxy membranes after complete curing process

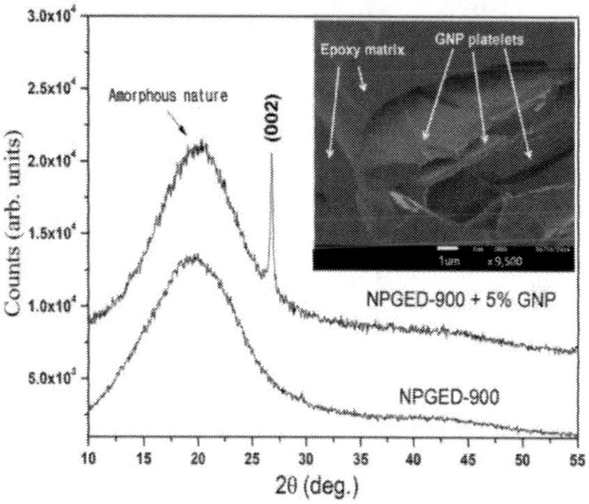

Figure 2. XRD patterns of pristine and composite membranes and SEM image of GNP composite.

Journal of Physics: Conference Series **618** (2015) 012036 doi:10.1088/1742-6596/618/1/012036

reported method [9] and the values are mentioned in Table 1. The effect of crosslinking density on the T_g, attributed to the thermal relaxations in molecular chains, is one of the fundamental aspects in the structure-property relationships [2]. The lower T_g for NPGED-900 (227 K) compared with the T_g of NPGD-230 (273 K) shows that the NPGD-230 sample presents higher crosslinking density (χ = 1.67 sites/nm^3) and more rigid molecular chain packings than NPGED-900.

Table 1. Free volume and CO_2 gas transport parameters for pure epoxy membranes.

Sample at 313 K	Cross-linking density (χ), sites/nm^3	Free volume parameter				CO_2 Gas transport	
		o-Ps lifetime (τ_3), ns (± 0.007)	o-Ps lifetime distribution (σ_3), ns	Free volume radius (R), nm	o-Ps Intensity (I_3), %	Permeability (Φ), mol/cm s Pa	Diffusivity (D), cm^2/s
NPGD-230	1.67	2.03	0.42 ± 0.03	0.29	23.73 ± 0.28	(8.36±0.20)X10^{-18}	(3.59±0.40)X10^{-8}
NPGED-900	0.90	2.56	0.57 ± 0.04	0.33	24.25 ± 0.27	(2.51±0.38)X10^{-16}	(6.67±1.00)X10^{-7}

CO_2 gas permeability and diffusivity in NPGD-230 and NPGED-900 membranes are reported in Table 1. From the results it is observed that CO_2 permeability diminishes with increasing crosslinking density of the epoxy membranes. NPGD-230 shows poor CO_2 gas permeability and diffusivity compared to NPGED-900. It seems that the rigid molecular chain packing in NPGD-230 membrane creates less diffusion channels for CO_2 gas molecules. In order to gain further insight into the effect of molecular chain packing on the CO_2 gas transport, we analyzed the free volume holes in the membrane microstructure, where the gas molecules are considered to jump for diffusion [10,11]. The free volume parameters (o-Ps lifetime, distribution and intensity) measured by PALS are reported in Table 1. The free volume radius (R) has been quantified using the semi-empirical Tau-Eldrup model [12,13]. It can be seen that free volume parameters are very well in aggreement with the microstructure of the membranes. NPGD-230 shows lower free volume radius and narrower o-Ps lifetime distribution (σ_3) compared to NPGED-900 due to its compact molecular chain networks. Therefore, it is understood that such compact molecular packings in the membrane provide less diffusion channels for gas molecules compared to the flexible networks present in the NPGED-900 membrane. It is proven that the NPGED-900 sample has ideal microstructure of polymer networks that facilitates the CO_2 gas permeabilty and diffusivity. There is no distinguishable change in the o-Ps intensity (I_3) which suggests the similar chemical nature in both membranes. It is attributed as the membrane chemical structure has no longer influence in gas transport results.

Polymer nanocomposites are interesting materials due to their enhanced properties with low amount of nanometer sized fillers. It is reported that the gas properties have been tailored by adding inorganic fillers in the polymeric membranes [3,14]. So we have prepared and studied a NPGED-900-5 nanocomposite membrane consisting of GNPs. In figure 2, XRD patterns of pristine and composite epoxy membranes confirm the amorphous nature of the materials with broad peak. The peak (002) at $2\Theta = 26.7°$ (d-spacing, 0.34 nm) is related to the distance between two GNP layers. The presence of only (002) reflection indicates the GNPs surface is parallel to the surface of the membrane [3]. Such GNPs morphology has been also confirmed by SEM images as shown in figure 2 (inset). Although, the 5 wt% of GNPs incorporated inside the membrane, the free volume parameter and gas permeabilty values measured at 313 K for NPGED-900 and NPGED-900-5 membranes show no significant change. To understand such anomalous behaviour, we measure the free volume radius and CO_2 permeabilty as a function of temperature, as shown in figure 3 and 4, respectively. It is attributed that the present concentration of GNP nanostructures inside the membranes is no longer responsible for any microstructural changes in the membrane matrix and also have no physical or chemical interactions with CO_2 gas permeant molecules as seen no drastic change in the permeability parameter. This suggests that higher wt% of filler would be necessary to modify the CO_2 transport properties in the present membrane.

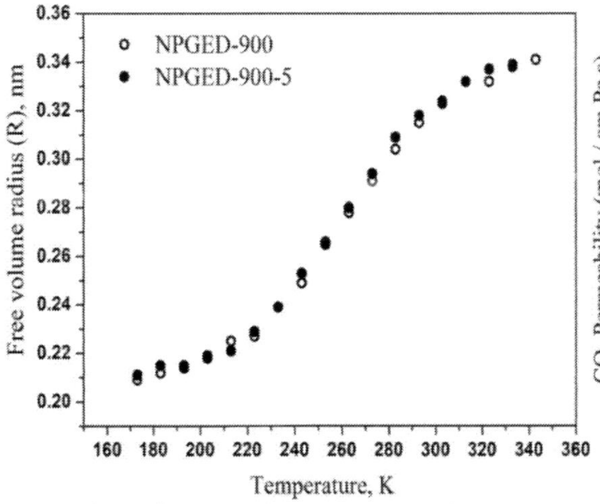

Figure 3. Free volume radius as a function of temperatures from 173 K to 343 K

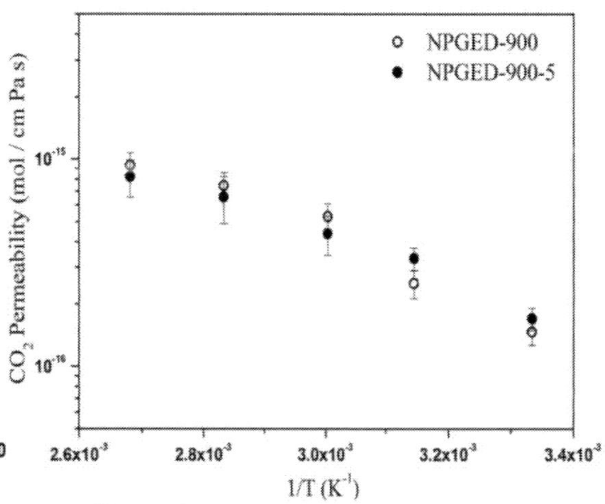

Figure 4. CO_2 permeability as a function of inverse temperatures from 293 K to 373 K

4. Conclusion

A good correlation between the average free volume properties measured by PALS and CO_2 gas permeability and diffusivity was found. The present study clearly reveals the influence of nanolevel molecular packing on the CO_2 gas diffusion. The membrane with low crosslinking density presents larger free volume spaces which acts like a diffusion channels for CO_2 gas permeant molecule. Composite membrane with 5wt% GNPs shows no significant change in CO_2 permeabilty which is in aggreement with the free volume results from PALS at different temperatures. Thus, the gas transport mechanism in polymeric membranes can be predicited by their free volume parameters.

Acknowledgement

This work was partially supported by the project ENAM (project No. 220068/s116).

References

[1] Powell C E and Qiao G G 2006 *J. Membr. Sci.* **279** 1-49.
[2] Patil P N, Rath S K, Sharma S K, Sudarshan K, Maheshwari P, Patri M, Praveen S, Khandelwal P and Pujari P K 2013 *Soft Matter* **9** 3589-3599.
[3] Checchetto R, Miotello A, Nicolais L and Carotenuto G 2014 *J. Membr. Sci.* **463** 196-204.
[4] Patil P N, Sudarshan K, Sharma S K, Maheshwari P, Rath S K and Pujari P K 2012 *Chem. Phys. Chem.* **13** 3916-3922.
[5] Baker R W 2006 *Membrane Technology and Applications* (New York: John Wiley & Sons)
[6] Kirkegaard P, Eldrup M, Morgensen O E and Pedersen N J 1981 *Comp. Phys. Comm.* **23** 307.
[7] Kansy J 1996 *Nucl. Instrum. Methods Phys. Res. A* **374** 235-244.
[8] Checchetto R, Bazzanella N, Patton B and Miotello A 2004 *Surf. Coat. Technol.* **177-178** 73.
[9] Wang X and Zhou W 2002 *Macromolecules* **35** 6747-6750.
[10] Belov N A, Zharov A A, Shashkin A V, Shaikh M Q, Raetzke K and Yampolskii Yu P 2011 *J. Membr. Sci.* **383** 70-77.
[11] Nagel C, Günther-Schade K, Fritsch D, Strunskus T and Faupel F 2002 *Macromolecules* **35** 2071-2077.
[12] Tao S J 1972 *J. Chem. Phys.* **56** 5499–5510.
[13] Eldrup M, Lightbody D and Sherwood J N 1981 *Chem. Phys.* **63** 51–58.
[14] Sanip S M, Ismail A F, Goh P S, Soga T, Tanemura M and Yasuhiko H 2011 *Sep. Purif. Technol.* **78** 208-213.

Feasibility of classification of clay minerals by using PAS

Y Honda[1] , Y Yoshida[2], Y Akiyama[2], S Nishijima[2]

[1] Institute of Scientific and Industrial Research, Osaka University, 8-1 Mihogaoka, Ibaraki, Osaka 567-0047, Japan
[2] Graduate school of Engineering, Osaka University, 2-1 Yamadaoka, Suita, Osaka 565-0871, Japan

E-mail: honda@sanken.osaka-u.ac.jp

Abstract. After the nuclear power plant disaster, the evaluation of radioactive Cs kept in soil, especially in clay minerals and the elucidation of its movement are urgent subjects to promote decontamination. It is known that the extractable level of Cs depends on the sort of clay minerals. We tried to find the characteristics of clay minerals belonging to phillosilicate group using positron annihilation spectroscopy (PAS) and the relationship between the results of PAS and the amounts of substantially extracted Cs from the clay minerals. The results showed that each clay mineral was found to be distinguishable from other clay minerals by PAS and the extraction rate of Cs was different among those clay minerals, however the direct correlation between the results of PAS and the extraction rates of Cs was not found.

1. Introduction

Radioactive cesium is known as a major species of fallout due to the Fukushima nuclear power plant accident and it has been kept in soil, especially in clay minerals. The trapping mechanism of Cs for different kinds of minerals have been studied [1,2]. Phillosilicate is a group of silicate having periodic laminar structure composed of tetrahedral sheet silicate and octahedral sheet silicate minerals, and is divided into two groups. One group has a unit consisting of one octahedral sheet such as aluminosilicate and one tetrahedral silicate sheet (1:1 clay mineral). Another group is consisting of an octahedral sheet silicate minerals sandwiched with tetrahedral sheet silicates (2:1 clay mineral). Cs is hard to be leached out of such 2:1 clay even with strong acid rather than 1:1 clay, because of local strong electric field due to the exchange of trivalent aluminium with divalent cations, leading to the localization of Cs at preferable spaces such as hexagons consisting of oxygen on the silicate layer, defects called Frayed Edge Site (FES), which is a wedged defect appearing at the edge of octahedral sheet, and interlayer between silicate sheets. Condensation of Cs would progress through cation exchange process [3]. An identification of the sort of soil or clay is crucial to decide substantial decontamination process. Positron annihilation spectroscopy (PAS) seems to be a good method to detect defects such as FES, and exchanged cations with Al or Si, since positron has a high mobility and is charge sensitive. Indeed the electronic state of FES in saponite was investigated by positron annihilation spectroscopy [4]. In such a study, the samples were treated to eliminate contaminated particles prior to the measurements, however such procedure would bring about a structural change due to reduction or exchange of cations. In this work, a classification of several sorts of clay minerals was tried by using PAS and a comparison between the results of PAS and the substantial extraction rate of Cs from those clay minerals is shown without making any pre-treatments for the samples.

Journal of Physics: Conference Series **618** (2015) 012037 doi:10.1088/1742-6596/618/1/012037

2. Experimental setup

The investigated samples were as follows; kaolinite which was synthesized with wet style, sintered vermiculite which was sold in a garden store, natural vermiculite which was brought from TOMOE Engineering Co., Ltd., montmorillonite which was purchased from Aquavitae Co., Ltd. and illite which was purchased from AromaFrance Co. Ltd. This illite was mixture of 65% illite and 25% kaolinite. Soil sold for the use of agriculture and indeed included much in the field of Fukushima was also prepared. No pre-treatment was made to avoid structural change. Many silver stings were included in natural vermiculite, which was observed by TEM-EDS (JEM-300F, JEOL), however there were no silver stings in the sintered vermiculite. The samples were also investigated with X-ray diffraction and the diffraction patterns were compared with the data in the PDF (Powder Diffraction File) of the ICDD (International Center for Diffraction Data). The diffraction pattern is so influenced by the presence of impurities that it is impossible to find the equivalent pattern to the measured one. However, the pattern of vermiculite was quite similar to the reference data both in peak angle and intensity profile. The measured major diffraction peaks of kaolinite and montmorillonite had almost the same peak angles as the reference data, though the intensity profiles were a little different. As the measured illite was a mixture of illite and kaolinite, the difference was larger than other samples, however the angles of the observed major diffraction peaks appeared also in the corresponding reference data of illite. Lifetime measurement of positron/positronium was carried out with a sealed Na-22 source covered with Kapton films. Its intensity was ca. 420 kBq and the time resolution was 220-240 ps. Coincidence Doppler broadening measurement (CDB), which is a simultaneous measurement of γ-rays emitted through 2 photon annihilation, was also carried out. To compare the results of PAS and the degree of Cs trapped in and extracted from the clays, all the samples were soaked in Cs solution with a concentration of 100 mg/l for 1 hour with shaking. The centrifugally separated contaminated samples were soaked in ammonium acetate solution with two different pH, 0.97 and 6.93 to extract Cs for 1 hour with shaking. The control of pH was made with nitric acid solution. The extracted Cs in solution from the sample was evaluated by an ICP-MS (Agilent7700 ;Agilent Technologies).

3. Results and discussions

3.1. Positron annihilation spectroscopy

3.1.1. Lifetime Measurements. Only kaolinite showed a small long-lived component (63 ns with ca. 2% relative intensity). Settling this lifetime, the other components were evaluated with 3 component analysis using PATFIT program as well as for other samples, where the lifetime spectrum was regarded as $N(t) = \sum_{i=1}^{3}(I_i/\tau_i)exp(-t/\tau_i)$. The τ_3s of vermiculite and montmorillonite were similar, that of kaolinite was slightly shorter and those of illite and soil were slightly longer (Figure1). This may reflect the structural difference among 1:1 group, 2:1 group and others. Indeed in 2:1 clay minerals, the interlayer distance varies with species of captured cation [1]. The size of hexagon on tetrahedral sheet is similar to the dimension evaluated by a lifetime of *o*-Ps. The I_3s of kaolinite, illite and soil were slightly higher than others, but

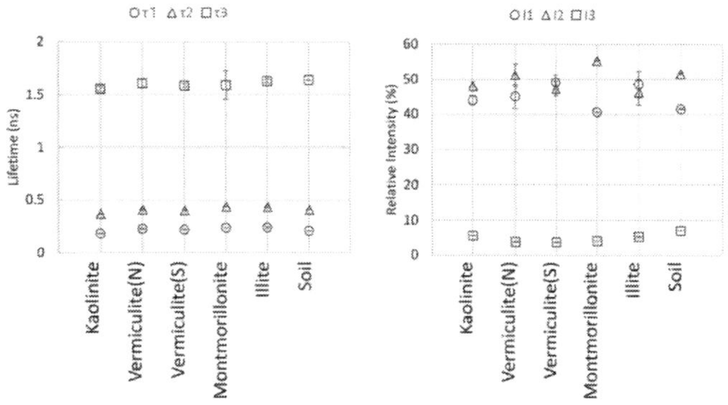

Figure1(a). Lifetimes and their relative intensities.

those values were very small. Why the I_3s were so small can be explained for 2:1 minerals by the fact that the tetrahedral sheets tend to stack as the result of Coulomb interaction between the sheet and interposed cation, reducing the interlayer gap. For 1:1 mineral, the unit layer is bound each other via hydrogen bond and the space of interlayer could be large, however the formation of *o*-Ps would be less. This is probably because the energy of electron relating to hydrogen bond is low and the space size formed with hydrated cations and the tetrahedral sheets is not enough to form positronium. In this analysis, as the source term correction of the sealed Na-22 could not be made, I_3 included the source components, however the relative intensity still have a meaning. As the strong electric field exists between layers, positron would approach the negative site, such as hexagon mentioned above, and annihilate without forming *o*-Ps. Concerning τ_1 and τ_2, kaolinite showed slightly shorter lifetimes. This would be due to the presence of hydrogen bond in the interlayer. Those for naturel and sintered vermiculite were almost equivalent, meaning that positron annihilated in an intrinsic structure of vermiculite without suffering from contaminants. The τ_1 and τ_2 of montmorillonite and illite were slightly longer than vermiculites and the soil. This would mean the interlayer gaps of montmorillonite and illite were wider than those of vermiculites.

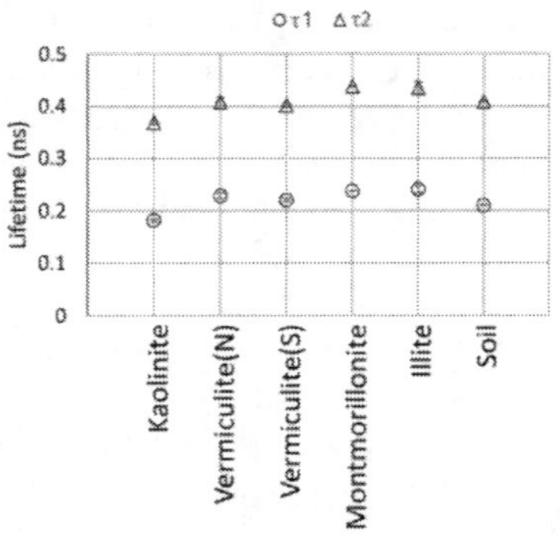

Figure1(b). Expanded τ_1 and τ_2.

3.1.2. CDB measurements. The profiles shown in Figure 2 are the ratios of the energy spectrum with respect to kaolinite, which was commonly measured in the different sets of measurement. The ratios for silver showed completely different profile in a higher momentum range. Kaolinite was found to have a narrower momentum distribution than other 2:1 clay minerals and silver. There was no evidence of silver in the spectrum of natural vermiculite and no significant difference between natural and sintered vermiculites. These results mean that positron approached some intrinsic structure of vermiculite and annihilated there, which was also deduced from the results of lifetime spectroscopy in the previous section. The ratio was larger in illite than those of montmorillonite and vermiculite at around normalized momentum of 0.03. Energy spread of vermiculite was slightly larger at around 0.02 than that of montmorillonite. The profile of ratio for the measured soil was quite similar to that of montmorillonite. It is said that the hydration of cation is an important process for trapping of cation at FES and diffusion of cation into interlayer [1]. The injected positron would come to the interlayer due to local electric field and annihilate with the electron relating to oxygen of tetrahedral sheet or water molecule attached to the cation or hydrogen bonding. Thus, those results would reflect the electronic status of interlayer.

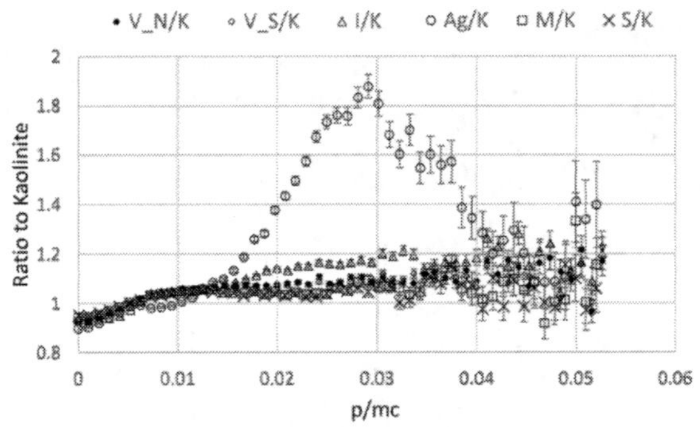

Figure 2. Energy spectrum ratios with respect to kaolinite. V_N:natural vermiculite, V_S:sintered vermiculite, I: illite, Ag: silver, M: montmorillonite, S:

3.2. Cesium extraction

The trapping ability and extraction of Cs for each clay mineral in neutral solution is shown in Figure 3. The absorption of Cs would be made at cation exchange site (CES), defects such as FES and interlayer including surfaces. Cs would be extracted easily from CES rather than from FES and interlayer. Absorption was not saturated in these experiments. The absorption rate of Cs for vermiculite and montmorillonite was rather higher than that of illite and kaolinite. Montmorillonite released about 70 % Cs, whereas vermiculite released only about 10% Cs in neutral solution regardless of whether it was sintered or not. In highly acid solution, the extraction rate for vermiculite and montmorillonite was increased a little but that for the other minerals and the soil showed no significant difference, meaning that the role of CES is not high for the sorption of Cs. Instead, the FES or fixing capacity of Cs in interlayer plays more important role for stable Cs trapping [2]. The results showed that the most stably adsorbed site of Cs was the FES and/or interlayer of vermiculite among the investigated clay minerals. The behaviour of Cs for the measured soil was similar to that of kaolinite and illite, though the ratio of CDB spectra showed quite similar profile to montmorillonite.

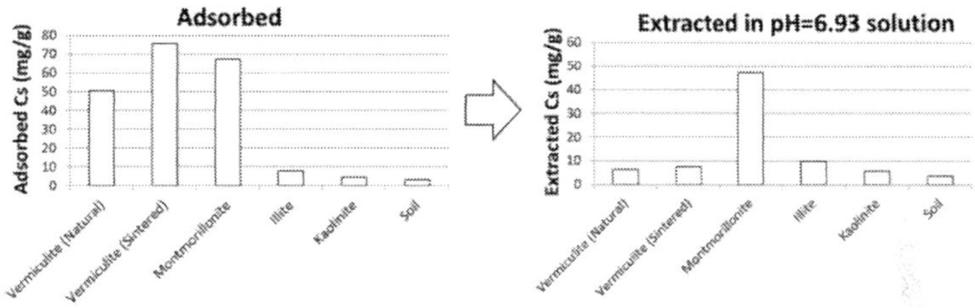

Figure 3. Sorption and extraction of Cs from clay minerals in neutral solution.

4. Summary

The difference between kaolinite (1:1 clay mineral) and other 2:1 clay minerals was detected by PAS. Taking into account of the results for natural and sintered vermiculite, it can be said that positron can sense its intrinsic structure. PAS could not detect the FES of 2:1 group with *o*-Ps. This is probably because the sites were filled with cations, and the space size was not enough for the formation of long-lived positronium. The annihilation process of positron was strongly influenced by the local electric field in the vicinity of interlayer. The annihilation γ-rays would convey the information such as the degree of hydration, binding energy of layers, hydrogen bonding, etc., leading to the difference in the CDB spectral ratio. However, there are no distinct correlation between the profile of the ratio and the extraction rate of Cs. Elucidation of the mechanism of Cs trapping in clay minerals, especially in vermiculite, seemed difficult based only on the results of PAS, however, it can be predicted that Cs cation would be preferentially exchanged at FES rather than other cations as is shown in elsewhere [1,2], and the difference of cation exchange capacity of Cs for different clay minerals might reflect on a difference around 0.02 of the normalized momentum in CDB spectral ratio. The soil bore resemblance to montmorillonite with respect to CDB results, but the behaviour of Cs in the soil through extraction process was different from that in montmorillonite and similar to kaolinite and illite, indicating that more information must be added to characterize substantial soil.

Acknowledgment

Authors acknowledge financial support from Ministry of Environment (Environment Research and Technology Development Fund (5Z-1301)).

References

[1] Sawhney B.L. 1992 *Clay Clay Miner.* **20** 93
[2] Nakao A, Thiry Y, Funakawa S, Kosaki T 2008 *Soil Sci. Plant Nutr.* **54** 479
[3] Kogure T, Morimoto K, Tamura K, Sato H, Yamagishi A, 2012 *Chem. Lett.* **41** 380
[4] Sato K, Fujimoto K, ai W, Munger M 2013 *J. Phys. Chem. C* **117** 14075

Accounting for the lack of nano-effect in a thermoset/clay nanocomposite: A positron annihilation study

S K Rath[1*], K Sudarshan[2], M Patri[1], P K Pujari[2]
[1]Polymer Division, Naval Materials Research Laboratory, Shil-Badlapur Road, Ambernath - 421506, Maharashtra, India
[2]Radiochemistry Division, Bhabha Atomic Research Centre, Mumbai - 400085, India

*Email: sangramkrath@yahoo.co.in

Abstract: The effect of nanoclay dispersion on the thermo-mechanical properties of an unsaturated polyester thermoset resin was studied by flexural and dynamic mechanical property measurements. Transmission electron microscopy studies revealed intercalated clay dispersion morphology in the nanocomposites. The thermomechanical measurements showed a steady decrease in the flexural strength and α relaxation temperature, with only moderate increase in the storage modulus at 1% clay loading, followed by a drop at higher clay loadings. In order to understand the absence of nano-effect in this case, free volume measurements were carried out by using positron annihilation lifetime spectroscopy. A bimodal distribution of o-Ps life times was observed. Nanoclay loading resulted in the increase of the o-Ps intensity corresponding to the longest life time as well as free volume fraction suggesting diminished chain packing efficiency in the nanocomposites. We posit that nanoclay induced decreased chain packing efficiency and the presence of higher free volume size elements might cause deterioration in mechanical properties of the nanocomposites.

1. Introduction

Recently, there has been a considerable focus on polymer nanocomposites, as incorporation of very small weight fractions of nanoparticles significantly alters the bulk thermal and mechanical properties [1, 2]. Among various nanofillers, clay minerals are one of the most widely used because of their unique layered structure, high aspect ratio and surface area as compared to other nanoparticles [1]. However, the understanding of structure/property relations in such composites is still an important area of research as it can enable designing materials with desired properties. There have been instances of deteriorating mechanical and thermal properties of polymers upon incorporation of nanoclay i.e. the absence of "nano-effect" [3]. In this study, we report such counterintuitive deterioration in mechanical properties of a thermoset/clay nanomposite and attempt to account for the reasons for the same by using positron annihilation studies.

2. Experimental

2.1 Materials and methods

A P-444 isophthalic unsaturated polyester resin (Pliogrip Resins and Chemicals Ltd, India) with 39 ± 3% dissolved styrene was used. The curing agents were 1.5% of 6% cobalt naphthenate as a catalyst, 0.1% of N, N-dimethyl aniline as an accelerator and 1.5% methyl ethyl ketone peroxide with 9% of active oxygen as an initiator. Closite-15A from Southern Clay Products was used as the nanoclay. Closite-15A was well mixed in the unsaturated polyester resin at different amounts (1, 3, 5 wt. %) and the curatives were added to the mixture and poured into aluminium moulds. The mixture was allowed to cure at room temperature for 24 hours, and subsequently post cured in an oven for 5 hours at 80°C. The 1,3,5 wt% clay loaded samples are referred to as USP-1, UPS-3 and USP-5, respectively in further discussion. For TEM, thin sections (45 nm) of the nanocomposites were obtained using a RMC MT-X ultramicrotome. These cryotomed sections were then transferred to copper grids. Unstained TEM lamellae were observed with a Philips CM 200 containing a LaB_6 filament and imaged using a Gatan digital camera.

The flexural tests were carried out on rectangular shaped samples according to ASTM D790, using a Hounsfield 50KN (UK) universal testing machine. The α relaxation temperature (T_g), storage modulus (E′), and the dissipation factor (tan δ) of the nanocomposites were determined in tensile mode using a Gabo, Eplexor 100 N, dynamic mechanical analyzer over a temperature range of 40-150°C at a frequency of 1 Hz, a ramp rate of 2° C min^{-1} and an initial strain of 0.2 %.

2.2. Positron annihilation lifetime spectroscopy (PALS)
PALS measurements of the nanocomposite samples were carried out under ambient conditions using a lifetime spectrometer based fast-fast coincidence and time resolution of 260 ps for gamma-rays of ^{60}Co was. A 15 μCi ^{22}Na positron source was sandwiched between two pieces of the same samples. The total area under each lifetime spectra was about 10^6 counts. The lifetime distributions were extracted from the lifetime spectra using CONTIN [4]. The probability density of the annihilation rates were converted to free volume radius distributions as described in ref [5].

3. Results and Discussion

3.1 Clay dispersion
Figure-1 shows the TEM micrograph of the USP/clay nanocomposites containing 3 and 5 wt % of nanoclay. It is seen that intercalated clay domains of silicate layers consisting of a few nearly parallel layers with basal spacing of 3-5 nm are uniformly distributed in the polymer matrix. Only a very few exfoliated single clay layers with a thickness of ≈ 1 nm thick are present, implying a dominant intercalated clay dispersion morphology.

3.2. Thermomechanical properties
The flexural strength of the nanocomposites as compared to the samples made of pure synthesized resin is shown in Figure 2. A monotonic decrease in the flexural strength with increased clay loading is observed despite the intercalation of the clay platelets in the resin. The DMA scans in the form of variation of the dissipation factor, tan δ vs. temperature for the neat USP resin and its nanocomposites are shown in Figure 3. The tan δ peak position for the polyester resin is located at 95°C and is observed to shift towards lower temperatures with clay loading, implying decreasing glass transition temperatures for the nanocomposites.

Figure 3B shows the variation of storage modulus with temperature. Contrasting effect of nanoclay on the dynamic storage modulus of USP resin is observed in the glassy and rubbery region. At 30°C (glassy region) the E□ trend is USP = USP-1 >USP-3 >USP-5, post T_g at 120°C (rubbery region) the trend is found to be USP-1 > USP =USP-3 >USP-5. This trend reversal is in accordance with the fact that the reinforcing effect of the clay is as expected larger at temperatures above T_g as compared to below T_g due to the decrease of the modulus of the polymer matrix when transitioning from the glassy to the rubbery state [6]. From the above thermmechanical results it is evident that there is a clear lack of nano-effect in the present case despite the evidences of intercalation of clay layers in the thermoset matrix. Reduced flexural strength, T_g and storage modulus with clay addition may be attributed to a number of factors, such as clay induced alteration in the curing kinetics of the USP resin leading to decreased cross-link density.

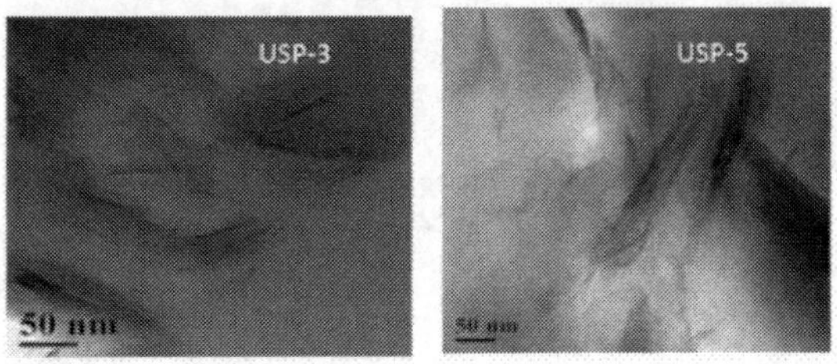

Figure 1. TEM micrographs of USP-3 and USP-5

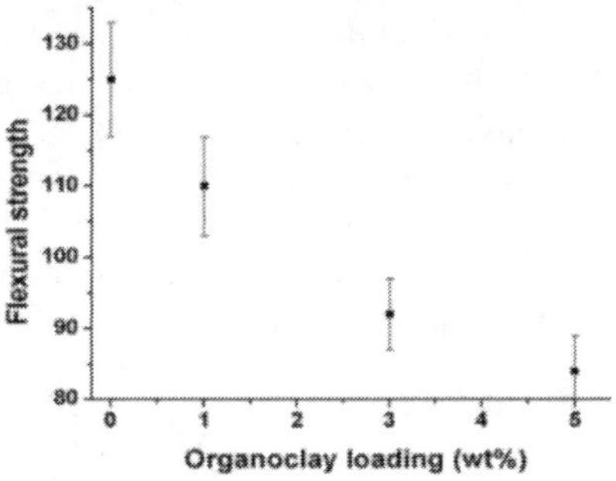

Figure 2. Variation of flexural strength of USP matrix with clay loading

This is one of the plausible reasons as nanoclay incorporation may alter the conductivity of the system [7]. The second probable mechanism is the possible network disruption as the USP network in the vicinity of the clay particles may be disrupted compared to the bulk matrix, which in turn decreases the effective cross-link density and *Tg*. In order to substantiate the above mentioned mechanistic origins of the altered thermomemchanical properties free volume measurements were carried out using PALS.

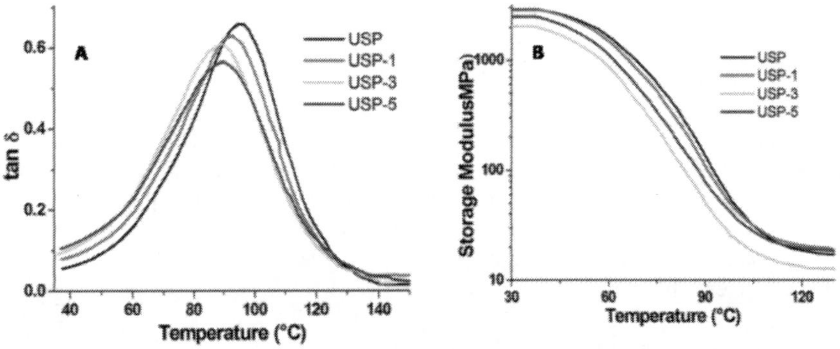

Figure 3. Dynamic mechanical analysis results of USP and its nanocomposites. (A) tan δ vs. temperature profiles and (B) storage modulus vs. temperature profiles

3.3. Free volume characterization

The positron annihilation lifetime spectra were fitted to lifetime distributions using CONTIN. The spectra could be best fitted to bimodal distribution of o-Ps lifetimes. The free volume hole radii distributions obtained from the o-Ps lifetimes (τ_3, τ_4) are shown in figure 4. While the PALS spectra of most polymers are described by a single o-Ps lifetime, τ_3, some high-free-volume polymers possess a second, longer o-Ps lifetime [6, 8, and 9]. It has been suggested that τ_4 corresponds to large, possibly interconnected free volume elements in such materials [6]. As seem from figure 4, the average of the shorter o-Ps lifetimes (τ_3) has marginally decreased with clay loading except in USP-3 while the intensity corresponding to larges hole sizes increased monotonically with clay loading. To get a clearer picture, the free volume fractions in the samples are calculated (Figure 5) based on average hole radius estimated using Tao-Eldrup model and o-Ps intensities. It is clearly seen from figure 5 that the free volume fraction increased with increase in clay loading and the increase is primarily due to the increase in the number of larger free volume holes. This evolution of higher free volume defects with increased clay loading imply diminished chain packing efficiency of the nanocomposites with clay loading. We posit that the inclusion of nanoclay disrupts the molecular packing efficiency of the USP matrix, leading to the observed deterioration in mechanical properties of the nanocomposites compared to the neat polymer matrix. Thus the ability of nanoparticles to significantly alter the physical and thermal properties of the polymer is significantly decreased in highly cross linked thermoset matrices. This is consistent with the observations of Putz *et al.* that increases cross-link density leads to increased network disruption by incorporation of nanoparticles in an otherwise densely connected system [7].

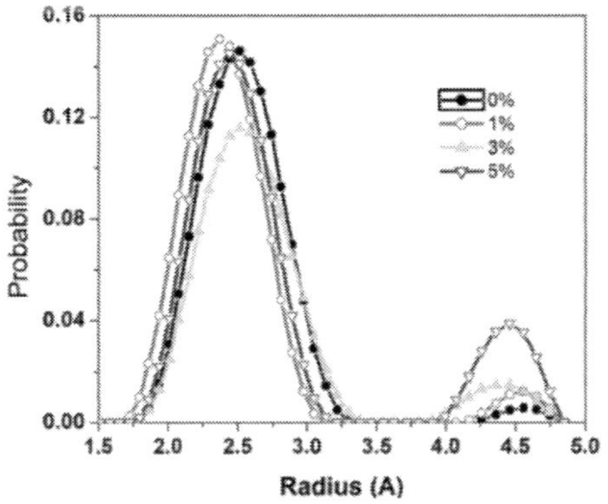

Figure 4. Free volume size distribution of USP and its nanocomposites

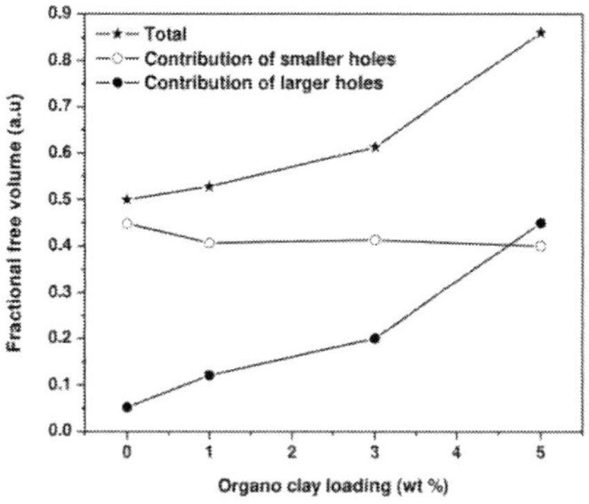

Figure 5. Free volume fraction in clay loaded samples and the contributions from different hole sizes

4. Conclusions

A clear absence of the nano-effect in unsatured polyester resin/clay nanocomposites through thermomechanical property measurements is observed even though the clay particles seemed to be well intercalated in the thermosets matrix. PALS measurements showed that clay loading caused an increase of the free volume fraction, suggesting a lower chain packing efficiency in these intercalated USP/clay nanocomposites. Furthermore, the o-Ps formation probability corresponding to the longer lifetime component increased with clay loading, implying nanoclay induced voids or defects in the USP matrix, which in turn may result in the observed lack of nano effect in the present case.

References

[1] Paul D R and Robeson L M 2008 *Polymer* **49** 3187.
[2] Rath S K, Aswal V K, Sharma C, Joshi K, Patri M, Harikrishnan G and Khakhar D V 2014 *Polymer* **55** 5198.
[3] Schaefer D W and Justice R S 2007 *Macromolecules* **40** 8501.
[4] Gregory R B and Zhu Y 1990 *Nucl. Instr. Meth.* **A 290** 172.
[5] Jean Y C, Yuan J P, Liu J, Deng Q and Yang H J 1995 *Polym. Sci. B: Polym. Phys.* **33** 2365.
[6] Winberg P, Eldrup M, Pedersen N J, van Es M A and Maurer F H J 2005 *Polymer* **46** 8239.
[7] Putz K W, Palmeriu M J, Cohn R B, Andrews R and Brinson L C 2008 *Macromolecules* **41** 6752.
[8] Merkel T C, Freeman B D, Spontak R J, He Z, Pinnau I, Meakin P and Hill A J 2002 *Science* **296** 519.
[9] Patil P N, Rath S K, Sharma S K, Sudarshan K, Maheshwari P, Patri M, Praveen S, Khandelwal P and Pujari P K 2013 *Soft Matter* **9** 3589.

Journal of Physics: Conference Series **618** (2015) 012039 doi:10.1088/1742-6596/618/1/012039

Collisional cooled Ps emitted into vacuum from silica-based porous materials: experiment to measure the Ps cooling time

S Mariazzi[1], **L Di Noto**[2,4], **G Nebbia**[3] **and R S Brusa**[2]
[1] Stefan-Meyer-Institut für subatomare Physik, Boltzmanngasse 3, 1090 Vienna, Austria
[2]Department of Physics, University of Trento and INFN-TIFPA, Via Sommarive 14, 38123 Povo, Trento, Italy
[3]INFN Padova, Via Marzolo 8, 35131 Padova, Italy
[4]Present address: Dipartimento di Fisica, Università degli Studi e INFN Sez. Genova, Via Dodecanneso 33, 16145 Genova, Italy.

E-mail: brusa@science.unitn.it

Abstract. In recent experiments on positronium time of flight (Ps-TOF) we have studied emission of cooled and thermalized Ps into vacuum from oxidized nanochannels synthetized in silicon. Ps cools down through collisions with the walls of the channels before exiting into vacuum. An important unknown parameter in the Ps-TOF measurements is the permanence time in the medium, i.e. the Ps cooling time before emission into vacuum. In this paper we describe an experiment that allows us to estimate the cooling time of Ps by analyzing the Ps-TOF spectra of cool Ps at three different distances from the sample.

1. Introduction

There is a demand for cool Ps in vacuum for spectroscopy experiments [1, 2] and for the production of anti-hydrogen beams through the charge exchange reaction between antiprotons and Ps excited in Rydberg states [3]. An efficient way of cooling Ps is to form it by injecting e^+ with energy (E_+) of some keV in porous silica materials with porosities open towards the vacuum [4]. Ps is emitted from the bulk into the pores with energy from ~1 eV to ~3 eV, then it reduces its kinetic energy by collisions with the walls of the pores and a cooled fraction escapes into vacuum [4]. In porous materials, with 2.7 nm pores size, a Ps kinetic energy of 42 meV was measured via Doppler broadening of the 1^3S-2^3P transition in the laser direction [5]. The incomplete thermalization was consistent with the energy of Ps quantum confined in the pores [6]. The limit imposed by quantum confinement was circumvented forming Ps in oxidized nanochannels with lateral dimensions of few nanometers [7] and Ps was found to escape with thermal energy at sample temperature in the 150-300 K range [8]. Knowledge of the time delay between e^+ implantation and cooled Ps emission (Ps cooling time) is important both for a correct interpretation of the Ps-TOF measurements [9] and for an efficient laser excitation of the Ps bunch [3].

The aim of this paper is to present an experimental method for measuring the Ps cooling time. The Ps-TOF measurements were done at room temperature (RT) in oxidized nanochannels. Only a rough estimation of the cooling time will be given because the poor statistics.

2. Experimental

Oxidized nanochannels with diameters in the 5-8 nm range were obtained by electrochemical etching as described in Ref. 7, 8. Nanochannels were produced in the first 2 μm of the silicon sample. Ps-TOF measurement were carried out with a slow continuous positron beam [10] keeping the sample at 300±1 K. The Ps energy component perpendicular to the target surface $E_\perp = (m_0 z_0^2)/t^2$, where m_0 is the e^+ rest mass and z_0 the distance of the slit center from the target, was derived measuring the time t between e^+ implantation on the target and the γ-rays of Ps annihilation in flight. The start signal was given by detecting the secondary electrons with two channeltrons, while the stop signals were given by 5 NaI(Tl) detectors arranged in an arc at a radial distance of 130 mm from the sample axis. A 100 mm thick lead shield, with a 5 mm slit was placed in front of the detectors (figure 1). Spectra were measured at three different distances z_0 from the target.

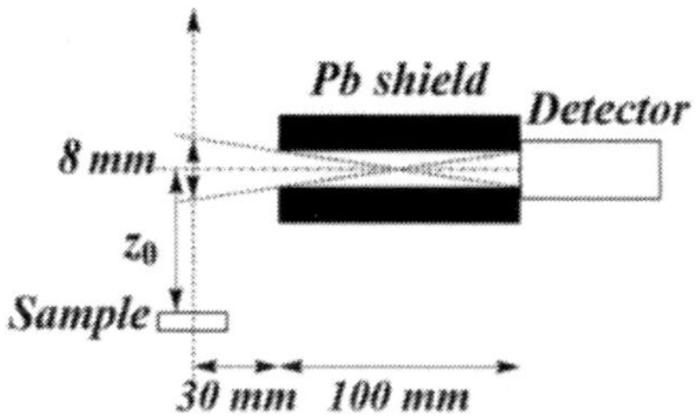

Figure 1. Sketch of the Ps-TOF set up.

Both start and stop signals, were combined in a fan-in/fan-out module and $\sim 10^4$ coincidence events were collected for each spectrum using a time-to-amplitude converter. Details of the Ps-TOF set up can be found in Ref. 11.

In order to evaluate the system resolution and the background a calibration run was performed on a sample made of a Si single-crystal. The resolution function was found to be a Gaussian with 16 ns FWHM by fitting the positron annihilation prompt peak. The Ps-TOF spectra were smoothed through a moving average filtering on 21 channels (0.405 ns/channel). The background due to prompt e+ annihilations was subtracted and the effect of Ps finite lifetime and permanence in front of the slit was taken into account by the multiplication factor $e^{t/142ns}/t$ [12]. The TOF spectra were multiplied by t^3 in order to obtain the energy spectra [12].

3. Results and discussion

With the aim of measuring the Ps cooling time when Ps approaches thermal energies, the e^+ implantation energy was increased up to the value of E_+ at which a first fraction of near thermal Ps is emitted. The average positron implantation depth $\bar{z}$ is related to the positron implantation energy by $\bar{z}[\text{nm}] = 40/\rho\ [\text{g/cm}^3](E_+\ [\text{keV}])^{1.6}$ where ρ is the sample density [5]. At low E_+ Ps is formed near the surface therefore it does not undergo enough collisions with the nanochannel walls to reach thermal energies. On the contrary, at higher E_+ Ps is formed deeper in the sample and the emission time becomes longer than the thermalization time. The energy spectra obtained by analyzing Ps-TOF spectra measured at $z_0 = 1$ cm and with $E_+ = 6$ keV and $E_+ = 7$ keV are shown in figure 2a-2b (in different scales) and in figure 2c respectively. In figure 2c the spectra measured at $E_+ = 7$ keV for $z_0 = 2$ cm and $z_0 = 3$ cm, are also reported. The energy spectra, in the following, are analyzed up to energies of 150-200 meV since for higher energies the tail of the prompt peak cannot be distinguished from the emission of fast Ps atoms.

The logarithmic spectra were fitted by superimposing two maxwellian-type distributions in which the exponents n and m were varied from -1/2 to 1 in steps of 1/2:

$$\ln\left(\frac{dN}{dE_\perp}\right) = \ln\left[\alpha \cdot E_\perp^m A(m, T_1) e^{\frac{E_\perp}{k_B T_1}} + \beta \cdot E_\perp^n B(n, T_2) e^{\frac{E_\perp}{k_B T_2}}\right] \qquad (1)$$

Where $A(m,T_1)$ and $B(n,T_2)$ are normalization factors such that $\int_0^\infty E_\perp^m A(m,T_1) e^{-\frac{E_\perp}{k_B T_1}} dE_\perp = 1$, $\int_0^\infty E_\perp^n B(n,T_2) e^{-\frac{E_\perp}{k_B T_2}} dE_\perp = 1$ and $\alpha + \beta = 1$.
The temperatures T_1, T_2 of the two distributions and the factors α and β, were free parameters in the fitting procedure, k_B is the Boltzman constant.
The best fits to all the energy spectra were obtained with m= - 1/2 which corresponds to a one-dimensional maxwellian distribution describing the coolest part of the spectra:

$$F(E_{c\perp}) = \frac{E_\perp^{-1/2}}{\sqrt{\pi k_B T_1}} e^{-\frac{E_\perp}{k_B T_1}} \qquad (2)$$

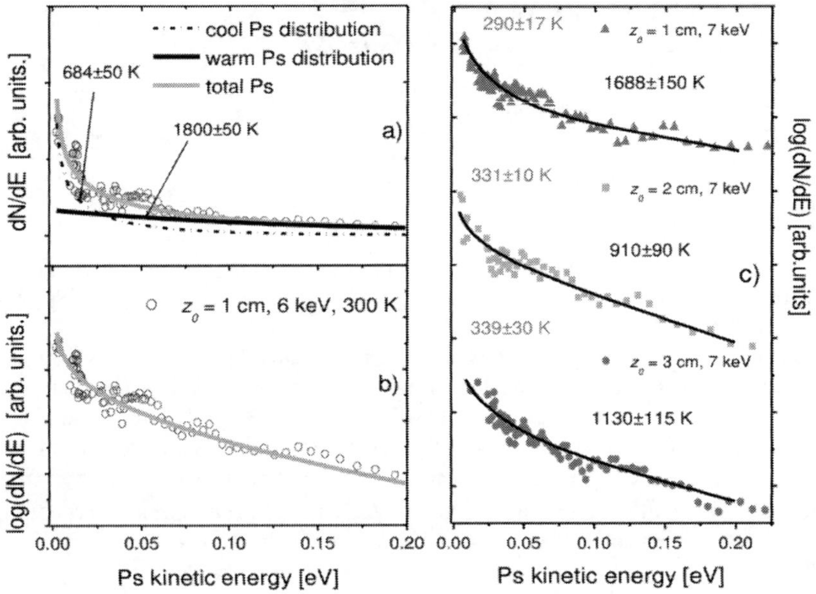

Figure 2. a) and b), Ps energy spectrum measured at $E_+ = 6$ keV and $z_0 = 1$ cm in linear and semilog scale, respectively. The light gray curves are best fits that result from the sum of a one-dimensional maxwellian and a beam maxwellian distribution as in equation (1). The one-dimensional maxwellian distribution corresponding to the cool Ps fraction and the beam maxwellian distribution corresponding to the warm Ps fraction are reported in a) as dash-dot and continuous black lines respectively. Their characteristic temperatures are also indicated. c) Ps energy spectra, in semilog scale, measured at $E_+ =$ 7 keV and with $z_0 = 1$ cm, 2 cm, 3 cm. The lines are best fits to the data obtained with equation (1). The characteristic temperatures of the two distributions are extracted from the fits.

whose average kinetic energy is given by $\langle E_{c\perp} \rangle = \frac{T_1 k_B}{2}$ and with *n*=0 which corresponds to a beam maxwellian distribution (i.e. a simple exponential function) describing the warm part of the spectra:

$$G(E_{w\perp}) = \frac{1}{k_B T_2} e^{-\frac{E_\perp}{k_B T_2}} \qquad (3)$$

whose average kinetic energy is given by $\langle E_{w\perp} \rangle = T_2 k_B$. From now on, we will call these distributions, cool distribution and warm distribution of the spectra, respectively.
As the fitting procedure does not support $\alpha = 0$ or $\beta = 0$, a single distribution is not sufficient to describe the energy spectra. A change in the slope of the energy spectra can be appreciated by observing the spectrum measured with $E_+ = 6$ keV and reported in linear scale in figure 2a. The lines in figure 2 are best fits to the experimental data. In figure 2a, one can notice how the beam maxwellian dominates at higher energy while the one-dimensional maxwellian is crucial to describe the steep slope of the data at lower energy. The fitted temperature T_1 and T_2 characterizing the two distributions are reported.
Increasing the positron implantation energy from 6 to 7 keV, the temperature T_1 of the cool distribution strongly decreases from 684 ± 50 K to 290 ± 17 K, while the temperature of the warm distribution is only reduced from 1800 ± 50 K to 1688 ± 150 K (figure 2c). The spectra measured at $E_+ = 7$ keV and $z_0 = 1$ cm, show that a fraction of Ps atoms approaches thermal equilibrium with the sample and is emitted into vacuum at near thermal energies. A more precise evaluation of the temperature requires taking into account the permanence time of Ps inside the sample before being emitted, i.e. the cooling time.
The blue and red curves in figure 3a represent the time distributions relative to the cool and warm energy distributions (equations (2) and (3) respectively) with the weights α and β, and the temperatures T_1 and T_2 extracted from the fits. In order to determine each time distribution, the corresponding energy distribution is divided by t^3 and multiplied by a normalization coefficient. The gray curves, sum of the red and blue ones, reproduce fairly well the Ps-TOF spectra.
The average time of flight $\langle t_m \rangle$ was calculated for the cool and the warm component of each time distribution measured at $z_0 = 1$, 2 and 3 cm. This time is the sum of two factors: $\langle t_m \rangle = \langle t_p \rangle + \langle t_f \rangle$ where $\langle t_p \rangle$ is the Ps average permanence time in the nanochannels before emission and $\langle t_f \rangle$ is the real average time of flight $\langle t_f \rangle = z_o/v_\perp$ with $v_\perp$ being the Ps perpendicular velocity not affected by the permanence time. $\langle t_m \rangle$ values as a function of z_0 were linearly fitted with the equation $\langle t_m \rangle = \langle t_p \rangle + z_o/v_\perp$ and are shown in fig 3b. The intercepts of the two sets of values give $\langle t_p \rangle = \langle t_{c1} \rangle$ and $\langle t_p \rangle = \langle t_{c2} \rangle$, the average Ps cooling time of the cool and warm distribution respectively. The slopes of the two straight lines give the reciprocal of the Ps perpendicular velocities $1/v_\perp$. The error on $\langle t_m \rangle$ arises from the the statistical error on the temperature of the two distributions (see figure 2c). The error on z_0 is due to the spatial resolution of ~8 mm at the target axis as determined by the geometry of the set-up (see figure 1). It was evaluated as represented by the standard deviation, $8/\sqrt{3}$ mm, associated to a uniform distribution.
The fit gives an average cooling time $\langle t_{c1} \rangle = 18 \pm 6$ ns for the Ps emitted with the one-dimensional maxwellian distribution. For the Ps of the warm distribution we can only conclude that the average time $\langle t_{c2} \rangle$ is estimated to be less than 7 ns. The Ps perpendicular velocities result $v_{c\perp} = 4.8\text{x}10^4 \pm 2\text{x}10^3$ m/s and $v_{w\perp} = 7.4\text{x}10^4 \pm 3\text{x}10^3$ m/s for the cool and warm distributions respectively.
The real average time of flight, $\langle t_{fc} \rangle$ and $\langle t_{fw} \rangle$, of each spectra are correlated to the Ps perpendicular velocities by the relation:

$$v_{c\perp} = \frac{z_0}{\langle t_{fc} \rangle}, \; v_{w\perp} = \frac{z_0}{\langle t_{fw} \rangle} \qquad (4)$$

In order to estimate the Ps temperature and average kinetic energy we assume that the shape of the t_f distribution is not altered by the permanence time compared to the measured time of flight t_m distribution. Thus the average time $\langle t_{fc} \rangle$ and $\langle t_{fw} \rangle$ of the two time distributions can be calculated as $\langle t_{fc} \rangle = 2.67\sqrt{\frac{2m_0}{2\pi k_B T_c}}\, z_0$ and $\langle t_{fw} \rangle = \sqrt{\frac{\pi 2 m_0}{2 k_B T_w}}\, z_0$. Substituting these expressions in equation (4), the best estimation of temperatures can be achieved as $T_c = (2.67)^2 \frac{(v_{c\perp})^2\, 2\, m_o}{2\pi k_B}$ for the cool one-dimensional maxwellian distribution [equation (2)], and $T_w = \frac{(v_{w\perp})^2 \pi\, 2\, m_o}{2 k_B}$ for the warm beam

maxwellian distribution [equation (3)]. The temperatures result to be T_c =342±25 K, and T_w = 1230±250 K. From equations (2) and (3) the corresponding average kinetic energies result to be $\langle E_{c\perp}\rangle$ = 15 ± 1 meV and $\langle E_{w\perp}\rangle$ =107 ± 22 meV. Usually when a single Ps-TOF spectrum at a distance z_0 is recorded, it is implicitly assumed that the permanence time t_p is negligible compared to t_f. Nevertheless this simple assumption could lead to an incorrect Ps energy evaluation when Ps is formed deep in a porous sample with small or tortuous pores.
The central value of the cool distribution temperature, evaluated in the spectra measured at different distance from the sample surface, rises from 290 K to 339 K approaching the real

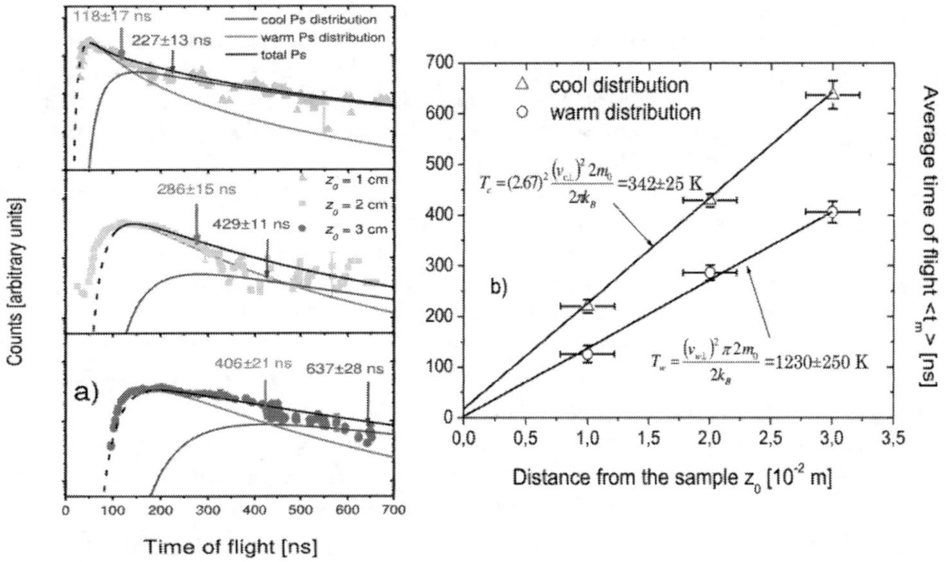

Figure 3.a) Ps-TOF spectra measured at E_+=7 and at $z_0 = 1, 2$ and 3 cm. The black curves are the time distributions corresponding to the energy spectra best fits. The blue curves are the time distributions of the cool Ps while the red curves are the time distributions of the warm Ps. The average time of flight $\langle t_m\rangle$ of each distribution is reported and its position marked by an arrow.
b) Average time of flight $\langle t_m\rangle$ calculated from the time distributions of figure 2c, as a function of the distance z_0 from the sample surface. Triangles for cool distributions and circles for warm distributions. The intercepts of the two best fit lines give the mean permanence time $\langle t_p\rangle$ of Ps before being emitted into vacuum. The slopes of the lines give the perpendicular flying velocity $v_{c\perp}$ and $v_{w\perp}$ of the cool and warm Ps distributions.

temperature T_c =342 K (see figure 2c). As a matter of facts, as the distance z_0 increases, the permanence time t_p becomes progressively negligible compared to the time of flight t_f.
From the measured Ps-TOF spectra we can observe that the cool distributions measured at a distance $z_0 \geq 2$ cm are characterized by a temperature compatible with T_c =342 K within the errors and are well fitted by a one-dimensional maxwellian function. This is an a posteriori confirmation that, for the present sample, the shape of the distribution is not (or slightly) affected by the permanence time t_p, as we have previously assumed evaluating T_c.
The measured $\langle t_p\rangle = \langle t_{c1}\rangle$ can be compared with a theoretical estimation of permanence time in similar nanochannels calculated in Ref.8. In the framework of the diffusion model, the mean

permanence time of diffusing Ps in 5-8 nm regular nanochannels after e^+ implantation at 7 KeV was calculated to be 16 ± 1 ns [8], in very good agreement with our experimental value of 18 ± 6 ns.

4. Conclusions

We have shown that measuring Ps-TOF spectra at different distances from the surface of a porous sample allows one to extract the cooling time of Ps. From very preliminary measurements done in nanochannelled samples we have estimated a mean Ps cooling time, corresponding to the mean permanence time $\langle t_p \rangle$ in the nanochannels before emission, of 18 ± 6 ns. In this time Ps reaches quasi thermal kinetic energies $\langle E_{c\perp} \rangle \cong 15$ meV. The method used to extract $\langle t_p \rangle$ will allow us to study in detail the cooling process of Ps by measuring $\langle t_p \rangle$ at different positron implantation energy and at different sample temperature. At present, we have installed an improved version of the Ps-TOF apparatus at the high intensity positron source NEPOMUC inside the FRMII in Munich. First tests with high positron fluxes performed by using a pair of large area MCPs as secondary electron detectors have shown that high signal to noise ratio can be obtained thus largely improving the quality of the data. With this set up more precise results can be obtained by acquiring five spectra at five different distances, in order to validate the linear fitting procedure. In the meanwhile a MonteCarlo simulation is under development in order to understand the origin of the two distributions.

References

[1] Castelli F, Boscolo I, Cialdi S, Giammarchi M G and Comparat D 2008 *Phys. Rev. A* **78** 052512
[2] Cassidy D B, Hisakado TH, Tom H W K and Mills A P Jr. 2012*Phys. Rev. Lett.***108** 043401
[3] Doser M et al., 2012 *Classical and Quantum Gravity* **29** 184009
[4] Brusa R S and Dupasquier A 2010 *Pysics with many positrons* ed A Dupasquier, A P Mills, Jr. and R S Brusa (IOS Press, Amsterdam, Oxford, Tokio, Washington DC), p. 245.
[5] Cassidy D B, Crivelli P, Hisakado T H, Liszkay L, Meligne V E, Perez P, Tom H W K and Mills A P Jr. 2010 *Phys. Rev. A* **81** 012715
[6] Mariazzi S, Salemi A and Brusa R S 2008 *Phys. Rev. B* **78** 085428
[7] Mariazzi S, Bettotti P, Larcheri S, Toniutti L and Brusa R S 2010 *Phys. Rev. B* **81** 235418
[8] Mariazzi S, Bettotti P and Brusa R S 2010 *Phys. Rev. Lett.* **104** 243401
[9] Cassidy D B, HisakadoT H, Meligne V E, Tom H W K and Mills A P Jr. 2010 *Phys. Rev. A***82**052511
[10] Macchi C, Mariazzi S, Karwasz G P and Brusa R S, Folegati P, Frabboni S and Ottaviani G 2006 *Phys. Rev. B***74** 174120
[11] Di Noto L, Mariazzi S, Bettonte M, Nebbia G and Brusa R S 2012 *Eur. Phys. Jour. D* **66** 118
[12] Howell R H, RosenbergI J, Fluss M J, Goldberg R E and Laughlin R B 1987 *Phys. Rev. B* **35** 5303

Journal of Physics: Conference Series **618** (2015) 012040 doi:10.1088/1742-6596/618/1/012040

N-heptane adsorption and desorption in mesoporous materials

R. Zaleski[a,1], M. Gorgol[a], A. Błażewicz[a], A. Kierys[b], J. Goworek[b]

[a] Maria Curie-Sklodowska University, Institute of Physics, Department of Nuclear Methods pl. M. Curie-Sklodowskiej 1, 20-031 Lublin, Poland

[b] Maria Curie-Sklodowska University, Faculty of Chemistry, Department of Adsorption, pl. M. Curie-Sklodowskiej 3, 20-031 Lublin, Poland

[1] E-mail: radek@zaleski.umcs.pl

Abstract. Positron Annihilation Lifetime Spectroscopy (PALS) was used for an *in situ* monitoring of adsorption and desorption processes. The disordered and ordered porous silica as well as the porous polymer were used as adsorbents, while an adsorbate in all the cases was *n*-heptane. The lifetimes and particularly the intensities of the ortho-positronium (o-Ps) components depend strongly on the adsorbate pressure. The analysis of these dependencies allows us to identify several processes, which are taking place during sorption. At low pressure, an island-like growth of the first layers of the adsorbate on the silica, in a contrary to a swelling of the polymer, is observed. A size of the pores, which remain empty, is estimated at the subsequent stages of the adsorbate condensation and evaporation. The adsorbate thrusting into micropores is deduced at $p/p_0 > 0.6$ whilst the mesopores are still not completely filled.

1. Introduction

Numerous applications of porous solids require a knowledge of their pore structure. The most popular methods of pore characterization base on an analysis of physical adsorption/desorption of a gas (usually nitrogen at 77 K). In order to determine a pore size distribution, the sequence of calculations like Barrett-Joyner-Halenda (BJH) scheme [1] or non-localized density functional theory (NLDFT) [2], arising from models of sorption process has to be performed. The accuracy of results depends on the correct interpretation of adsorption/desorption isotherms. Most methods of an isotherm analysis were developed basing on measurements of macroscopic quantities (usually the volume of adsorbed liquid) and a theoretical modelling (conventionally, the Kelvin equation of the classical thermodynamics is used). Still, an interpretation of the results often requires a supposition that the ideal course of pore filling and emptying was distorted by some processes [3,4]. A cavitation, pore blocking, free volume formation at the liquid-solid interface and changes of the condensate density are among them. However, an experimental observation of these effects is a real challenge. This can be achieved only by alternative methods for the adsorption porosimetry, such as small-angle X-ray scattering (SAXS), small-angle neutron scattering (SANS) or PALS. The last technique is the most promising one, because it is sensitive directly to the free volumes, such as pores, which are not filled by an adsorbate or subnanometer free spaces present among others on the liquid-solid interface. Moreover, introducing an adsorbate to a sample does not hinder the PALS measurement in any way.

The PALS studies of adsorption and desorption and the accompanying processes were already performed for *n*-heptane adsorbate and several porous silica adsorbents [5-9]. The conclusions derived from comparison of the most representative results and the results of the measurement performed for porous polymer adsorbent are presented in this paper.

2. Experimental

2.1. Materials

Three porous materials with different pore structures and compositions, but the comparable porosity (pore size, specific surface and total volume) were used as adsorbents (Table 1). The ordered porous silica SBA-15 was prepared following the procedure reported in [7,10]. The tetraethyl orthosilicate (Sigma-Aldrich) as the silica source and the amphiphilic triblock copolymer Pluronic P123 (BASF) as the structure-directing agent were used in the synthesis. The silica gel Si-A was prepared according to the procedure described in [8,11] with the use of Amberlite XAD7HP (ROHM & HAAS, now Dow Chemical Co.) as a porous polymer template. The porous polymer Amberlite XAD4 is commercially supplied by (ROHM & HAAS). *N*-heptane of purity 99.5% (Fluka) was used as an adsorbate.

Table 1. Parameters characterizing the porosity of the investigated samples obtained from nitrogen adsorption/desorption at 77K: specific surface of the pores S_{BET}, total pore volume V_p, average pore diameter from BJH for a desorption isotherm D_{av}. An arrangement of pores was determined from TEM images.

Sample	S_{BET} (m^2/g)	V_p (cm^3/g)	D_{av} (nm)	Ordered	Material
SBA-15	1029	1.24	5.4	yes	silica
Si-A	1135	1.06	3.5	no	silica
XAD4	898	1.27	5.5	no	polymer

2.2. Methods

The powder samples were poured to a loosely closed container fastened inside a vacuum chamber, equipped in a gas dosing system. Prior to the adsorption experiment the samples were kept in vacuum (~10^{-4} Pa) for about 24 h. A 0.5 MBq ^{22}Na positron source sealed in an 8 μm Kapton envelope was placed in the middle of each sample. Before *n*-heptane (Fluka, 99.5% purity) was allowed to permeate the sample, it was degassed by freezing in liquid nitrogen followed by pumping out impurities while n-heptane was slowly melting. This procedure was repeated three times. Series of PALS measurements at various *n*-heptane pressures from zero to vapor pressure (increasing and then decreasing) was conducted at room temperature.

A standard 'fast-slow' delayed coincidence spectrometer was used to collect the positron annihilation lifetime spectra. The spectrometer was equipped with BaF_2 scintillators. Each spectrum has been measured for 8 hours, which was enough to collect 1.5-2.0 × 10^7 counts. A high counting rate was obtained at the expense of the resolution, which had to be approximated by two Gaussians (typically with *FWHM* of about 290 ps and 450 ps, and a contribution of 80% and 20% respectively).

The PALS spectra were analyzed using the LT 9.2 program [12]. In order to decrease a number of fitted parameters, the lifetime of para-positronium was fixed at 125 ps. The free positrons lifetime was varied from 350 to 480 ps depending on a material investigated and whether its pores were filled with *n*-heptane or not. In addition to these short-lived components, three ortho-positronium components were fitted during the analysis. The lognormal distribution of lifetimes was assumed for the longest-lived component. A width of the distribution was described by a dispersion parameter σ.

3. Results and discussion

Before *n*-heptane introduction to the samples three o-Ps components can be attributed to the structure of the pore walls (τ_3 ~ 1.5-2 ns), micropores (τ_4 ~ 6-10 ns) and mesopores (τ_5 > 80 ns) respectively. Almost all o-Ps in the studied porous materials annihilate in mesopores (Fig.1). The intensities of two short-lived components in both silica are approximately 1%. Only in XAD4 they reach about 3% each. This situation changes distinctly when *n*-heptane fills the pores. Reaching the vapor pressure results in a disappearance of the mesopore related component ('mesopore component') in all cases except SBA-15, but even there it decreases to 1%. The major origin of the short-lived components becomes different, i.e. *n*-heptane – pore wall interface ('interface component', still $\tau_3 \sim 1.5$-2 ns) and bubbles created by o-Ps [13] in liquid *n*-heptane ('liquid component', $\tau_4 \approx 3.7$ ns).

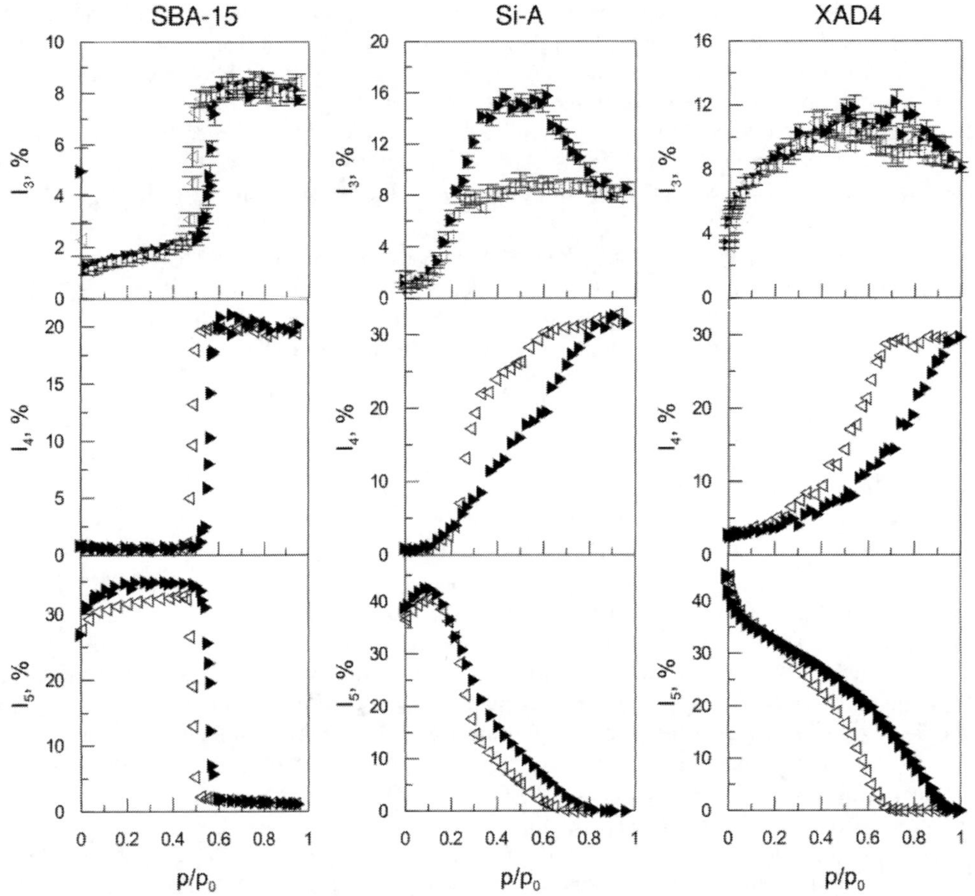

Figure 1. The intensities of the o-Ps components related to *n*-heptane – pore wall interface (I_3), liquid *n*-heptane (I_4) and mesopore (I_5) in SBA-15 (left) Si-A (middle) and XAD4 (right) as a function of *n*-heptane pressure relative to its vapor pressure during adsorption (full triangles) and desorption (open triangles).

The intensities of o-Ps components are much more sensitive to *n*-heptane adsorption comparing to the lifetimes. An interchange of intensities between the mesopore component (I_5) and the liquid component (I_4) is visible in Fig.1 as almost mirror reflection of these intensities dependency on relative pressure. This is a clear confirmation of filling initially empty pores (giving contribution to the mesopore component) by *n*-heptane (contributing to the liquid component). The shape of a $I_4(p/p_0)$ dependency is different for each material. This is caused by a differences in their pore size distributions (the larger the pore, the greater pressure is required for an adsorbate condensation inside of it). Almost vertical I_4 increase between 0.5 and 0.6 in SBA-15 proves practically a uniform pore size in this material. On the contrary, in Si-A I_4 starts to change at a very low pressure (~0.1) and continues almost linear increase until 0.8 is reached. Such a dependency suggests a wide distribution of pore sizes. Similarly, there is a whole range wide increase of I_4 in XAD4. However, in this case the steepest increase is visible at $p/p_0 > 0.6$.

In each porous material desorption of *n*-heptane from the pores of a particular size occurs at a different pressure than adsorption until the pressure decreases to about 0.2. It is visible as hysteresis loops on the dependencies of I_4 and I_5 in Fig. 1. This effect is primarily caused by a different mechanism of adsorption (condensation at roughly cylindrical surface parallel to pore walls) and desorption (evaporation from spherical meniscus formed at pore ends). This single effect is observed in SBA-15, which is consisted of parallel cylindrical pores forming honeycomb-like hexagonal structure. Still, the hysteresis is too large in Si-A and XAD-4 to be explained this way. In these two samples the pressure at which desorption begins is probably determined by the size of pore entrances, which are narrower than the pore internal size ('ink-bottle' pores). This hypothesis is consistent with what is expected to be a structure of pores in XAD4: free spaces left between polymer filled spheres. The Si-A structure can be similar, because it is synthesized using the polymer template similar to XAD4. However, in this case a predominance of very small entrances is expected, because the greatest I_4 decrease during desorption is observed at $p/p_0 = 0.2\text{-}0.3$.

Although I_5 dependencies on relative pressure are almost mirrored of I_4, there is a distinct difference at the relative pressure below 0.2 (0.1 in Si-A). No significant change of I_4 is observed in this range, while I_5 increases in both silica or decreases in polymer by several percent. There can be several causes of a change of the positronium formation probability, among them the most probable is a development of the pore surface. A greater surface roughness of a pore wall leads to a higher probability that a positron-electron pair correlated in a spur or blob [14] encounters the surface during its diffusion-like movement. In a consequence, the probability of a bulk leaving by the pair, followed by Ps formation in a pore, increases. The surface would develop in silica if the adsorbate, instead of forming layers, builds island-like structures. On the other hand, such structures on a polymer surface are not expected, because *n*-heptane initially penetrate between a polymer chains causing the material swelling. In this case most of the gaps between polymer chains close and the surface becomes smoother.

Similarly, a presence of small hysteresis loops between the adsorption and desorption run of I_5 at low pressure in the silica can be explained on the basis of a surface roughness. This is unlikely that island-like structures formed during adsorption appear also during desorption. More probable is that during evaporation the surface of the adsorbate covering pore walls is pretty smooth. Thus I_5 is smaller at low pressure during desorption comparing to adsorption, when the surface is more rough.

The shortest-lived component the interface component is related to an annihilation site, which is competitive to bubbles in *n*-heptane, i.e. both the interface component and the liquid component increase when the mesopore component decreases. This site was identified a priori as an *n*-heptane – pore wall interface. An argument for such identification is the I_3 increase at pressure lower than pressure of the I_4 increase. This is in agreement with an unquestioned scenario, where an *n*-heptane layer (discontinuous isle-like during adsorption) forms before condensation during adsorption or remains after evaporation during desorption. However, this straightforward interpretation of the interface component is most likely simplified. It should be noticed, that o-Ps annihilates not necessarily in the same place, where it was formed. If there is a connection between free volumes, o-Ps can migrate from one free space to another [15]. The direction of positronium migration is mostly from a smaller to larger free volume, because it is energetically favorable. Keeping in mind this effect we can suppose, that o-Ps formed in pore walls (mostly in micropores) gives its contribution to the interface component instead of the mesopore component, when the walls become at least partially covered by an adsorbate. This explains the slope of the I_3 pressure dependency in SBA-15 in the pressure range below the adsorbate condensation.

The origin of the hysteresis loop in the I_3 pressure dependency in Si-A can be also explained by taking into account the o-Ps migration. The high I_3 value (~15%) during adsorption is most likely the result of an annihilation of o-Ps formed both directly in the interface and in micropores connected with

Journal of Physics: Conference Series **618** (2015) 012040 doi:10.1088/1742-6596/618/1/012040

the interface. Because o-Ps formed in the micropores cannot migrate to mesopores, which was the case without adsorbate, they annihilate with the lifetime specific for the interface component. However, above $p/p_0 = 0.6$ *n*-heptane is pushed into the micropores, what results in decrease of I_3 to about 8%. During desorption the micropores remain filled until n-heptane is removed from the pore surface at $p/p_0 < 0.2$, so there is no contribution from them and I_3 remains constant. The origin of I_3 changes in XAD4 is partly different. Already relatively high I_3 increases mostly due to a change of the concentration of a free spaces between polymer chains caused by a swelling. Only about 2% increase above $p/p_0 = 0.1$ can be attributed to the origin similar as in the silica.

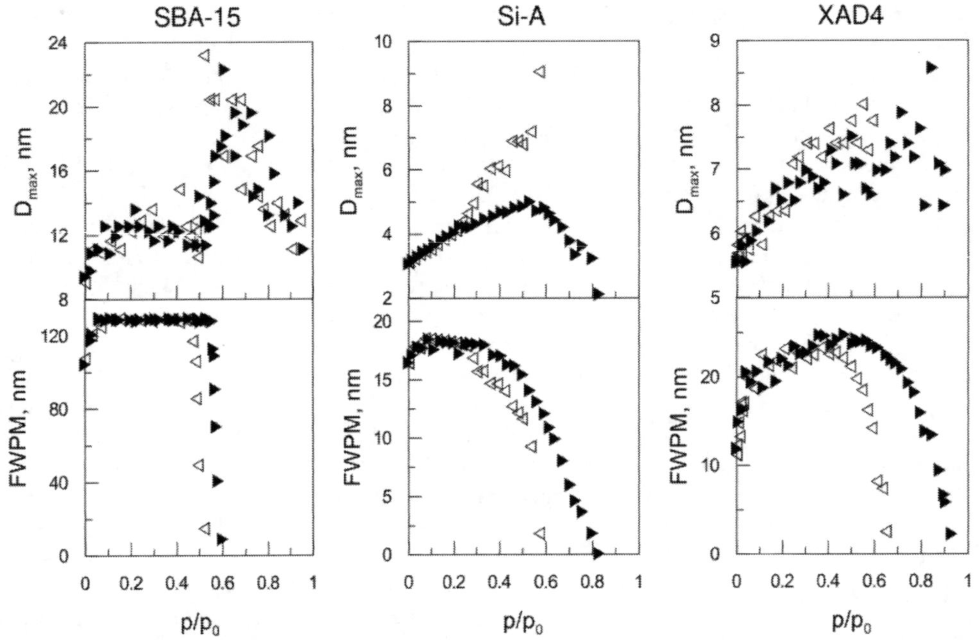

Figure 2. The diameter at a maximum of mesopore size distribution (D_{max}) and the full width at 1% of the maximum of the highest distribution (*FWPM*) in SBA-15 (left) Si-A (middle) and XAD4 (right) as a function of the relative *n*-heptane pressure during adsorption (full triangles) and desorption (open triangles).

The complementary information to the one provided by the intensities can be obtained from an analysis of the lifetimes, especially the longest-lived one proportional to the mesopore sizes. The relation given by the Extended Tao-Eldrup model [16] can be used to calculate pore sizes from lifetimes, but in order to obtain full information provided by positron lifetime spectra both an average lifetime and a dispersion have to be taken into account. Therefore, the lognormal shape of a lifetime distribution defined by an average lifetime and a dispersion was transformed into a pore size distribution according to the calculation scheme described in ref. [5]. Finally, the position of the maximum of a mesopore size distribution (D_{max}) and the full width at 1% of the maximum of the highest distribution for a particular material (*FWPM*) were calculated (Fig.2). The *FWPM* was used to obtain a width of a size distribution at a constant level for the whole pressure range.

The pore sizes D_{max} calculated from PALS data for empty pores are in a fairly good agreement with the nitrogen adsorption/desorption results (Table 1) excluding SBA-15. Most likely in SBA-15 the lifetime specific for inter-particle spaces (i.e. the spaces within a conglomerate of randomly oriented structures of hexagonally arranged pores, which forms a powder grain) instead of primary cylindrical pores is observed. This is the result of o-Ps migration outside the straight and relatively wide primary pores. Also very large *FWPM* is inadequate for pores of a uniform size. Thus no useful information can be obtained from these data. In other samples an increase of D_{max} with pressure is observed. It is the expected dependency, which confirms filling of the smallest pores in the first during adsorption and emptying them in the last during desorption. The increase is not preserved at relative pressure above 0.55 during adsorption in Si-A. It indicates a predominance of the pore narrowing over the closure of the smallest ones. This effect is weaker in XAD4, where D_{max} stabilizes only. Such a change of the slope of the D_{max} against pressure dependency is visible during desorption in any sample. This is another confirmation of different shape of *n*-heptane meniscus during evaporation in comparison to condensation.

A width of the distribution (*FWPM*) shows some increase at low pressure. This can be related to a change of the pore structure due to the swelling in the polymer (*FWPM* almost doubled), while its origin in the silica is uncertain (but in this case the increase is much smaller). Above the low pressure increase a plateau in *FWPM* is observed. A pressure increase in this range results in a shift of the pore size distribution towards larger sizes without a change of its width. A closure of the smallest pores is expected as it was already mentioned, but a simultaneous appearance of large ones is difficult to explain. It may be a result of fixing the lognormal shape of lifetime distributions. If a real shape is different (e.g. its maximum is shifted towards greater diameters comparing to the maximum of the lognormal distribution), such a distortion may cause an artificial broadening of the distribution. Nevertheless, above the pressure at which I_5 decreases to about a half of its highest value, *FWPM* begins to decrease as expected. This indicates the closure of consecutive groups of pores of a certain (the smallest ones still remain open) size. The changes of *FWPM* during desorption are shifted towards a lower pressure following the hysteresis observed in I_4 and I_5.

4. Conclusions

A PALS adsorption study provides numerous additional information about the studied material in comparison to a classic adsorption experiment. A pore structure determination based on the shape of the I_4 pressure dependency and the hysteresis loop provides practically the same information as a volume of adsorbed *n*-heptane in the classic experiment. Additional information about the initial growth or the final decay of the adsorbate layers on adsorbent surface is provided by I_5. A presence of a hysteresis in a I_3 pressure dependency can indicate presence of micropores. Probably an adjustment of an adsorbate molecule size may allow to detect micropores of various minimal sizes. The strength of the adsorbent-adsorbate interaction may also be important in a study of this effect. A direct measurement of a size of the free space, which is still free of adsorbate, allows us to verify the adsorption/desorption models.

Acknowledgments

The research was supported by The Polish Ministry of Science and Higher Education through the grant no. 2013/09/D/ST2/03712.

References

[1] Barrett EP, Joyner LG, Halenda PP 1951 *J. Am. Chem. Soc.* **73** 373
[2] Ravikovitch PI, Neimark AV 2001 *J. Phys. Chem. B* **105** 6817
[3] Ravikovitch PI, Neimark AV 2002 *Langmuir* **18** 9830
[4] Coasne B, Galarneau A, Di Renzo F, Pellenq RJM 2006 *Langmuir* **22** 11097
[5] Goworek J, Zaleski R, Buda W, Kierys A 2010 *Appl. Surf. Sci.* **256** 5316
[6] Zaleski R, Dolecki W, Kierys A, Goworek J 2012 *Microporous Mesoporous Mater.* **154** 142
[7] Kierys A, Zaleski R, Gorgol M, Goworek J 2013 *Microporous Mesoporous Mater.* **179** 104
[8] Kierys A, Zaleski R, Tydda M, Goworek J 2013 *Adsorption* **19** 529
[9] Zaleski R, Dolecki W, Goworek J 2013 *Mater. Sci. Forum* **733** 207
[10] Hartmann M, Vinu A 2002 *Langmuir* **18** 8010
[11] Zaleski R, Kierys A, Dziadosz M, Goworek J, Halasz I 2012 *RSC Advances* **2** 3729

Journal of Physics: Conference Series **618** (2015) 012040 doi:10.1088/1742-6596/618/1/012040

[12] Kansy J 1996 *Nucl. Instrum. Methods Phys. Res., Sect. A* **374** 235
[13] Ferrell RA 1957 *Phys. Rev.* **108** 167
[14] Stepanov SV, Byakov VM 2002 *J. Chem. Phys.* **116** 6178
[15] Zaleski R, Błażewicz A, Kierys A 2013 *Nukleonika* **58** 233–238
[16] Goworek T, Ciesielski K, Jasinska B, Wawryszczuk J 1998 *Chem. Phys.* **230** 305

Journal of Physics: Conference Series **618** (2015) 012041 doi:10.1088/1742-6596/618/1/012041

Confined water in controlled pore glass CPG-10-120 studied by positron annihilation lifetime spectroscopy and differential scanning calorimetry

O Šauša[1], I Maťko[1], E Illeková[1], E Macová[2] and D Berek[2]

[1] Institute of Physics, Slovak Academy of Sciences, Dúbravská cesta 9, 845 11 Bratislava, Slovakia
[2] Polymer Institute, Slovak Academy of Sciences, Dúbravská cesta 9, 845 41 Bratislava, Slovakia

E-mail: ondrej.sausa@savba.sk

Abstract. The solidification and melting of water confined in the controlled pore glass (CPG) with average pore size 12.6 nm has been studied by differential scanning calorimetry (DSC) and positron annihilation lifetime spectroscopy (PALS). The fully-filled sample of CPG by water as well as the samples of CPG with different content of water were used. The measurements show the presence of amorphous and crystalline phases of water in this type and size of pores, freezing point depression of a confined liquid and presence of certain transitions at lower temperatures, which could be detected only for cooling regime. The localization of confined water in the partially filled pores of CPG at room temperature was studied.

1. Introduction

Porous materials with nanometer size pores are extensively exploited in science and technology. They are used as adsorbents, catalyst support and chromatographic packings. Such materials can suppress the crystallization of confined materials and allow to prepare different structures in comparison with bulk state [1]. Water belongs to the most important substances in nature and is essential for the life. The knowledge of water properties in confined systems is very important, because water exhibits complicated behaviour and many anomalies [2,3].

Systematic research of confined systems associating differential scanning calorimetry (DSC) and positron annihilation lifetime spectroscopy (PALS) methods has been launched [4-6]. The localization of confined liquid in the pores of the matrix in the case of partially-filled pores is also interesting information [6]. In the present study, the thermal behaviour of water solidified in the controlled pore glass CPG-10-120 of mesoporous category (12.6 nm average pore size) as well as the localization of water in the partially-filled pores at room temperature are investigated. The results are compared with the study of the bulk water and ice by positron annihilation method [7].

2. Experimental

For this study CPG-10-120 obtained from BDH Chemicals Ltd., England with mean pore diameter 12.6 nm, pore volume 0.87 mL g^{-1} and pore surface area 187.1 m^2 g^{-1} has been used. This mesoporous material prepared by specific thermal treatment of borosilicate glass [5] with narrow size pore distribution (± 7.8 %, information from producer) is characterised by sponge-like morphology. The distilled water (standard laboratory quality) was used. The process of water filling was based on the action of capillary forces, which pulled defined quantity of water into the pores (filled in air). The partial mass of water (or coefficient of filling) in each sample is defined as $k = m_{H2O}/(m_{H2O} + m_{CPG})$, where m_{H2O} and m_{CPG} is mass of water and CPG (± 0.1 %), respectively, in the prepared sample.

The DSC measurements were performed using power-compensation DSC8500 (Perkin-Elmer) with automatic intracooler (200 K). Samples of 15 – 30 mg encapsulated in air and measured in dynamic nitrogen atmosphere were used. Continuous cooling and heating regimes with the rates of 2 K min^{-1} were applied. The precision of the measured data is ± 0.4 K and ± 2 J g^{-1}.

The low-temperature dependence of o-Ps lifetimes in the CPG/water composite for the $k = 0.522$ in the temperature region 60-300 K (closed cycle He-refrigerator, temperature stability better then 1K) was studied by PALS. At room temperature the samples with different content of water (sequential evaporation of water from the primary sample with $k = 0.522$) were investigated also. Lifetime spectra were measured by the conventional fast-fast coincidence time spectrometer with the time resolution about 320 ps FWHM determined by Al defect-free sample. PATFIT-88/POSITRONFIT [8] code was used for the analysis. The correction on the positron annihilation in the source as well as in the sample container Kapton foil window was taken into account. The free volume pore sizes were estimated from the o-Ps lifetime using EELViS routine [9] for the spherical cavity.

3. Results and discussion

PALS results for o-Ps lifetimes and their relative intensities for the different k at room temperature in air are in Figs. 1-3 (four-component analysis used). The τ_4 (k) dependence shows uniform value and I_4 increases with decreasing of k in large range of k (0.12-0.45, area II on figures). Water gradually escapes from large pores without creating of thick layer on the CPG inner pore surface. In fully-filled and over-filled samples I_4 is negligible (τ_4 with large errors, k>0.45, area III) and τ_3 is similar to the value of liquid bulk water. For low content of water (k<0.12, area I) τ_3 increases more evident with decreasing k. This effect is connected with an influence of similar value of o-Ps lifetime in small pores in pure matrix (k=0) and creation of additional small free-volumes, when monolayer of water on pore surface is probably crushing whenever water escapes from pore surfaces.

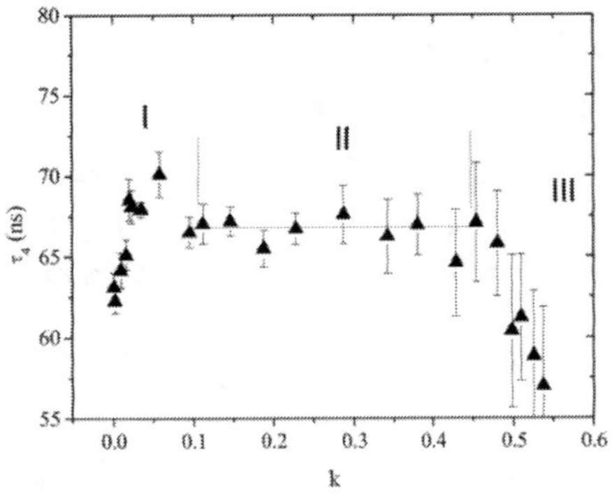

Figure 1. Dependence of o-Ps lifetime τ_4 (connected with large CPG pores) on the coefficient of filling k.

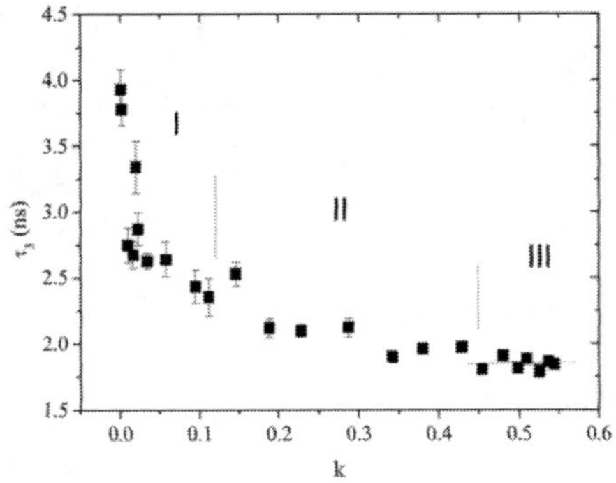

Figure 2. Dependence of o-Ps lifetime τ_3 on the coefficient of filling k.

The temperature dependence of o-Ps lifetime τ_3 (Fig. 5, three-component fit was used) for the region 260-300 K is very similar to that of bulk water. The sharp change of the τ_3 at phase transition of water to ice presented at the bulk state [7] is absent in confined system. The global trends in the temperature dependence change here gradually, indicating a less ordered structure and being similar to amorphous systems. Thence τ_3 is so high at low temperature compared with bulk water value.

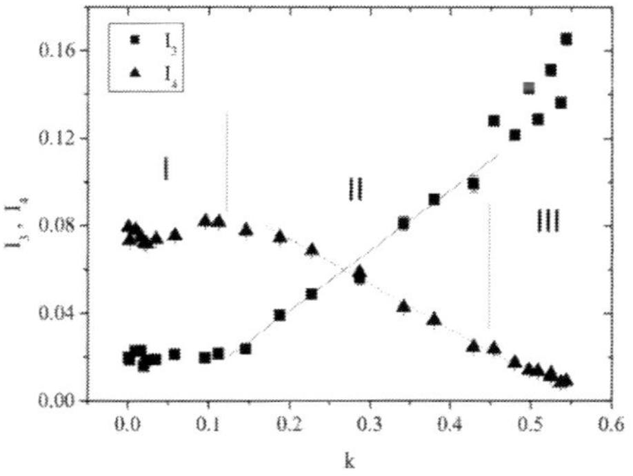

Figure 3. Dependence of relative intensity I_3 and I_4 on the coefficient of filling k.

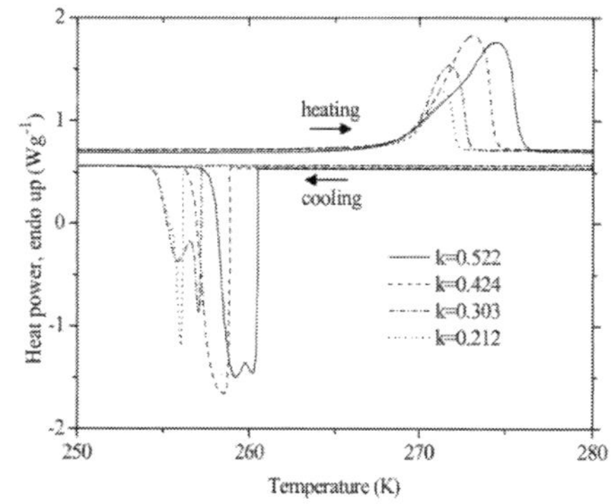

Figure 4. DSC curves exhibit a shift of solidification peaks towards lower temperatures with decreasing water content.

The glass transition temperature T_{gPALS} is determined as the intersection of lines for low temperature region and for heating regime in the region 200-240 K. This estimated value 186 K is in good agreement with the measurements on the water confined in matrices with smaller pore sizes [1]. The transformation area at T_{gPALS} is wide and more complex (mixture of glassy state and temperature dependence of τ_3 connected with creation of vacancies in crystalline ice, similar to bulk state [10]). Accordingly PALS measuremnts exhibit local changes of τ_3 (visible mainly in the cooling regime) and I_3 (see Fig. 6) at temperature below 231 K (solidification by homogeneous nucleation) as well as changes of I_3 in the temperature region of 230-273 K during different processes of freezing and melting in CPG pores.

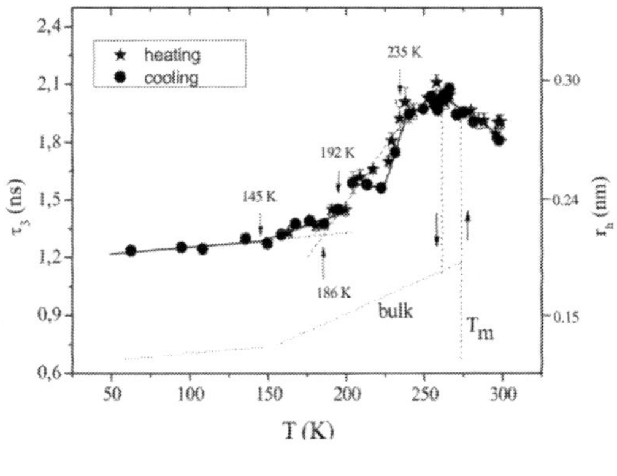

Figure 5. Dependence of o-Ps lifetime τ_3 (or free-volume cavity size r_h) on the temperature T for confined ($k = 0.522$) and bulk water (dashed, from ref. [7]).

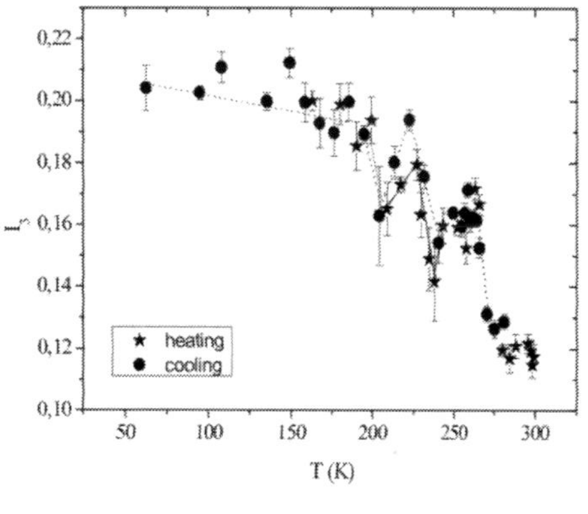

Figure 6. Dependence of relative intensity I_3 on the temperature T and $k = 0.522$.

Characteristic DSC curves exhibit hysteretic behaviour, i.e. crystallization in the pores is detected at lower temperatures compared to melting (Fig. 4), as observed in many previous studies [11].

Journal of Physics: Conference Series **618** (2015) 012041 doi:10.1088/1742-6596/618/1/012041

Especially for the cooling regime dual thermal response is evident. While a scattering of onset of solidification peaks for individual measurement cycles of one sample usually reaches ~1 K, the difference between onset of freezing peaks for lowest and highest filling reaches at least 5 K. DSC thermogram shows a weak peak at 231 K in cooling cycle for all *k*. This peak is related to crystallization of confined water in the spherical cavities by homogenous nucleation [12,13].

The estimation of amorphous and crystallized fraction of water in CPG was based on the difference between the total enthalpy determined from DSC thermogram for cooling cycle relative to 1 g water and total enthalpy of pure water. Results are in Tab.1. From these results it follows that larger amorphous fraction is observed if water is absent in the central part of large pores of CPG.

Table 1. Content of water in the CPG samples expressed by coefficient *k* and estimated fraction of crystalline water from DSC measurements.

k	Crystalline fraction (wt%)
0.52	73
0.424	72
0.212	43

4. Conclusions

PALS temperature dependences confirm the presence of amorphous and crystalline phases of water confined in 12.6 nanometer pore size of CPG for the fully-filled pores. From the $\tau_3(T)$ dependence the glass transition temperature T_{gPALS} was estimated to the value 186 K. Confined water in CPG of the mesoporous category exhibits an evident dependence of thermal behaviour on the filling of the porous structure. This effect manifests itself by shift of DSC solidification peaks towards lower temperatures with decreasing water content. Proposed interpretation, based on the influence of microstructural properties on ice front percolation, is well supported also by the detected presence of solidification by homogeneous nucleation at typical temperature (231 K). The fraction of amorphous and crystallized water was estimated for different pore filling. The amorphous fraction is larger for the lower content of water in the CPG pores. Water escapes by evaporation from large pores with increasing of number of "empty" large pores (I_4 increase) with no significant changes of τ_4 (no thick layer on the inner pore surface) for $k > 0.1$.

Acknowledgements

The authors wish to thank the Agency VEGA Slovakia, projects no. 2/0001/12, 2/0189/14, as well as APVV, projects 0492-11, 0109-10 and CEX FUN-MAT.

References

[1] Rousenova M, Alam M A, Townrow S, Kilburn D, Sokol P E, Guillet-Nicolas R and Kleitz F 2014 *New. J. Phys.* **16** 103030
[2] http://www.lsbu.ac.uk/water/
[3] Malenkov G 2009 *J. Phys.: Condens. Matter* **21** 283101
[4] Iskrová M, Majerník V, Illeková E, Šauša O, Berek D and Krištiak J 2009 *Mat. Sci. Forum* **607** 235
[5] Illeková E, Krištiak J, Macová E, Maťko I and Šauša O 2013 *J. Therm. Anal. Calorim.* **113** 1187
[6] Šauša O, Illeková E, Krištiak J, Berek D and E Macová 2013 *J. Phys. Conf. Ser.* **443** 012059
[7] Šauša O, Iskrová M, Sláviková B, Majerník V and Krištiak J 2011 *Mat. Sci. Forum* **666** 115
[8] Kirkegaard P, Pedersen N J and Eldrup M 1989 *PATFIT-88*, Riso Report – M2740
[9] Zaleski R 2009 EELViS, http://eelvis.sourceforge.net. Accessed 15.06.2011
[10] Eldrup M and Mogensen O 1972 *J. Chem. Phys.* **57** 495
[11] Riikonen J, Salonen J and Lehto V P 2011 *J. Therm. Anal. Calorim.* **105** 811
[12] Morishige K, Yasunaga H, Denoyel R and Wernert V 2007 *J. Phys. Chem. C* **111** 9488
[13] Janssen A H, Talsma H, van Steenbergen M J and de Jong K P 2004 *Langmuir* **20** 41

Journal of Physics: Conference Series **618** (2015) 012042 doi:10.1088/1742-6596/618/1/012042

Positron-Annihilation Lifetime Spectroscopy using Electron Bremsstrahlung

A Wagner[1], W Anwand[1], M Butterling[1], T E Cowan[1,2], F Fiedler[1], F Fritz[1,2], M Kempe[1,2,†], R Krause-Rehberg[3]
[1] Helmholtz-Zentrum Dresden-Rossendorf, Institute of Radiation Physics, Bautzner Landstr. 400, 01328 Dresden, Germany
[2]Technische Universität Dresden, 01062 Dresden, Germany
[3]Martin-Luther-Universität Halle-Wittenberg, Institute of Physics, 06099 Halle, Germany
†deceased

E-mail: a.wagner@hzdr.de

Abstract. A new type of an intense source of positrons for materials research has been set up at the superconducting electron linear. The source employs hard X-rays from electron-bremsstrahlung production generating energetic electron-positron pairs inside the sample under investigation. CW-operation allows performing experiments with significantly reduced pile-up artefacts in the detectors compared to pulsed mode operation in conventional accelerators. The high-resolution timing of the accelerator with bunch lengths below 10 ps full width at half maximum (FWHM) allows positron annihilation lifetime spectroscopy (PALS) measurements with high time resolution. A single-component annihilation lifetime of Kapton has been measured as (381.3 ± 0.3) ps. Employing segmented detectors for the detection of both annihilation photons allows for the first time to perform a 4D tomographic reconstruction of the annihilation sites including the annihilation lifetime.

1. Introduction

Positron annihilation lifetime spectroscopy (PALS) serves as a suitable tool for the characterization of nanometer-sized lattice defects in crystals at low defect concentrations. Moreover, the production of Positronium allows studying open volumes in polymers and nanoscale porosities. Several complementary techniques have been developed in the past which make use of either kinematical observables of the annihilation radiation of positrons with electrons from the sample materials and of the annihilation lifetime of positrons after injection into the sample material. Long positron annihilation lifetimes beyond a few ns are caused by voids and porous structures inside the sample which may be of high relevance for material failure and/or structural failure. The new source is especially suited for extended bulk samples, or samples cannot be exposed to external sources of positrons (low-energy positron beams, radioactive sources), e.g. because they are imposing hazardous conditions (high pressure, high temperature, intrinsic radioactivity) to the source, or if the sample handling imposes difficulties (fluids, gases, organic samples). Here, we also report about a new method which allows correlating positron annihilation lifetime studies with a three-dimensional tomographic analysis of a bulk sample. The well-established PALS technique makes use of radioactive positron emitters (e.g. ^{22}Na) where the positron emission is accompanied by electromagnetic transitions from excited states in the daughter nuclei. Because photon emission happens promptly with respect to the positron emission (the lifetime of the 2^+ state in ^{22}Ne is only 3.6 ps) the time resolution of the additional photon detector adds to the overall achieved accuracy in lifetime measurements. Both, the contribution of the lifetime of the excited state and the contribution of the detector resolution can be avoided using an accelerator-based source of positrons. Furthermore, both the source strength and the time structure can be matched to the sample conditions under study. Here, we present a new versatile source of positrons using a superconducting electron accelerator for position-resolved positron annihilation lifetime spectroscopy.

2. The superconducting electron linear accelerator

The Helmholtz-Zentrum Dresden-Rossendorf operates the superconducting electron accelerator ELBE [1] which delivers electron beam energies up to 40 MeV and average beam currents up to 1.6 mA in continuous-wave mode. The electron beam serves as a source of secondary radiation like coherent infrared light from free-electron lasers, THz radiation from undulators and dipole magnets, photo-neutrons produced inside a liquid-lead target, hard bremsstrahlung, and positrons. Due to individually accessible caves, the setup of experiments is possible while beams are delivered to other caves. The layout of the facility and the various end stations are shown in Fig. 1. The superconducting technology allows adjusting the time structure for time-of-flight experiments (neutrons), pump-probe experiments, or annihilation lifetime experiments as described in this contribution. Specifically, the micro-pulse repetition rate can be selected as $1/2^n \times 26$ MHz with $n = 0\ldots8$. Typically, the micro-pulse repetition rate for positron annihilation lifetime experiments is chosen to be 26 MHz or 13 MHz with micro-pulse intervals of 38 ns or 77 ns, respectively, depending on the lifetime components expected in the sample under investigation. The micro-pulse width has been measured to be less than 5 ps using electro-optical sampling [2]. The ELBE installation operates as a 24/7 user facility while the free-of-charge access is granted through an external advisory committee.

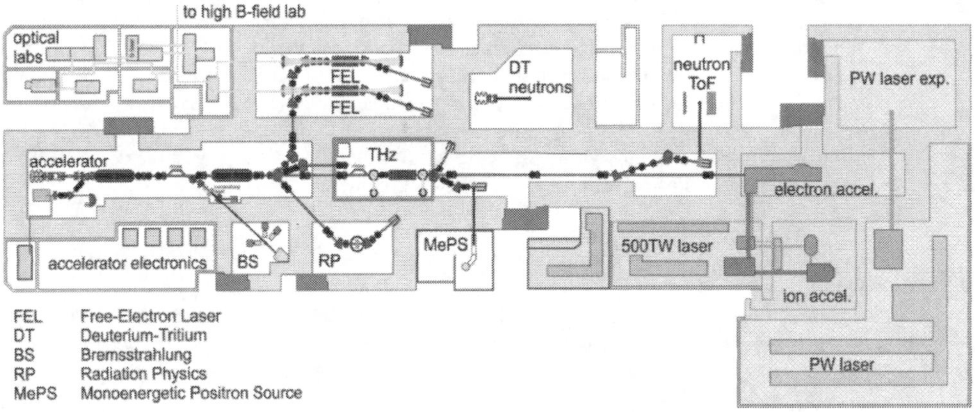

Figure 1. Layout of the ELBE centre for high power radiation sources. The overall width is about 100 m. Experiments discussed here took place at the bremsstrahlung facility labelled BS. Colour-shaded areas indicate high-power laser setups which are currently being developed for electron beam-laser interactions [3] and laser plasma ion acceleration.

3. Positron Annihilation Lifetime Spectroscopy of Bulk Samples

A new kind of positron annihilation lifetime spectroscopy setup has been realized at the bremsstrahlung facility [4] (see Fig. 2) at the ELBE LINAC. The pulsed electron beam with energies up to 16 MeV is converted into hard photons using thin Niobium radiator foils with thicknesses around 10^{-3} radiation lengths. Shaped by a 2600 mm long collimator the bremsstrahlung beam impinges onto a sample under study where it generates secondary positrons via pair production. These positrons thermalize in a few ps inside the sample. Characteristic annihilation quanta are being registered in coincidence by 4 sets of pairs of high-purity Germanium semiconductor detectors (HPGE) for accurate energy determination and BaF_2 scintillation detectors for accurate timing. Coincidence conditions and list mode storage of both energy and timing signals help efficiently reduce background. Furthermore, annihilation lifetime and electron momentum correlation studies, aka age-momentum correlation (AMOC) [5], can be performed with ease. Time walk effects of the timing discriminators can be minimized by offline corrections using the event-by-event recorded list mode data. In comparison to pioneering works at a normal conducting electron machine [6], the new installation makes use of continuous-wave (CW) beams which are only available at a superconducting (SC) accelerator. The CW–mode prevents detector pile-up effects which are present in macro-pulse bunched machines. Moreover, due to higher bunch charges the bunch lengths in normal conducting machines are significantly higher than in SC-machines. With typical ns-bunch widths positron annihilation lifetime measurements of metals become impossible.

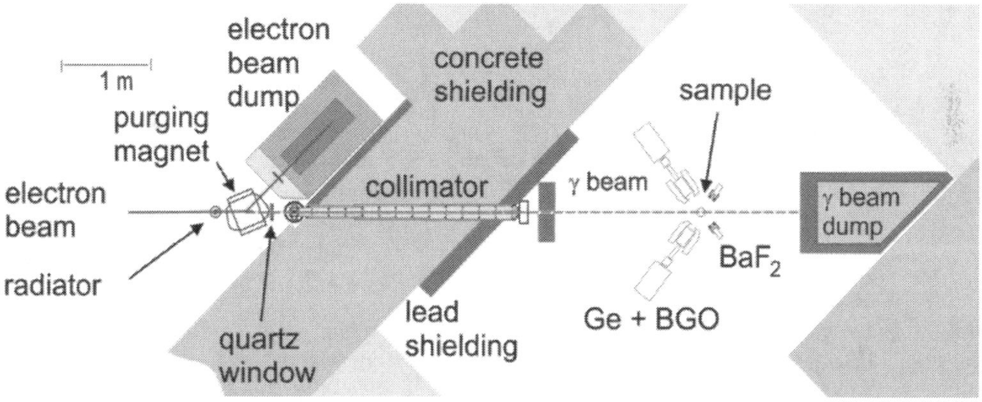

Figure 2. Layout of the bremsstrahlung facility at ELBE. The electron beam (red line) impinges from the left onto a radiator foil producing bremsstrahlung (blue line).

The facility has already been used in various positron lifetime studies and age-momentum studies (e.g [7, 8, 9]. As a benchmark example, we present here annihilation lifetime studies for bulk Kapton© (DuPont, Wilmington, NC), a polyimide widely used as cover material for radioactive sources in annihilation lifetime studies. Since positrons annihilate not only inside the sample material, the contribution inside the cover material has to be subtracted from the spectrum of the investigated sample material. Therefore, the source correction requires exact knowledge of the positron lifetime of the material. In contrast to other foil materials like Mylar, the annihilation parameters of Kapton© are not sensitive to temperature, making it an excellent candidate for PALS measurements. Up to now, there is a discussion about the positron lifetimes of Kapton©. The presence of just a single component with a lifetime of 382 ps was accepted for long time (for example [10]). On the other hand, a bimodal distribution with two components (280 ps with 30 % intensity and 410 ps) was reported in [11]. In order to help clarifying the discussing a measurement of the positron annihilation lifetime in Kapton© has been performed with positrons from pair production. The bulk Kapton© sample had dimensions of 7x7x7 mm^3. The beam repetition rate has been selected to be 77 ns in order to avoid pile-up effects from annihilation events with long lifetimes, e.g. from ortho-Positronium formation. From the combined analysis of the four independent lifetime measurements (see Fig. 3) one derives a lifetime of (381.3 ± 0.3) ps. Parameterizations with more than one lifetime component resulted in a worse χ^2 value. In order to check for the significance of Positronium formation at the Kapton© surfaces, a stack of foils has been studied, as well, but no indication for an extended lifetime component has been found. Fig. 3 shows the annihilation lifetime distributions measured by one of the four HPGe/BaF_2 combinations and the result of a parameterization with a single-component exponential and a constant value for beam uncorrelated background events. The structure observed at about 8 ns stems from background by scattered photons behind the sample.

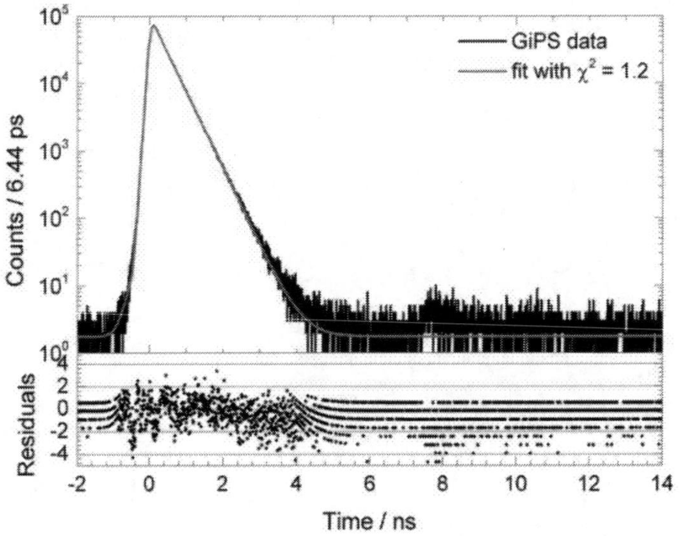

Figure 3. Positron annihilation lifetime distribution of bulk Kapton© for one of the four sets of HPGe/BaF2 detectors and result of the single-component exponential fit.

4. Tomographic Positron Annihilation Lifetime Spectroscopy

The positron annihilation lifetime facility discussed before has been extended by a set of position-sensitive photon detectors with the aim to reconstruct a three-dimensional image of the distribution of positron lifetimes inside bulk samples [3]. Two pixelated photon detector modules, each made from 13 × 13 crystals of Lu_2SiO_5:Y (LYSO) of 4 × 4 × 20 mm^3 volume (Siemens AG, Munich: Biograph® PET scanner), have been used. Each module is equipped with 4 photomultiplier tubes. Dedicated preamplifiers and timing discriminators have been developed resulting in a time resolution of 530 ps (FWHM) for the combined mean-time of all channels. The photon energy deposition is calculated using the sum of all four charge-integrated signals and individually calibrated crystal responses. The obtained energy resolution is 11.4% (FWHM) at 511 keV photon energy for both detectors, respectively. Signal partitioning between the four photomultiplier tubes of one detector allows identification of the crystal in which the photon interacted. Figure 4 shows a sketch of the setup with both detectors fixed perpendicular to the direction of the incoming photon beam and rotational stage with the mounted sample. A three-dimensionally structured sample has been employed. It consists of a 25 mm diameter PTFE (Teflon) cylinder with embedded slabs made from Copper, Iron, and Aluminium having the same volume (12 mm^2 × 25 mm) but different geometrical shapes, see Fig. 4. During the course of the measurement, the sample is rotated by 360° in steps of 2°.

In the case of two single-pixel events between the two detectors the line-of-response is calculated as the 3-dimensional connection between the two pixel positions. The simplest image reconstruction is performed for three-dimensional distributions by overlaying all lines-of-response while rotating the sample around an axis perpendicular to the beam. Gating on positron lifetimes in excess of 225 ps (1 σ of the time resolution) discriminates for regions inside the sample with enhanced formation of o-Ps, namely regions with increased porosity (here: PTFE). Taking the mean timing of both detectors as the annihilation lifetime the inaccuracy due to different positron production sites perpendicular to the beam cancels. From previous studies [3] a lateral position resolution of 2.8 mm has been obtained.

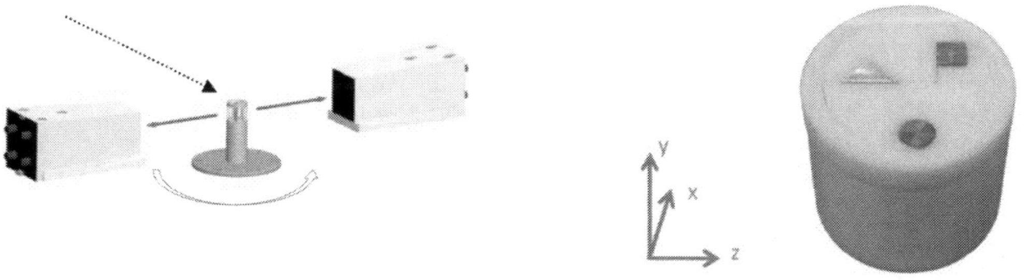

Figure 4. Left side shows the setup of two pixelated LYSO photon detectors employed in the tomographic system. Right side shows a PTFE cylinder of 2.5 cm diameter with inserts made of Al (triangle), Cu (disk), and Fe (square).

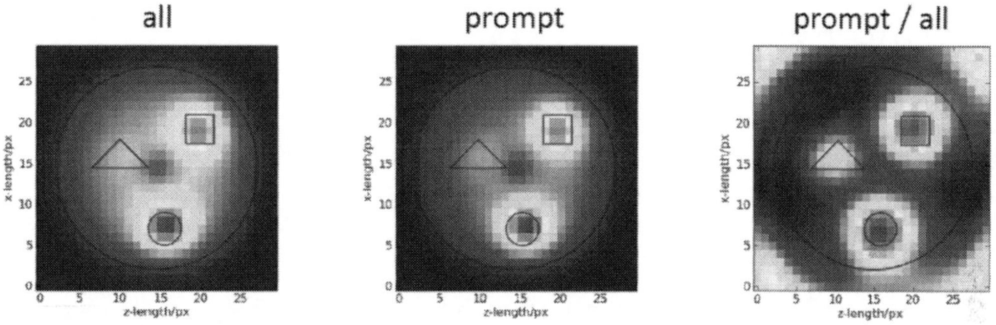

Figure 5. Reconstructed distribution of annihilation events for different regions of annihilation lifetimes after thirty iterations of the MLEM algorithm. Left picture shows all measured annihilations while the central picture is gated on a time interval between –1 ns to 2 ns (prompt). The right picture shows the ratio between prompt and all events discriminating for short annihilation lifetimes. The boundaries of the sample materials are indicated as thin lines.

In order to obtain a reconstructed distribution of annihilation events, the Maximum Likelihood Expectation Maximization (MLEM) algorithm [12] has been employed. The algorithm allows reconstructing iteratively a 3-D distribution of events from the set of all measured annihilation events and the system matrix which comprises all possible combinations of detector pixels (169 × 169), rotation angles (180), image voxels (30 × 30 × 30) resulting in an overall size of 138 billion matrix cells. Two reconstructed distributions which are obtained from full (Fig. 5, left) and gated on only prompt annihilation events (Fig. 5, centre) are again used to generate a contrast enhanced annihilation-lifetime sensitive source distribution shown in Fig. 5, right side.

The newly developed system complements earlier developments of a positron annihilation microprobe which enabled high-resolution two-dimensional defect analysis at surfaces [13, 14] or a high-energy positron beam for PALS studies in bulk material [15]. Further developments of the presented system aim at improving the position resolution by using smaller scintillator crystals and improved time resolution by employing digital silicon photomultiplier from Philips Digital Photon Counting which showed an improved time resolution already [16].

Acknowledgements
The authors like to thank M. Sobiella, K. Heidel, A. Hartmann for the experimental support and the ELBE crew for stable beams.

References

[1] Gabriel F, Gippner P, Grosse E, Janssen D, Michel P, Prade H, Schamlott A, Seidel W, Wolf A, Wünsch R 2000 *Nucl. Inst. Meth. Phys. Res.* **B 161** 1143.
[2] Kaya C, Schneider C, Al Shemmary A, Seidel W, Kuntzsch M, Bhattacharyya J, Mittendorff M, Evtushenko P, Winnerl S, Staats G, Helm M, Stojanovic N, Michel P, Gensch M 2012 *Appl. Phys. Lett.* **100** 141103.
[3] Jochmann A, Irman A, Bussmann M, Couperus JP, Cowan TE, Debus AD, Kuntzsch M, Ledingham KWD, Lehnert U, Sauerbrey R, Schlenvoigt HP, Seipt D, Stöhlker T, Thorn DB, Trotsenko S, Wagner A, Schramm U 2013 *Phys. Rev. Lett.* **111** 114803.
[4] Schwengner R, Beyer R, Dönau F, Grosse E, Hartmann A, Junghans AR, Mallion S, Rusev G, Schilling KD, Schulze W, Wagner A 2005 *Nucl. Inst. Meth. Phys. Res.* **A 555** 21.
[5] Wagner A, Anwand W, Butterling M, Cowan TE, Fiedler F, Kempe M, Krause-Rehberg R 2012 *Defect and Diffusion Forum* **331** 41.
[6] Selim FA, Wells, Harmon JF, Williams J, J. Appl. Phys. **97** (2005) 113539
[7] Zvezhinskiy D, Butterling M, Wagner A, Krause-Rehberg R, Stepanov SV 2013 *J. Phys.: Conf. Ser.* **443** 012057
[8] Zvezhinskiy D, Butterling M, Wagner A, Krause-Rehberg R, Stepanov SV 2014 *Acta Physica Polonica* **A 125** 821.
[9] Butterling M, Anwand W, Cowan TE, Hartmann A, Jungmann M, Krause-Rehberg R, Krille A, Wagner A 2011 *Nucl. Inst. Meth. Phys. Res.* **B 269** 2623.
[10] MacKenzie IK, Fabian J 1980 *Nuovo Cimento* B 58 162.
[11] McGuire S, Keeble DJ 2006 *J. Phys. D* 39 3388.
[12] Shepp LA, Vardi Y 1982 *IEEE Transactions on Medical Imaging* **1** 113.
[13] Haaks M, Bennewitz K, Bihr H, Männig U, Zamponi C, Maier K 1999 *Appl. Surf. Sci.* **149** 207.
[14] Oshima N, Suzuki R, Ohdaira T, Kinomura A, Narumi T, Uedono A, Fujinami 2009 *Appl. Phys. Lett.* **94** 194104
[15] Oshima N, Kuroda R, He C, Kinomura A, Ohdaira T, Toyokawa H, Suzuki R, Hayashizaki N, Hattori T 2007 *Phys. Status solidi* **C4** 4023.
[16] Hueso-González F, Golnik C, Berthel M, Dreyer A, Enghardt W, Fiedler F, Heidel K, Kormoll T, Rohling H, Schöne S, Schwengner R, Wagner A, Pausch G 2014 *J. Instr.* 9 P05002

Journal of Physics: Conference Series **618** (2015) 012043 doi:10.1088/1742-6596/618/1/012043

Spin polarized low-energy positron source

V N Petrov[1,2], S N Samarin[2], K Sudarshan[2], L Pravica[2], P Guagliardo[2], J F Williams[2]
[1] St. Petersburg State Polytechnical University, St. Petersburg 195251, Russia
[2] Centre for Atomic, Molecular and Surface Physics (CAMSP), School of Physics, The University of Western Australia, Perth WA 6009, Australia

E-mail: nikiforovich@hotmail.com

Abstract. This paper presents an investigation of spin polarization of positrons from a source based on the decay of ^{22}Na isotopes. Positrons are moderated by transmission through a tungsten film and electrostatically focussed and transported through a 90 deg deflector to produce a slow positron beam with polarization vector normal to the linear momentum. The polarization of the beam was determined to be about 10% by comparison with polarized electron scattering asymmetries from a thin Fe film on W(110) at 10^{-10} Torr. Low energy electron emission from Fe layer on W(100) surfaces under positron impact is explored. It is shown that the intensity asymmetry of the electron emission as a function of the incident positron energy can be used to estimate the polarization of the positron beam. Also several materials with long mean free paths for spin relaxation are considered as possible moderators with increased polarization of the emergent positrons.

1. Introduction

Comprehensive studies of the characteristics of surfaces and thin films frequently depend on effective sources [1] and detectors [2] of polarized electrons. Electron exchange and various spin-obit coupling phenomena are usually identified through angular momenta measurements and their symmetries. Correlations between quantum particles must concern both charge and spin degrees of freedom. Our research is directed towards improved understanding of quantum correlations and hence the development and use of sources and detectors of polarized electrons and positrons.

One major aspect of our research direction concerns the applicability of a positron beam as a tool for the detailed investigation of magnetic properties of material surfaces. This report concerns the design of a low-energy spin-polarized positron source and preliminary measurements. Positrons emitted from radioactive sources have preferred spin direction because of parity violation in the process of beta decay [3] and generally retain their spin orientation, even after the moderation process when passing through a thin (~ 1 micron) tungsten film. These features are the basis for a slow spin-polarized positron beam.

2. Experimental

The experiments were carried out at the University of Western Australia using a positron beam system fabricated at the University of Texas [4] and subsequently modified for working at UHV about 10^{-10} Torr. The positron beam system consist of two parts namely variable energy positron beam system (figure 1) and experimental scattering chamber (figure 2), fitted with surface science instrumentation. The positron beam-forming apparatus in figure 1 consists of a radioactive source of positrons, a

Journal of Physics: Conference Series **618** (2015) 012043 doi:10.1088/1742-6596/618/1/012043

positron moderator and positron optics. The positron source is based on the radioactive decay of ~20 mCi ^{22}Na. Polycrystalline tungsten foil of thickness of 1μm is used to moderate positrons which are electrostatically focused and transported into the experimental chamber. The potential between the moderator and the sample determines the beam energy. The use of an electrostatic (non-magnetic) transport system maintains the direction of spin polarization of the positron beam. Potentials on the various elements of the optical system were optimized using SIMION (www.simion.com) program. The experimental chamber contained position-sensitive microchannel plate detectors grids fitted with several planar grids to enable, for example, retarding potential energy analysis of either secondary electrons or positrons. A vacuum of low 10^{-10} mbar was maintained. Details of the system can be found in ref [5].

The direction of the spin polarization of the positron beam leaving the 90 degree deflector is horizontal in the plane of the paper of figure 1. Electrons ejected from the foil were detected by one of two MCP detectors fixed at 90 degrees one to another and at 45 degrees to the beam.

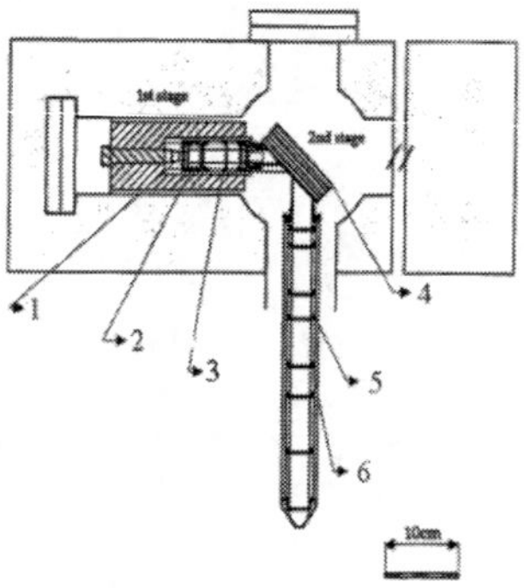

Figure 1. Schematic of positron beam system delivering positrons into the scattering chamber (figure 2). 1. A radioactive source of positrons 2. Moderator. 3. Input of positron optics, 4. The system of rotation of positrons 5. Focusing system 6. Output of the positron optics.

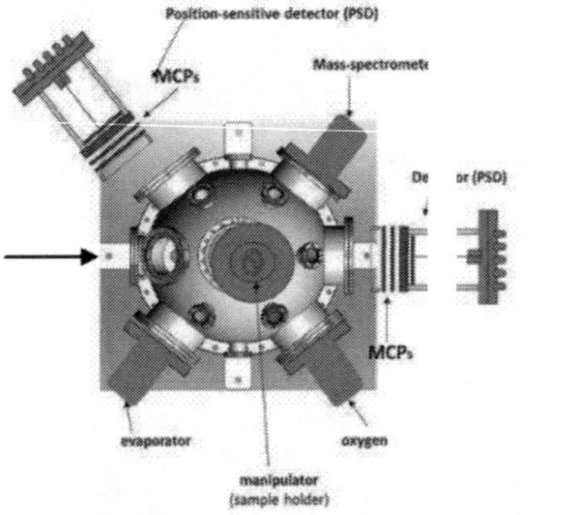

Figure 2. The experimental surface scattering chamber showing the location of detectors and attached surface characterization instruments. The positrons beam enters horizontally from the left.

3. Results and discussion

Samples of W (100), with or without Fe over layers, were studied by various methods of electron spectroscopy to determine their electronic and magnetic characteristics.

Figure 3 shows the dependence of the intensity of positrons and electrons as a function of the energy of the positrons incident (moderator voltage) on the tungsten surface. It is seen that at the energy of about 15 eV for positrons and electrons the peak is clearly observed that corresponds to the first diffraction maximum. Note that when using a position-sensitive microchannel detector for positrons, we observed a typical diffraction pattern as a set of corresponding reflections.

Figure 4 shows the energy spectrum of positrons emanating from the W(100) surface when the incident positron energy is set by the moderator potential at 15 V. The energy of the specularly scattered positrons is 17.5 eV It can be seen that the spectrum looks like a typical spectrum for secondary electrons in the case of surface excitation with an electron beam. The shift of the elastic peak by 2.5 eV with reference to moderator voltage is due to the negative work function of the moderator for positrons.

Figure 5 shows the spectrum of secondary electrons when 17.5 eV positrons from the positron beam are incident on the tungsten surface. It can be seen that there are no elastic peaks in the spectrum, and there is a significant part of low-energy electrons. Note also that the intensity of the electrons at the energy of 17.5 eV and above does not equal zero. This is due to the mechanism of formation of secondary electrons. Annihilation of positrons in the surface region leads to the formation of holes that are filled with electrons. The liberated energy was used for the appearance of secondary electrons, even with energy higher than the energy of the primary positrons.

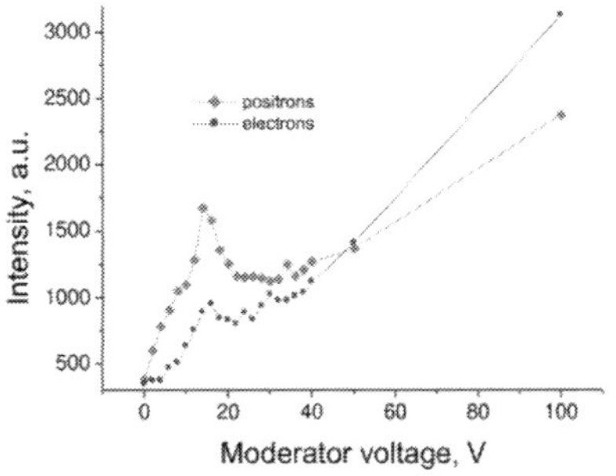

Figure 3. Intensity of positrons and electrons as a function of the energy of positrons incident on the tungsten surface observed at an angle of 22.5 degrees.

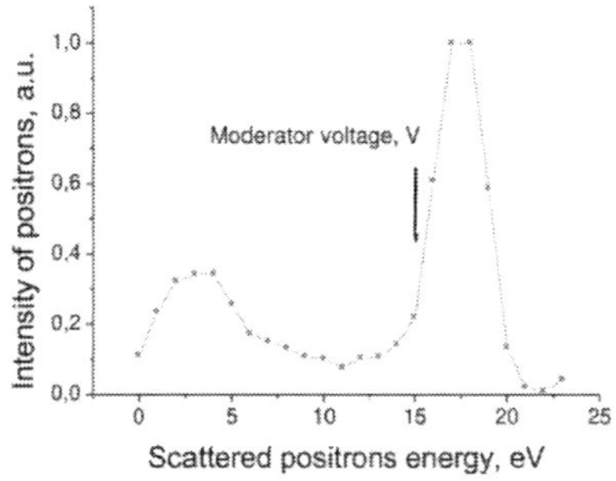

Figure 4. Spectrum of scattered positrons from W(100). Voltage of the moderator is 15 V.

In the next phase of work the magnetic properties of iron films deposited on the tungsten surface were investigated (figure 6). It was found that for both samples, there is a peak that appeared due to the surface resonance [6]. It allowed us to use the W (100)-Fe film system to measure spin polarization of the positron beam. Note that figure 3 shows all emitted electrons as a function of the moderator potential, whereas figure 6 shows only electrons with the energy close to the incident energy of positrons. In the latter case an appropriate potential was applied to the grid in front of the detector.

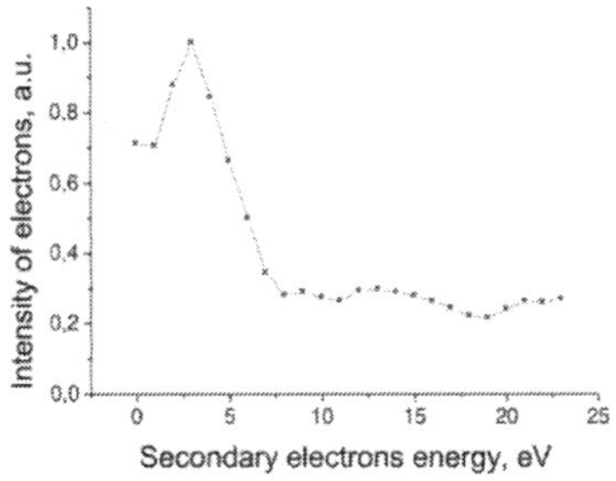

Figure 5. Spectrum of secondary electrons with the incident positron energy of 17.5 eV.

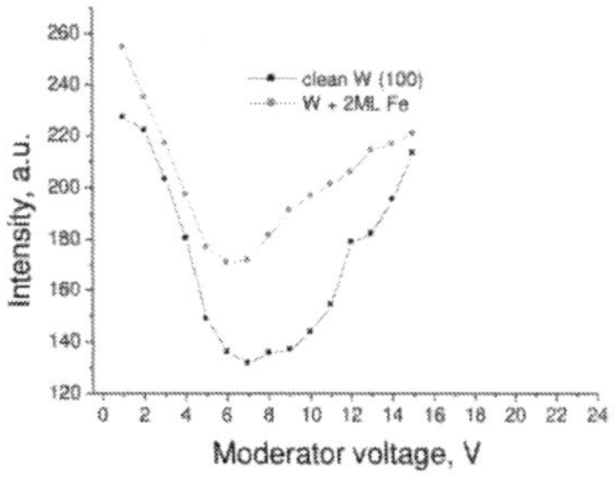

Figure 6. The intensity of the electrons (of energy approximately equal to the incident positron) as a function of incident positron energy from clean W(100) and W(100) deposited with over layers of Fe.

Subsequently, the iron film was sequentially magnetized in opposite directions and the electron emission intensity (as in figure 6) was measured as a function of primary positron energy. The asymmetry in the intensity of the electron emission is calculated by the formula $A=(I^+ - I^-)/(I^+ + I^-)$, where I^+ and I^- are the electron intensities for different directions of magnetization of the iron film, relative to the direction of polarization of the positron beam (figure 7). Below is a simple model that explains our result.

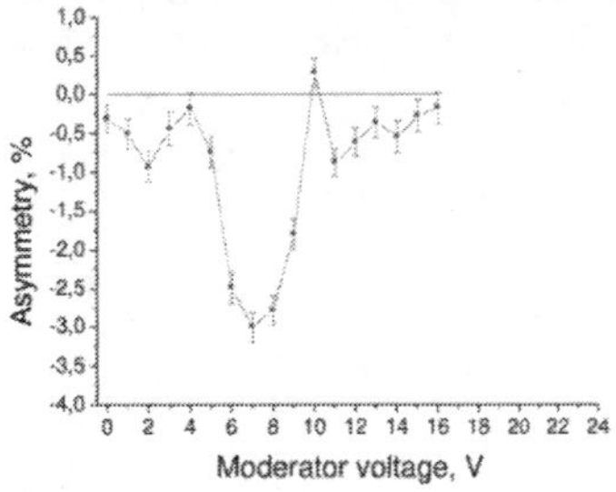

Figure 7. The asymmetry in the intensity of the emitted electrons as a function of energy of the incident positrons. The sample is the Fe film deposited on the W (100).

It can be assumed that the polarized positrons with a certain energy and polarization excite more electrons with a certain spin direction and less with another spin direction and excite them into resonance states. The electrons escaping from the resonance state can be detected at a certain energy and exit angle. In parallel, these resonant electrons interact with iron, which acts as a spin filter. As a result of this process there is an asymmetry of the intensity of the electrons leaving the surface due to the changing of the magnetization vector of the iron film.

One can also consider a model where the positrons are polarized while passing through the magnetic film. However, in the first model, the electrons due to resonance spend significantly longer time in the interface region than the positrons and the probability of processes involving spin is larger. Therefore, we believe that the first model is preferable.

It is seen that in the field of electron resonance the asymmetry reaches a significant value, so you can use this method to calculate the polarization of a positron beam. According to our calculations, it is about 10% in our beam. For this calculation, we used data on the scattering of polarized electrons on tungsten [7].

According to the research conducted in the laboratory of surface magnetism at St. Petersburg Polytechnic University the best material for the moderator that decelerates positrons is palladium, a material that has a very large mean free path of electrons with respect to the spin relaxation (Spin relaxation length Λ_S). Spin relaxation length Λ_S of palladium is hundreds of times greater than a similar property of tungsten.

4. Conclusion

We demonstrate the evidence that the low-energy positron beam generated by our positron gun is spin-polarized. To continue work on the improvement and creation of a truly high efficiency low-energy spin-polarized positron source it is desirable to use an alternative moderator and another scheme of spin detection. For the positron spin detection a classical Mott detector can be used.

Acknowledgement

This research was supported by the Australian Research Council and the University of Western Australia. V. Petrov is grateful to the UWA for the hospitality.

References

[1] Pierce D T et al 1976 *Phys. Rev. B* **13** 5484
[2] Petrov V N et al 2003 *Rev. Sci. Instrum* **74** 1278
[3] Rich A et al 1988 *Hyperfine Interactions* **44** 125
[4] Zhou H 1996 *PhD thesis The University of Texas at Arlington*
[5] Sudarshan K et al 2013 *J. Physics:Conf. Ser.* **443** 012070
[6] McRae E G et al 1981 *Phys. Rev. B* **24** 4230
[7] Petrov V N et al 1996 *Solid State Physics* **38** 1423

Triple coincidence PALS setup based on fast pulse digitizers

D. Bosnar[1], B. Đurđević[1], L. Pavelić[1] and S. Bosnar[2]

[1] Department of Physics, Faculty of Science, University of Zagreb, Zagreb, Croatia
[2] Division of Materials Chemistry, Ruđer Bošković Institute, Zagreb, Croatia

E-mail: bosnar@phy.hr

Abstract. We have assembled positron annihilation lifetime spectroscopy system that consists of three BaF_2 detectors with XP2020URQ PMTs which are directly coupled to DRS4 evaluation board. Both time and energy information of the detected gamma rays are reconstructed. In the off-line analysis, by using selected cuts on energies of gamma rays, double, as well as triple coincidence events can be extracted. The setup is very suitable for the determination of contributions of positronium three gamma annihilations which can be of interest in the investigations of various porous materials.

1. Introduction

Positron annihilation lifetime spectroscopy (PALS) systems are usually based on classical, analog, electronics, and they extract only time information from the registered gamma rays, see eg. [1]. One of the improvements of such setups, but still relaying on the analog electronics, includes simultaneous recording of time and energy information of the detected gamma rays and this additional information can be used in investigations of contributions of three gamma positronium annihilations [2]. A new approach to PALS measurements has been recently introduced by the employment of fast pulse digitizers, see eg. [3]. We have used a low-cost DRS4 evaluation board [4], similarly as in [5], in order to further investigate properties and possibilities of PALS measuring systems based on fast pulse digitizers. Our focus is on the determination of contributions of positronium three gamma annihilations. Our PALS setup comprises three BaF_2 detectors directly coupled to the DRS4 evaluation board. Both time and energy information of the detected gamma rays are reconstructed. In the off-line analysis double coincidence events can be selected by choosing 1.27 MeV gamma as a start signal in one detector and 0.511 MeV gamma as a stop signal in one of the other two detectors. Triple coincidence events can be selected by demanding coincidences among 1.27 MeV gamma in one detector and two annihilation gamma rays from positronium decay in the remaining detectors. The setup is now being used in the investigations of selected samples with large voids and for the determination of contributions of three gamma positronium annihilations. These contributions can be of special interest for various porous materials [6,7,8].

In the second section we give brief overview of the PALS setup, the third section is devoted to the energy and time determination from the data collected by the digitizer, the fourth section briefly describes ongoing measurements of the selected samples and the fifth section gives summary and conclusion.

Journal of Physics: Conference Series **618** (2015) 012044 doi:10.1088/1742-6596/618/1/012044

2. PALS setup

Our positron annihilation lifetime spectroscopy system comprises three BaF_2 scintillation detectors. The BaF_2 scintillators have cylindrical shape with the base diameter of 1 inch and height of 1 inch. All sides of the scintillators, except one base, are wrapped with aluminum foil. The scinitllators are placed inside aluminum housings and are coupled to XP2020URQ PMTs which are protected by mu-metal shieldings. All three detectors are in one plane and each detector is positioned on an aluminum holder that can be rotated around the fix central point at which sample with ^{22}Na source is placed. The angles among the detectors and the distances of the detectors from the source can be varied. Three Ortec 556 HV suppliers are used for the BaF_2 detectors. Anode outputs of the PMTs are directly coupled to the inputs of the DRS4 evaluation board. The board has four input channels and it is connected with the USB cable to the computer. Linux based control and data acquisition software supplied by the producer of DRS4 board is used for display of signals on the computer screen and for the data acquisition. The sampling rate can be varied in the interval 0.7-5 GS/s and it is possible to collect 1024 sampling points for each channel, covering the signal time range of 1.4 μs - 200 ns, respectively. The software allows selection of thresholds for the used channels. Also, selection of a combination of AND / OR conditions among the used channels for the data acquisition trigger is possible. In the measurements complete signals from the detectors in the selected time range and for selected trigger conditions are stored.

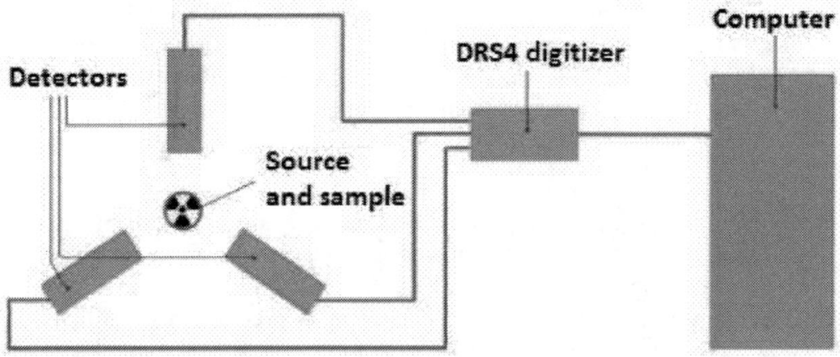

Figure 1. The PALS setup with three BaF_2 detectors and DRS4 evaluation board.

3. Energy and time determination

Algorithms in Root software [9] for timing and energy determination from the recorded signal pulses have been developed. For timing determination an emulation of constant fraction method was applied.

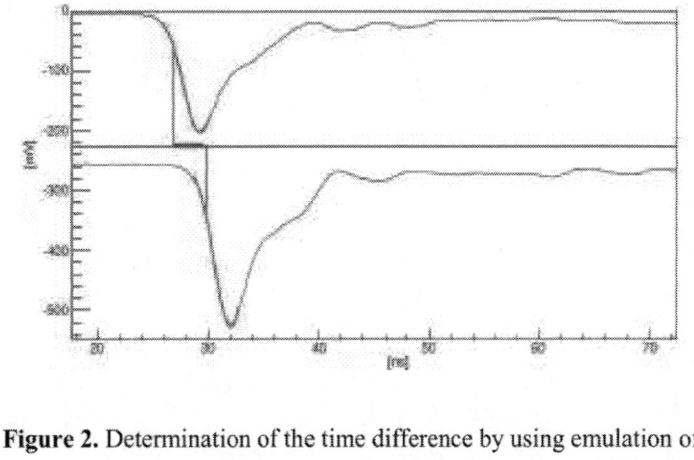

Figure 2. Determination of the time difference by using emulation of constant fraction method.

Journal of Physics: Conference Series **618** (2015) 012044 doi:10.1088/1742-6596/618/1/012044

A Gaussian fit on the rising edge of the signal was made and the hight of the signal was determined. The time point at a selected fraction of signal height was taken as the time of signal appearance and time differences for coincidence gammas were determined, Figure 2.The selected fraction had been varied until the best time resolution in measurements with ^{60}Co was obtained.

For energy determination the integration of the signal pulse in the whole range of recorded signal has been made, and it clearly allows separation of gammas in the photo-peak and Compton gamma rays, Figure 3.

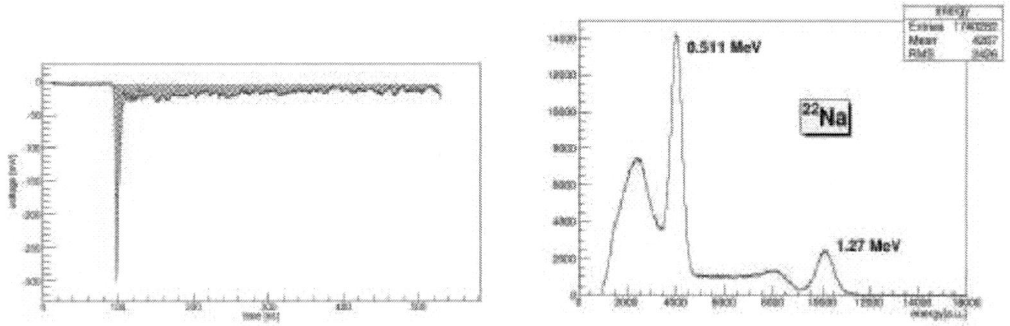

Figure 3. Energy reconstruction by integration of the recorded signal, left. Spectrum of ^{22}Na Gamma rays recorded by one of BaF_2 detectors, right.

In the measurements with ^{60}Co source and maximal sampling rate of 5GS/s by selecting only gamma rays from photo-peak the achieved time resolution was 145 ps (FWHM), and by using both Compton gamma rays and gamma rays from photo-peak time resolution was 153 ps. Achieved time resolution at the sampling rate of 2 GS/s with ^{60}Co source was 153 ps when we used gammas from photo-peak, and 185 ps when we used both Compton gamma rays and gamma rays from photo-peak, Figure 4.

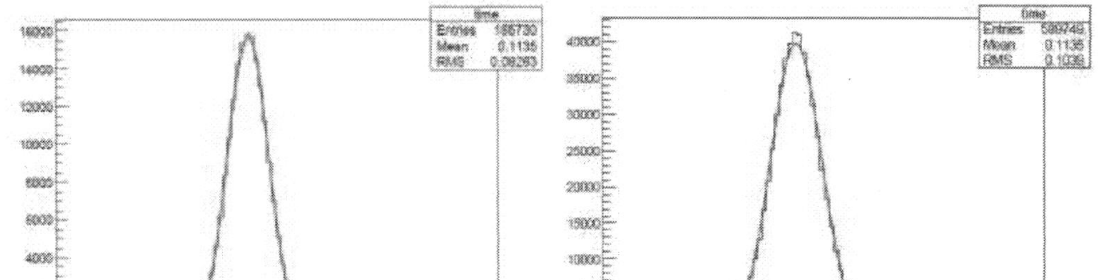

Figure 4. Time resolution determined in the measurements with ^{60}Co at the sampling rate of 2 GS/s: 153 ps, only gamma rays from photo-peak, left; 185 ps, both Compton gamma rays and gamma rays from photo-peak, right.

The time resolution at the minimal sampling rate of 0.7 GS/s becomes much worse and it is not suitable for PALS measurements. In our measurements, in order to also measure positronium three gamma decays, we use the sampling rate of 2 GS/s.

Journal of Physics: Conference Series **618** (2015) 012044 doi:10.1088/1742-6596/618/1/012044

4. Measurements of selected samples

The setup is currently being used for PALS measurements of some selected samples of porous materials, in particular hierarchical zeolites and mesoporous materials. In larger sample voids a portion of positronium can directly decay in three gammas. Due to these decays the intensities of longer lifetimes from positronium pick-up annihilations determined by standard PALS measurements are lower and eventually they suggest lower relative concentrations of such voids. In order to determine this portion we are currently investigating mesoporous materials in which lifetimes up to 41 ns have been observed, and zeolite samples with hierarchical porosity in which lifetimes up to 36 ns have been observed with our standard PALS setup [10]. In the measurements we are collecting data from which double and triple coincidences can be extracted. Procedures for the determination of contributions of three gamma positronium annihilations which take into account different efficiencies for two- and three- gamma detections are being developed. In these measurements we also are examining the stability of the employed evaluation board for the long term measurements.

5. Summary and conclusion

We have assembled PALS setup based on three BaF_2 detectors and DRS4 evaluation board and the software for timing and energy determination of recorded gamma rays has been developed. It can be used for double and triple coincidence measurements. Excellent properties regarding time resolution have been achieved. In the off-line analysis energy selection of the detected gamma rays can be done. These features of the setup can be used for the determination of contributions of positronium three gamma annihilations. On the hardware side this is very simple setup, but the complexity of the relevant information extraction has been shifted toward the software for the data reconstruction and analysis. Also, the amount of collected data substantially increases. The stability of this particular board in long term PALS measurements is being examined. The achieved results clearly demonstrate some of possible advantages of PALS system based on such kind of fast pulse digitizers.

Acknowledgments
This work has been supported by projects at the MSES of the Republic of Croatia 098-0982904-2953, 098-0982886-2893, 119-1191005-1021 and IAEA CRP contract 17251.

References

[1] Becvar F *et al* 2000 *Nucl. Instr. Meth.* A **443** 557
[2] Bosnar D *et al* 2007 *Nucl. Instr. Meth.* A **581** 91
[3] Becvar F *et al* 2005 *Nucl. Instr. Meth.* A **539** 372
[4] www.psi.ch/drs/evaluation-board
[5] Petriska M *et al* 2014 *J. Phys.: Conf. Ser.* **505** 012044
[6] Kobayshi Y *et al* 2007 *Rad. Phys. Chem.* **76** 224
[7] Kajcsos Zs *et al* 2007 *Rad. Phys. Chem.* **76** 231
[8] Huang S H *et al* 2012 *Rad. Phys. Chem.* **81** 791
[9] root.cern.ch
[10] Bosnar S *et al* 2014 *Acta Phys. Pol.* **125** 775

Design of a pulsed positron system at Trombay

S. Mukherjee[1],S. K. Sharma[1], Priya Maheshwari[1], S. K. Gupta[2] and P. K. Pujari[1]

[1] Radiochemistry Division,Bhabha Atomic Research Centre, Mumbai 400085.
[2] Ion Accelerator Development Division, Bhabha Atomic Research Centre, Mumbai 400085.

E-mail: pujari@barc.gov.in

Abstract. We present here the design of a pulsed beam setup to deliver narrow time width positron pulses. The major constituents of the setup include- ^{22}Na radioactive source and moderator assembly, ExB deflector for filtering out high energy positron and gamma rays, chopper-prebuncher-buncher assembly for time bunching of the slow positrons. In the ExB section, crossed electric and magnetic fields guide the slow positrons through an off-centered hole in a tungsten block. The initial beam will then be time bunched by using a reflection type chopper and a double gap prebuncher. The main buncher is designed as a quarter wave resonator with base frequency of 150 MHz. To prevent the sagging of the cantilevered inner tube of the resonator, we will support the inner conductor using an alumina post. There will be provision of tuning the frequency by using a tuner made of conducting material. The incident beam energy will be varied by biasing the sample.

1. Introduction

Energy tunable pulsed positron beam can be used to probe defect characteristics of surfaces, subsurfaces and buried interfaces in thin films. Such experiments provide a huge advantage over the conventional positron annihilation lifetime technique where the bulk of the sample is probed ($\sim$ 100 μm) owing to the wide energy range of the emitted positrons from the radioactive source. We present here a simulation and design of a ^{22}Na radioactive source based pulsed positron beam. Conventionally, positron pulsing techniques are derived from the ion bunching field because of the low intial current and hence higher need to minimize loss (ion beams tend to be in nanoamps while positron beams are in femtoamps). Also, the positron β values ($\sim 10^{-2}$) for materials science applications are comparable to that of the heavy ion beams[1]. Almost all of the existing pulsed positron beams are based on some variation of chopper-prebuncher-buncher arrangement. The Munich group [2] used a time dependent transverse electric field to kick away the non-time-synchronized positrons by increasing the Larmor radius of the positrons. For the prebuncher and buncher, they used the conventional design of resonant cavities. The pulsed beam developed by Suzuki et al. [3] used a reflection type chopper composed of three parallel grids to convert the DC positron beam to a pulsed beam of 2.5 ns FWHM. PALS setup at both these groups have a timing resolution ~250 - 300 ps which is sufficient for lifetime measurements.

2. Design and Simulations

In the following sections, we will discuss the various components of the spectrometer: ExB filter, chopper & prebuncher, buncher. The complete beam assembly is shown in Fig. 1.

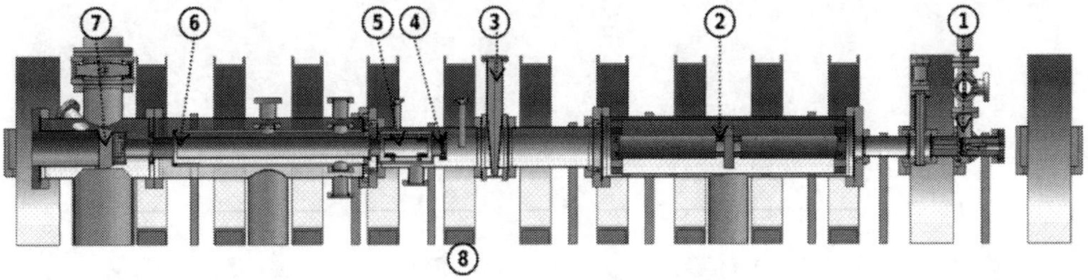

Figure 1. Design of the pulsed positron beam. 1. Source+Moderator Setup 2. ExB filter 3. Isolation Valve 4. Chopper 5. Prebuncher 6. Main buncher 7. Sample 8. Helmholtz coils.

2.1. Energy filtering of fast positrons

The positrons emitted from the ^{22}Na source have a wide energy distribution with end point being $\sim$ 540 keV and a mean of $\sim$ 190 keV. These high energy positrons are moderated using a 1 μm Tungsten foil which produces slow positrons with mean energy equal to the absolute value of the positron work function ($\sim$ 3 eV) with a spread of $\sim$ 0.5 eV and efficiency of $\sim 10^{-4}$. Most of the high energy positrons ($\sim$ 90%) actually pass through the moderator unaffected. Conventionally, a solenoidal bend in the beam is used to filter the high energy positron beam with the assumption that the change in the B field will not be adiabatic for the fast non-thermalized positrons [4]. But the drawback of such setups can be understood by noting that range of energy of the positrons allowed to pass through for a 90^0 bend with B= 90 G is 0-3 keV. Use of crossed magnetic fields (BxB) for fast positron filtration has been suggested as well which has same drawback as the "90^0 bend" [5]. On the other hand, ExB filtration provides excellent energy selection (10%)[6]. Conventionally, it is achieved by using a set of parallel plates where the electric field (E) is perpendicular to the magnetic (B) field. The positrons, which were initially following the magnetic field, are deflected up (direction perpendicular to both E and B fields) and are made to pass through an off centered hole. The deflection, y, is given by $y = \frac{EL/B}{\sqrt{2T/m}}$, where E is the electric field, B is the magnetic field, L is the length of the plate, T is particle's kinetic energy and m is the mass of the positron. This arrangement introduces distortion such that the initial circular cross section of the beam is transformed to an elliptical one. Hence we have used cylindrical plates to eliminate these distortions [7]. A disadvantage in using cylindrical plates for ExB filters is that such arrangement only works for one set of beam energy and plate voltage. Hence we have decided to float the ExB filter assembly. The SIMION[8] simulations indicate that with the dimensions of ExB plates (L=250 mm, B=40 G) with 30 mm vertical deflection will deliver an almost Gaussian beam of mean energy 250 eV having FWHM of 10 eV (see Fig. 2)[9].

2.2. Time bunching of positrons

The time bunching of the positrons is achieved by a combination of chopper, prebuncher and buncher. The chopper will be made of three tungsten mesh (2 μm, 100 mesh/in) which are separated by Macor insulators with a gap of 5 mm. The middle mesh is fed a square wave RF of 37.5 MHz with a DC bias of 260 V. The chopper has a duty cycle of 15 %. The prebuncher consists of two concentric cylinders and the RF pulse is fed to the inner tube. The acceleration gap at the entrance and exit of the outer cylinder is defined by the axial distance between the inner and the outer cylinder (10mm). The two bunching gaps are separated by 125 mm which is also the half wave length of the RF pulse (37.5 MHz) at the positron energy of 250 eV. The chopper and prebuncher combination was simulated in SIMION.

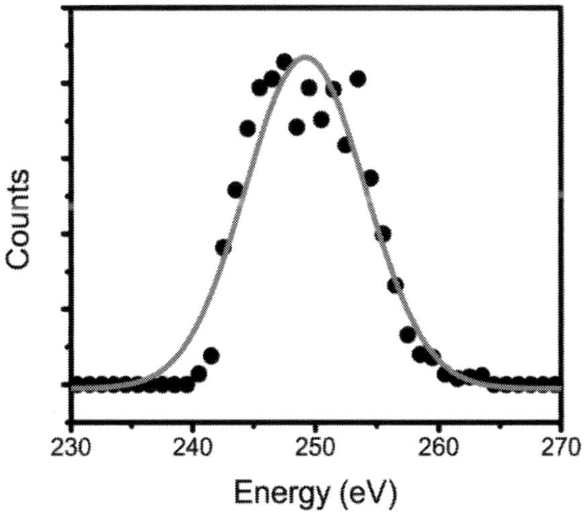

Figure 2. Transmission function of the ExB plates. The beam is centered at 250 eV with FWHM of 10 eV.

Table 1. Parameters of the quarter wave resonator(E_{max} is the value of maximum electric field).

Parameter	value	Parameter	value	Parameter	value
Length(inner)	469 mm	Gap Length	10 mm	Diameter(inner)	20 mm
Diameter(outer)	52 mm	E_{max}	9x10$^7 V/m$	Q	2200

2.3. Main Buncher

The main buncher is designed as a quarter wave resonator with a base frequency of 150 MHz and a Q value of 2200. The main parameters of the buncher are summarized in Table 1. The buncher has been simulated using CST Microwave software. Since the inner tube of the buncher is cantilevered at one end, it can sag at other end causing the change in resonant frequency. For this we have proposed to support it using a post made of alumina (Fig. 3). The change in frequency as a function of the alumina post position is shown in Figure 3(b).

Provision has been kept to vary the frequency of the resonant cavity using a copper tuner of 14 mm diameter and 30 mm length. The variation of the frequency is related to the reduction in the volume (achieved by the varying the plunger depth) by the Slater's perturbation theorem [10]

$$\frac{\delta f}{f} = \frac{\int_{\delta V}(\mu_0 H^2 - \epsilon_0 E^2)dV}{\int_V(\mu_0 H^2 + \epsilon_0 E^2)dV}$$

where δV is the change in cavity volume, V is the cavity without the tuner, E and H are the magnitude of the unperturbed electric and magnetic field. The simulated variation of the frequency with tuner depth is shown in Fig 3(b).

3. Conclusion

In conclusion, we have optimized preliminary design of a pulsed positron beam using SIMION and CST microwave studio. This design will provide 250 eV positron pulses of duration less than 200 ps. The distance between the moderator and the sample will be $\sim$ 2 m and the average beam intensity will be $\sim 10^3$. When combined with timing resolution of the gamma detectors

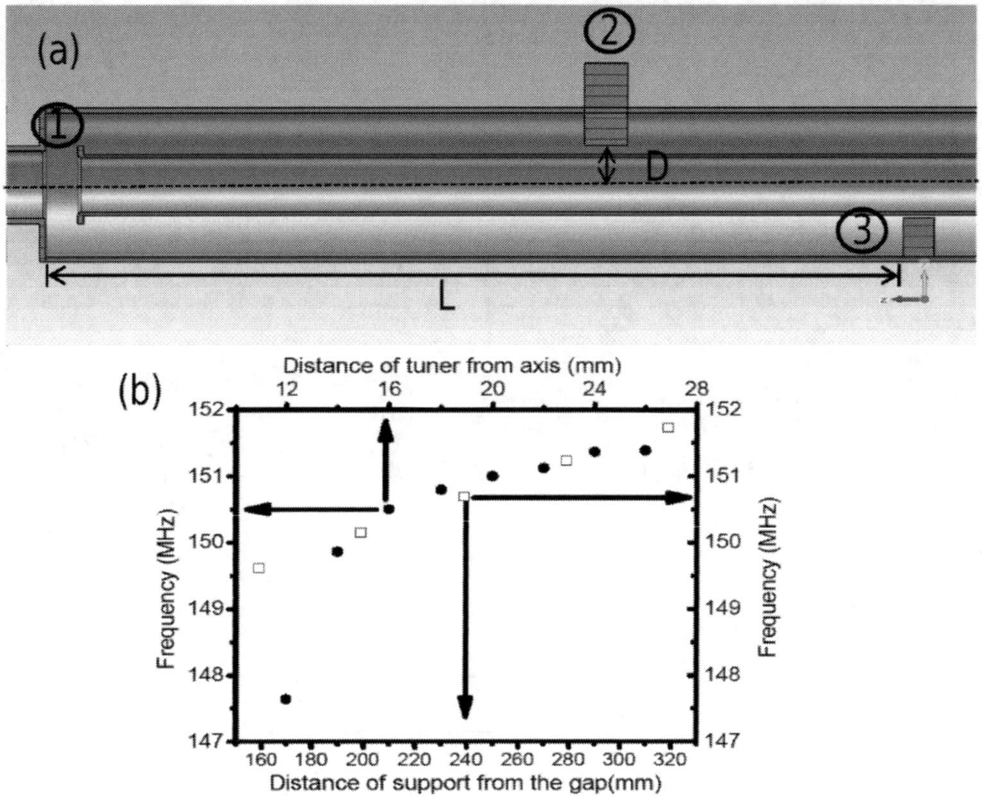

Figure 3. (a) CST setup of the main buncher. 1. Bunching gap. 2. Tuner 3. Alumina support. The distance of the tuner from the buncher axis is represented by D while the distance of the support from the gap is represented by L. (b) Variation of the resonant frequency as a function of the distance of support from the gap(L) and distance of tuner from axis (D).

(BaF_2, ~200 ps), it is expected that this spectrometer will provide time resolution of ≤300 psec for depth dependent PALS.

References

[1] Lynch F, Lewis R, Bollinger L, Henning W and Despe O 1979 *Nuclear Instruments and Methods* **159** 245
[2] Schödlbauer D, Sperr P, Kögel G and Triftshäuser W 1988 *Nuclear Instruments and Methods in Physics Research Section B: Beam Interactions with Materials and Atoms* **34** 258 – 268 ISSN 0168-583X
[3] Suzuki R, Kobayashi Y, Mikado T, Ohgaki H, Chiwaki M and Yamazaki T 1991 *Japanese Journal of Applied Physics* **30** L532
[4] Jackson J D 2003 *Classical Electrodynamics* (Wiley)
[5] Hathaway A 2011 Ph.D. thesis North Carolina State University
[6] Mills A P 1980 *Applied Physics Letters* **37** 667–668
[7] Hutchins M, Coleman P, Stone R and West R 1986 *Journal of Physics E* **19** 282
[8] Manura D and Dahl D 2008 Simion (r) 8.0 user manual
[9] Barr W and Perkins W 1966 *Review of Scientific Instruments* **37** 1354
[10] Wrangler T P 2008 *RF Linear Accerlators* (Wiley-VCH Verlag GmbH Co.)

Institute of Physics
Dirac House, Temple Back
Bristol BS1 6BE UK

CURRAN ASSOCIATES INC
proceedings.com